Ultrastructure of Human Gametogenesis and Early Embryogenesis

ELECTRON MICROSCOPY IN BIOLOGY AND MEDICINE

Current Topics in Ultrastructural Research

SERIES EDITOR: P. M. MOTTA

Already published in this series

Vol. 1 Motta, P. M. (ed.): Ultrastructure of Endocrine Cells and Tissues. ISBN: 0-89838-568-7.

Vol. 2 Van Blerkom, J. and Motta, P. M. (eds.): Ultrastructure of Reproduction: Gametogenesis, Fertilization, and Embryogenesis. ISBN: 0-89838-572-5.

Vol. 3 Ruggeri, A. and Motta, P. M. (eds): Ultrastructure of the Connective Tissue Matrix. ISBN: 0-89838-600-4

Vol. 4 Motta, P. M., Fujita, A. (eds.): Ultrastructure of the Digestive Tract. ISBN: 0-89838-893-7.

Ultrastructure of Human Gametogenesis and Early Embryogenesis

edited by

Jonathan Van Blerkom Ph.D.

Department of Molecular, Cellular and Developmental Biology, University of Colorado, Boulder, Colorado

and

Reproductive Genetics in Vitro, Denver, Colorado

and

Pietro M. Motta Ph.D., M.D.

Department of Anatomy, University of Rome, Rome, Italy

1989 **Kluwer Academic Publishers**

BOSTON DORDRECHT LONDON

Distributors

for the United States and Canada: Kluwer Academic Publishers 101 Philip Drive, Assinippi Park, Norwell, MA 02061 USA
for the UK and Ireland: Kluwer Academic Publishers. Falcon House, Queen Square, Lancaster LAI 1RN, UK
for all other countries: Kluwer Academic Publishers. Distribution Centre, P.O. Box 322, 3300 AH Dordrecht, The Netherlands

Library of Congress Cataloging in Publication Data

Ultrastructure of human gametogenesis and early embryogenesis / edited by Jonathan Van Blerkom, Pietro Motta.
p. cm. — (Electron microscopy in biology and medicine)
Includes bibliographies and index.
ISBN 0-89838-383-8
1. Gametogenesis. 2. Embryology. 3. Germ cells — Ultrastructure.
I. Van Blerkom, Jonathan, 1947– . II. Motta, Pietro M. III. Series.
[DNLM: 1. Embryo — ultrastruct. 2. Gametogenesis. 3. Oocytes-ultrastruct. QS 604 U47]
QM601.U45 1988
612′.6 — dc 19
DNLM/DLC
for Library of Congress 88-15890
CIP

PRINTED IN THE UNITED STATES

Preface

In the last few years, the adoption and worldwide proliferation of clinical procedures for medically assisted conception have been associated with the examination and analysis of spermatozoa, oocytes and early embryos under a variety of in vivo and in vitro conditions. These analyses have enabled correlations to be made between the behavior of gametes, the pattern of early embryonic development and the initiation of a normal pregnancy. Collectively, the findings have not only enormously increased our understanding of the process of early human development, but also have provided new insights into the origin and causes of reproductive failure in man.

The research presented in this volume describes recent results derived from the study of normal and abnormal patterns of human spermatogenesis, oogenesis and early embryogenesis. The chapters discuss aberrations in morphodynamic and morphophysiological processes that have clinical relevance in human infertility and conception. Two of the chapters describe, respectively, the basic research that allows the cryopreservation of human oocytes and embryos, and the development of in vitro systems that permit the study of cell differentiation and interaction during the peri-implantation period. When relevant, each chapter extrapolates findings from in vitro experimentation to the comparable situation that is observed in vivo.

The rapid clinical application of basic research findings in the treatment of human infertility underscores the necessity of having a comprehensive and fundamental understanding of the morphogenetic and morphophysiological bases of normal and abnormal gametogenesis and embryogenesis. This volume is intended to provide such information, especially as it relates to the causes and treatments of infertility. Because each chapter is a discussion of the current state of knowledge, the authors indicate both where information is still incomplete and what specific research is required to expand our understanding of the origins of infertility and early developmental failure.

J. VAN BLERKOM and P.M. MOTTA

Contents

Contributing authors

Dr. L. Accinni, Regina Elena Institute for Cancer Research, Viale R. Elena, Rome, Italy

Prof. C. De Martino, Regina Elena Institute for Cancer Research, Viale R. Elena, Rome, Italy

Prof. M. Fabbrini, Regina Elena Institute for Cancer Research, Viale R. Elena, Rome, Italy

Prof. G. Familiari, Department of Anatomy, University of Rome "LA SAPIENZA", Rome, Italy

Prof. S. Francavilla, Regina Elena Institute for Cancer Research, Viale R. Elena, Rome, Italy

Prof. P. Hytell, Dept. of Obstetrics and Gynecology, Rigshospitalet, University of Copenhagen, Blegdamsvej 9, DK-2100, Denmark

Dr. S. Lindenberg, Chromosome Lab 4041, Dept. of Obstetrics and Gynecology, Rigshospitalet, University of Copenhagen, Blegdamsung 9, Denmark

Dr. S. Makabe, Dept. of Obstetrics and Gynecology, Toho Univ. School of Medicine, 6-11-1 Ohmorinishu 6 Chome, Tokyo, 143 Japan

Prof. P.M. Motta, Dept. of Anatomy, University of Rome "LA SAPIENZA", Rome, Italy

Dr. S. Nottola, Department of Anatomy, University of Rome "LA SAPIENZA", Rome, Italy

Dr. A.H. Sathananthan, Lincoln School of Health Sciences, LaTrobe University, 625 Swanston Street, Carlton, Victoria, 3053 Australia

Prof. A. Trounson, Centre for Early Human Development, Monash Medical Centre, Clayton Road, Clayton, Victoria 3168, Australia

Prof. J. Van Blerkom, Dept. of Molecular, Cellular and Developmental Biol. University of Colorado, Boulder, Colorado USA

Prof. H. Wartenberg, Anatomisches Institut Der Rheinischen Friedrich–Wilhelms Universitat, Nussalle 10, D 5300 Bonn 1, Federal Republic of Germany

Ultrastructure of Human Gametogenesis and Early Embryogenesis

CHAPTER 1

Mammalian spermatogenesis and its disorders in man

CESARE DE MARTINO, SANDRO FRANCAVILLA, ALDO FABBRINI, & LIDIA ACCINNI

1. Introduction

In mammals, spermatogenesis begins at puberty and results in the formation of a unique population of haploid cells, the spermatozoa. This process continues throughout the life of the animal and involves the constant renewal of the stem cells, the spermatogonia. In contrast to oogenesis, spermatogenesis is a continuous process that proceeds without pauses at specific stages of sperm development. An exception to this rule is observed in mammals with seasonal cycles, such as the coyote, fox, ground squirrel, and free-ranging monkey. During the nonbreeding period for these mammals, the seminiferous tubules are almost exclusively composed of Sertoli cells and a single layer of spermatogonia.

2. Development of male germ cells

The development of male germ cells may be divided into three main periods. In the first, or *extragonadal period*, the so-called *primordial germ cells* are identified in the mammal in the endodermal yolk sac and in the hindgut epithelia (see as review, 1). These cells are capable of mitotic divisions and amoeboid movements and rapidly migrate through the primitive intestine toward the genital ridge (Figs. 1-1a and b). During their migration, the primordial germ cells often follow an aberrant path and acquire an ectopic location in extragonadal tissues and/or organs (Fig. 1-1c). During this period, the germ cells of both sexes appear to be alike in morphology with the exception of the chromosome set. The second period of development occurs when the gonads and sex begin to diverge. In the course of development, the primordial *sex cords* appear within the inner zone of the male gonad, whereas the outer zone thins out and gives rise to the *albuginea* (Fig. 1-1d). In the eight-week-old human fetus, the sex cords contain the *support cells*, which are somatic cell precursors of the *Sertoli cells*, and the primordial germ cells, now known as *gonocytes*. In man, during this period, which is often called *pre-spermatogenesis*, the gonocytes go through a mitotic activity intercalated with cell degeneration. In addition, a direct interaction between somatic and germ cells occurs. However, this second period is not always well-defined and probably does not exist in many animals in which the process of spermatogenesis begins almost immediately after germ cells become localized in the gonad (1). Even in those animals in which a pre-spermatogenic period occurs, it is difficult to define precisely when it ends and when the third period, *spermatogenesis proper*, begins.

3. Spermatogenesis

The interval between birth and the triggering of spermatogenesis varies among mammals. In the rat, for example, spermatogenesis begins on the fourth postnatal day with the appearance of type *A spermatogonia*, whereas in primates, it commences several years (three to seven) after birth.

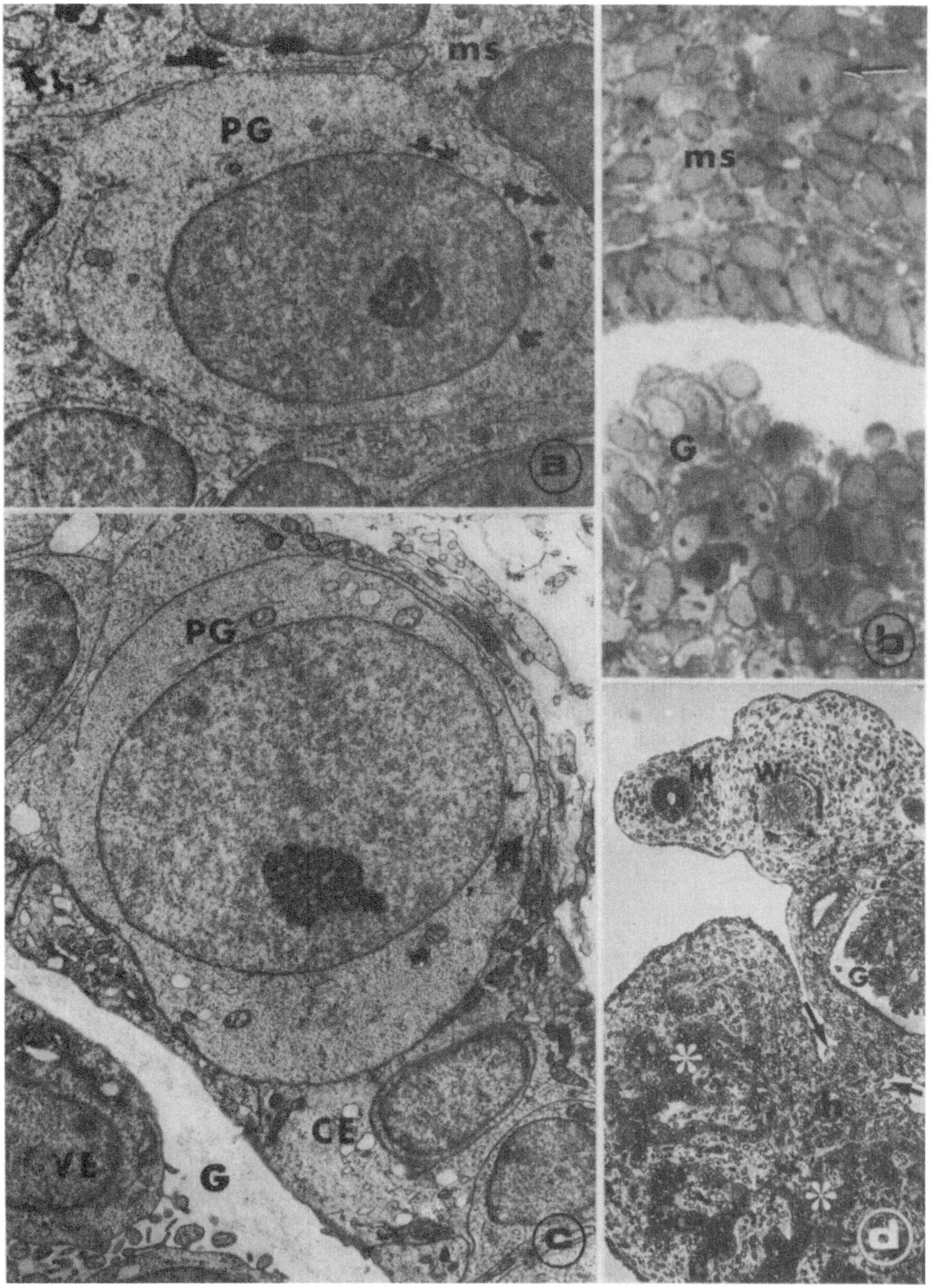
ms
PG
a
ms
G
b
PG
CE
VE
G
c
M
W
G
d

Also in man, a quiescent period that lasts 10–14 years is observed between birth and onset of testicular growth.

The germ cells never develop in isolation. They always grow in close aggregates, usually in synchronized *clones* syncytially connected by *intercellular bridges* (Fig. 1-2g) in such a way that a given germ cell influences and is influenced by other germ cells. Along the vertebrate phylogenetic scale, the model of clonal development is also present in those animals in which germ cell development and maturation do not recognize clear stages of the cycle of the seminiferous epithelium, i.e., fishes and amphibians (1). In addition, phylogenetically, clonal development seems to be present from the lowest animals, such as porifera, to the highest mammals. However, in the testes of *Schistosoma mansoni*, a dimorphic flatworm characterized by spermatogenesis so rapid and continuous as to allow the fertilization of an egg every five minutes during the parasitic stage, the germ cells appear to be randomly distributed and completely separated in all stages of their maturation (1), even though small zones or clumps of synchronous germ cells are occasionally observed (Figs. 1-3a and b).

In vertebrates, the pattern of clonal development differs according to the species studied. In fact, the synchronous clonal development of germ cells in casts or cysts (1) is typical of the spermatogenesis of fishes and amphibians (Figs. 1-3c, d, and e), while in Amniota (Reptilia, Aves, Mammalia), the gametes develop in several concentric layers representing successive generations (Figs. 1-2g and h). These evolve cyclically, in concert and in symbiosis with the somatic Sertoli cells, which are in contact with the germ cells at all stages (Fig. 1-2g). By contrast, in fishes and amphibians, the Sertoli cells, called *cyst cells*, are in close relationship with the spermatids, but run outside the spermatocyte casts (Fig. 1-3e) (for details see 1).

3.1. Mammalian spermatogenesis

Once activation of the gonocytes has occurred, development proceeds for the majority of mammals through a series of fairly well-known steps. However, because spermatogenesis has been most extensively studied in the mouse and rat, which represent animal models for comparison of the spermatogenic process with other mammals, an extensive description of the spermatogenetic stages that take place in the rat is described below.

3.2. Spermatocytogenesis or spermatogonial stage

Spermatogenesis includes all phases of development of the germinal epithelium from spermatogonia to testicular spermatozoa (Figs. 1-2, 1-4–1-9). The first step of spermatogenesis proper is *spermatocytogenesis* and the cells involved in this stage are called *spermatogonia*. The *primary spermatogonia*, which are relatively large cells provided with dispersed chromatin and one or two prominent nucleoli, undergo division. The daughter cells of primary spermatogonia and their progeny are called *secondary spermatogonia*. In mammals, the number of successive spermatogonial mitoses is, within narrow limits, apparently species-specific — usually three to five. Secondary spermatogonia are differentiated cells characterized by nuclear and cytoplasmic changes. Thus, the successive spermatogonial generations show characteristics that differ progressively from the former

←

Figure 1-1a–c. Six-week-old human embryo. Primordial germ cells (PG in Figures 1-1a and c, arrow in Figure 1-1b) are visible in the genital ridge between the cells of the subcoelomic and paramesonephric mesenchyme. They are characterized by a large round shape, rare scattered mitochondria and a prominent nucleolus with an evident nucleolonema. In Figure 1-1c, a PG is localized in an extragonadal position between the cells of the capsular epithelium (CE) of a mesonephric glomerulus (G). VE: glomerular visceral epithelium. Figures 1-1a and c: X 9,000; Figure 1-1b: X 800.

Figure 1-1d. Nine-week-old human embryo. Early stage of male gonadal differentiation. In cross section, the testis shows an ovoid shape and is connected to the mesonephric mesenchyme by a short mesorchium (arrows). The outer zone, anlage of tunica albuginea, is narrow while the inner zone shows the developing sex cords (asterisks). These cords (or future seminiferous tubules) are almost uniform in width and converge toward the hylum (h). M: Müllerian duct; W: Wolffian duct; G: mesonephric glomerulus. X 125.

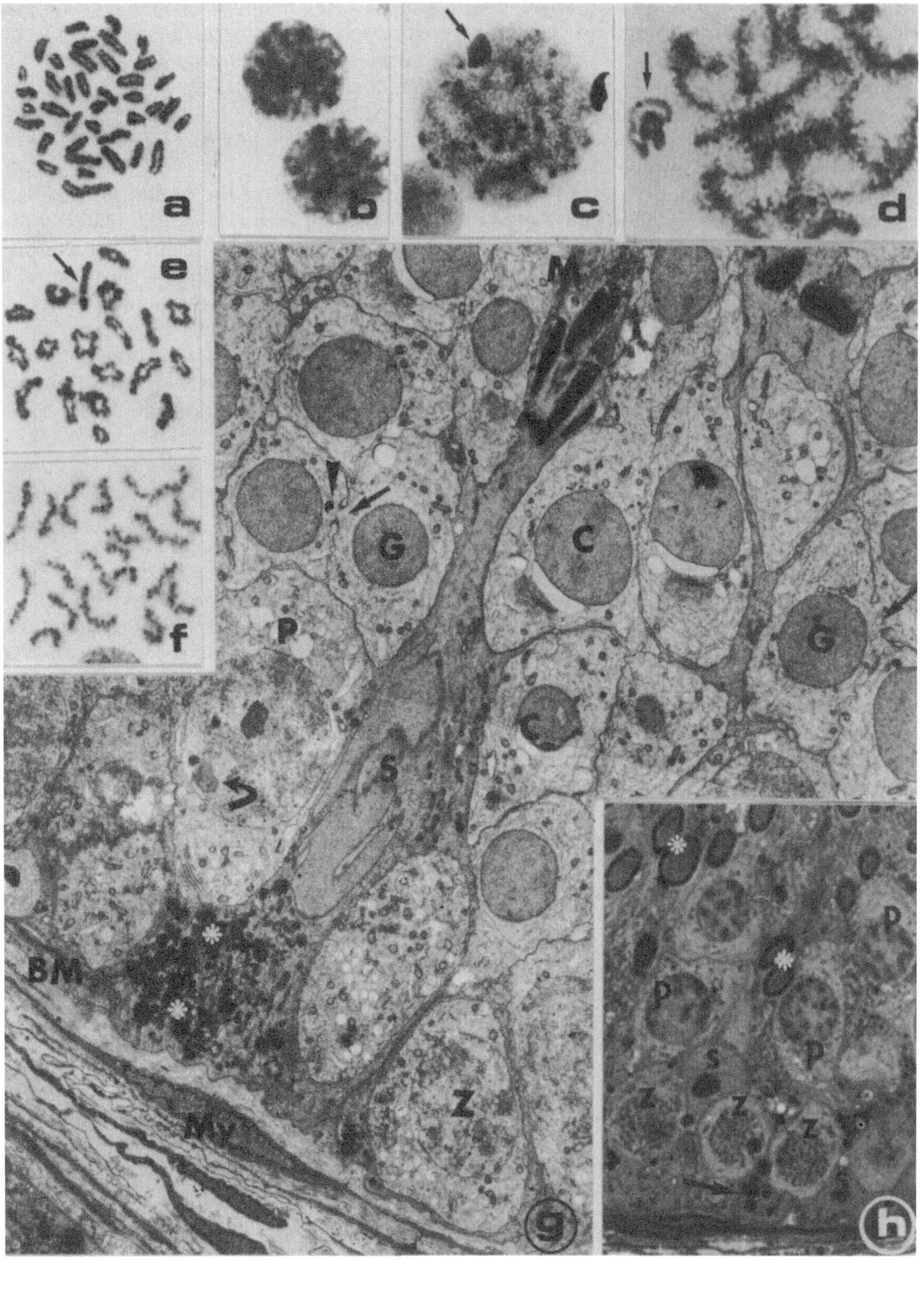
a
b
c
d
e
f
M
G
C
P
S
BM
My
Z
g
h

generation. Because germ cell proliferate almost continuously through adult life in the male, a constant number of spermatogonial divisions is essential for the provision of a supply of germ cell precursors, or *self-renewing stem cells*, and a constant number of spermatogonia committed to spermatogenesis (Fig. 1-4). The mechanisms responsible for this process have been thoroughly investigated, and many theories have been proposed. After the original description of rat stem cell renewal by Clermont and Leblond (2), it became apparent that there are several distinct types of spermatogonia, called *A spermatogonia*, with mitotic peaks at specific stages of the cycle of the seminiferous epithelium. These different types of A spermatogonia are designated progressively: $A_{1\text{-}2\text{-}3\text{-}4}$. Subsequently, two main hypotheses on stem cell renewal in the rat have been proposed by Clermont and Bustos-Obregon (3) and by Huckins (4). The Huckins model (Fig. 1-4) for the rat was described contemporaneously for the mouse by Oakberg (5). According to Clermont and Bustos-Obregon (3), A spermatogonia are divided into five classes: A_0–$A_{1\text{-}4}$. The A_0 spermatogonia normally do not divide and do not contribute to the replenishment of A_1 spermatogonia. These cells are considered *reserve stem cells* that become active only if the more mature spermatogonia are depleted by noxious agents such as x-rays. The $A_{1\text{-}4}$ spermatogonia are engaged in proliferative activity that leads both to the formation of more mature germ cells and to the renewal of stem cells. In fact, the majority of dividing A_4 spermatogonia give rise to the *intermediate* (In) and *B spermatogonia*, while only a few of them return to type A_1, which initiates a new cycle. A similar model of stem cell renewal has been proposed for both the monkey and man (6,7). In these species, however, only two classes of type A spermatogonia have been identified: in monkey, A_1 and A_2, the former considered as the reserve stem cell and the latter as the renewing stem cell, and in man the *dark* and the *pale* spermatogonia, the former considered as stem cell. In contrast, Huckins (4) and Oakberg (5) indicated that type A_0 spermatogonia incorporate ^{3}H-labeled thymidine and designated this type of spermatogonia A_s or *A stem cells* (Fig. 1-4). A_s spermatogonia can divide to form additional A_s spermatogonia, which are morphologically identical to the parental A_s spermatogonia and remain scattered singly or as paired cells. These additional A_s spermatogonia, after successive divisions resulting in chains of cells, stop dividing, transform into A_1–A_4 spermatogonia, and begin an irreversible development sequence during which A_1 spermatogonia transform into preleptotene spermatocytes through a series of mitotic steps (A_2–A_3–A_4–In–B spermatogonia).

3.3. Meiotic or spermatocytic stage

The next step in spermatogenesis is the *meiotic* or *spermatocytic stage*, which is characterized by two meiotic divisions (Fig. 1-2a–f). The cells in the

←

Figure 1-2a–f. Spermatogonial mitotic division, primary spermatocytic prophase and I–II meiotic divisions as appears in air-dried isolated mouse germ cells after in vivo treatment with 0.01% vinblastine sulfide for 20 min. Figure 1-2a: spermatogonial metaphase; Figure 1-2b: leptotene; Figure 1-2c: zygotene; Figure 1-2d: middle pachytene; Figure 1-2e: diakinesis; Figure 1-2f: second meiotic division. Note the reduction of chromosomes from the diploid to the haploid number in Figures 1-2a and f, respectively, and the chromosomic loops and chiasmata in Figure 1-2d. Arrows point to sex vesicles and sex chromosomes. X 1,500.

Figure 1-2g. Monkey testis (*Macaca mulatta*). Stage IV of the cycle of the seminiferous epithelium which, in monkeys, is divided in 12 stages. As opposed to Amphibia, mammalian spermatogenesis develops in concentric clones. Note how the germ cells appear interconnected by intercellular bridges (arrows). In this monkey stage, three types of differentiating spermatids, i.e. two generations of round spermatids — one at Golgi phase (G) with the acrosomic granule within the idiosome (arrowhead) and the other at cap phase (C) — appear associated with spermatids at maturation phase (M). Note the dendritic cytoplasmic ramifications of the Sertoli cell (S) surrounding spermatocytes and spermatids. Z: zygotene spermatocyte; P: pachytene spermatocytes; curved arrow: sex vesicle; asterisks: sertolian lipid bodies; BM: basement membrane; My: myoid cells. X 3,000.

Figure 1-2h. Testis of Beagle dog. The cycle of the seminiferous epithelium in this mammal is divided into eight stages. The stage shown in the figure is characterized by two concentric layers of spermatocytes at different prophasic periods, i.e. zygotene (Z) and pachytene (P), and one generation of spermatids at acrosomic phase (asterisks). S: Sertoli cells; arrow: sertolian lipid droplets. X 800.

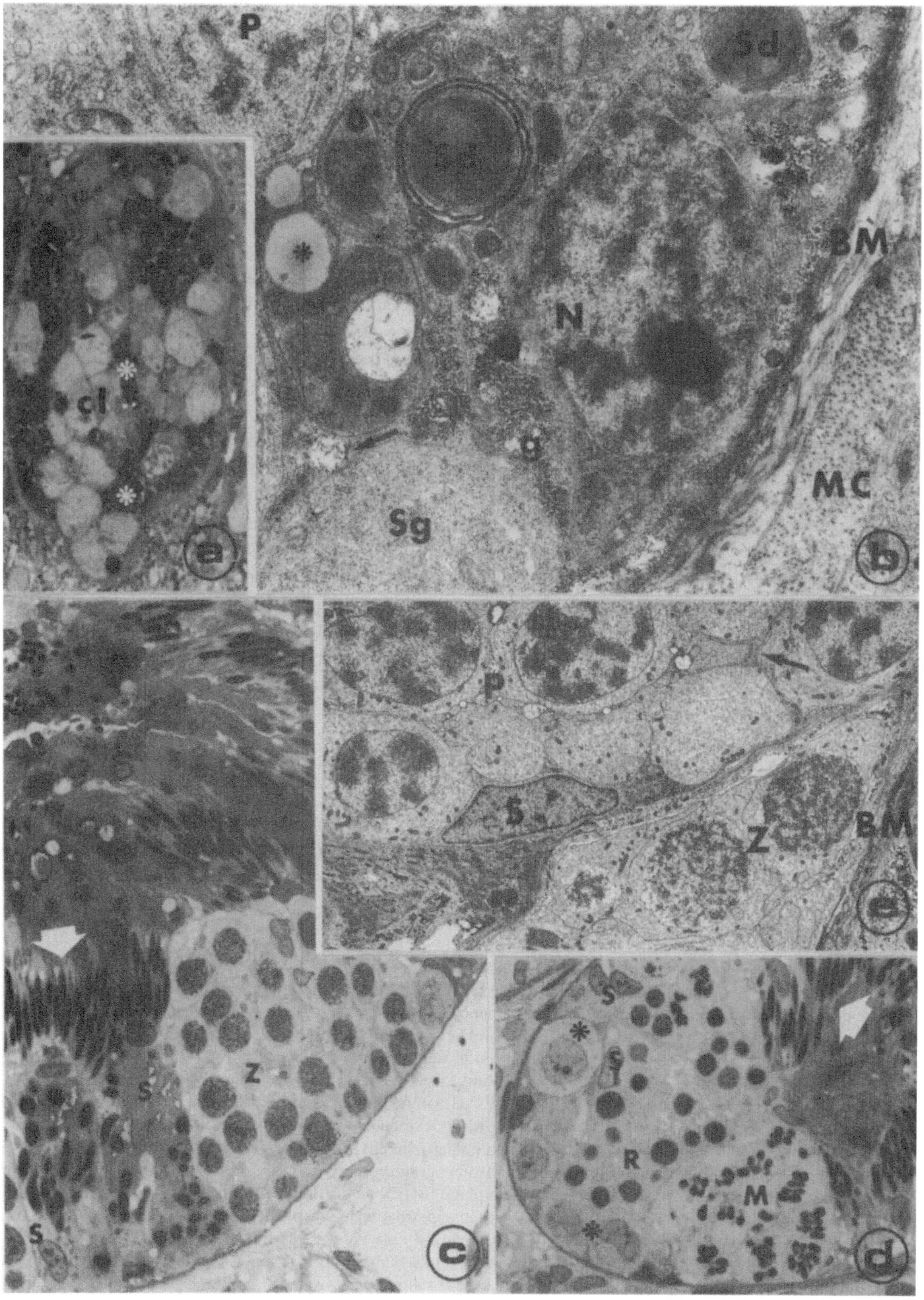
P
Sd
BM
N
g
Sg
MC
cl
a
b
P
S
Z
BM
e
S
Z
c
S
R
M
d

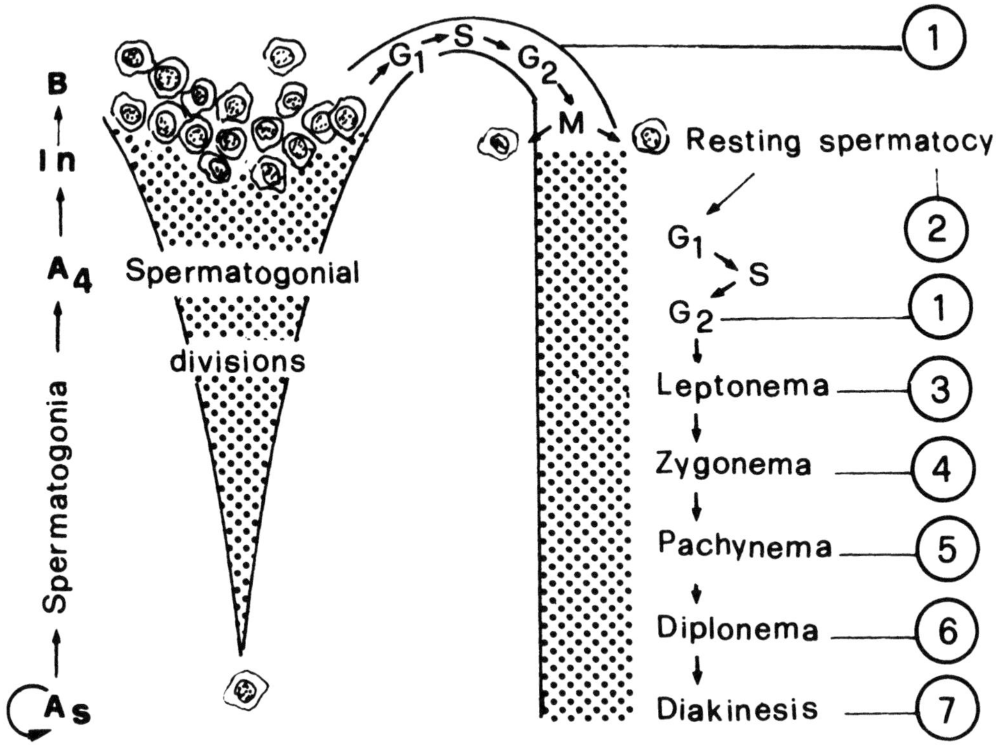

Figure 1-4. Composite diagram presenting the model of spermatogonial renewal according to Huckins (4) and Oakberg (5) on the right, and the prophase stages of the first meiotic division on the left.

←

Figure 1-3a–b. Testis of Platylhelminth *Schistosoma mansoni.* Figure 1-3a illustrates one of the seven aligned cranial testes typical of this species, showing irregular clumps of synchronous aggregation of germ cells which are separated by basal and central sustentacular cells (asterisks). Figure 1-3b illustrates the ultrastructural features of a testicular basal portion, showing a random maturation of germ cells with spermatids (Sd) close to the basement membrane (BM). The sustentacular cells project narrow cytoplasmic processes (arrow) between the developing germ cells; their basal cytoplasm contains abundant glycogen (g) and lipid droplets (asterisk), while the large nucleus (N) shows spots of heterochromatin. The tubule wall is formed by a basal lamina that is surrounded by fibrillar material and muscle cells (MC) provided with both thick and thin filaments. P: pachytene spermatocytes, Sg: spermatogonium. Figure 1-3a: X 800; Figure 1-3b: X 4,000.

Figure 1-3c–e. Testis of anuran *Rana esculenta.* This frog's spermatogenesis occurs in casts or cysts, separated from each other by cyst or Sertolilike cells that run along the periphery of the cast. Each cyst contains one clone of germ cells, i.e. single or paired spermatogonia and numerous primary or secondary spermatocytes. No cytoplasmic projections of supporting cells penetrate deeply into the cast (arrow in Figure 1-3e). However, during spermiogenesis the clonal germ cells, as round spermatids, are in close morphological relationship with the basal cytoplasm of the supporting cells, and as elongated spermatids or spermatozoa, with the apical cytoplasm. S: supporting cells; asterisks: spermatogonia; Z: zygotene spermatocytes; M: meiotic spermatocytes; P: pachytene spermatocytes; R: round spermatids; white arrows: elongated spermatids and spermatozoa; BM: basement membrane. Figures 1-3c and d: X 600; Figure 1-3e: X 3,000.

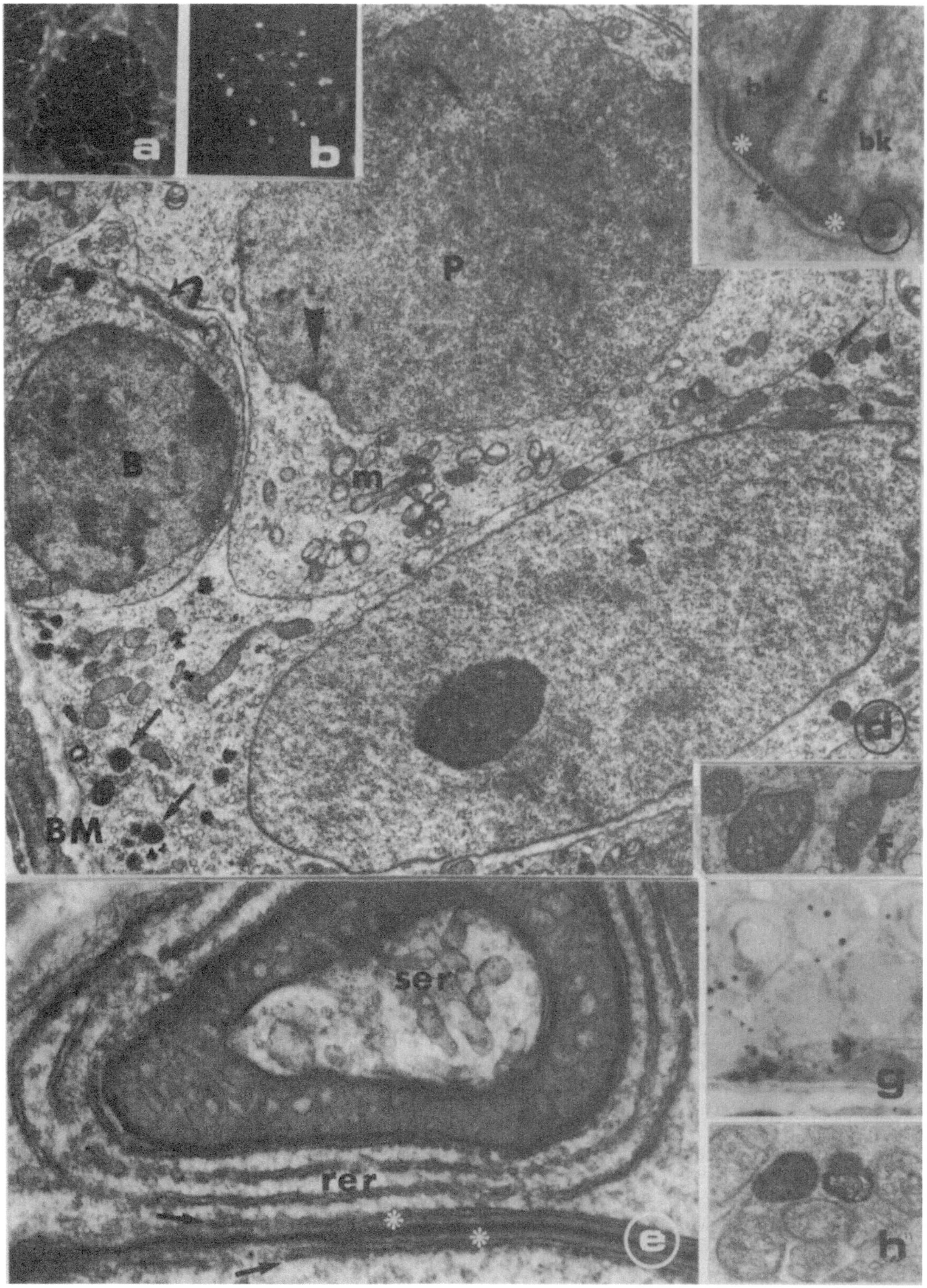
a
b
bk
c
bk
c
P
B
m
S
d
BM
f
ser
g
rer
e
h

first meiosis, called *primary spermatocytes* or *spermatocytes I*, are characterized by a short interphase period (resting or preleptotene spermatocytes) followed by a long, elaborate prophase (Fig. 1-4). After division, these cells give rise to the *secondary spermatocytes* or *spermatocytes II*, which in turn rapidly undergo a second meiotic division without a proper interphase. Thus, the final step of spermatogenesis begins. This stage, called spermiogenesis, indicates spermatid maturation, which, through important morphological modifications, gives rise to the testicular spermatozoa. These three stages of spermatogenesis are usually found in all mammals, including man, even though the spermatogonial population and the cycle of the seminiferous epithelium may differ among mammalian species.

The essential event of the meiotic stage is the reduction of the diploid set of chromosomes to the haploid complement. Morphological changes, chiasmata formation and segregation of homologous chromosomes occur in the primary spermatocytes during their elaborate prophase, which is divided into the following stages (Figs. 1-2a–f and 1-4):

Preleptotene (or *preleptonema*). This stage is difficult to identify. The cells presumably involved, also called *resting spermatocytes*, cannot be distinguished from type B spermatogonia (Fig. 1-5d) due to their similar nuclear characteristics, i.e., a spherical dark nucleus with coarse chromatin spots along the nuclear membrane, except when thin threads of chromatin are seen.

Leptotene (or *leptonema*). The leptotene nucleus is characterized by condensation of chromatin, which forms threadlike patches (Figs. 1-2b and 9b). The chromosome number is still diploid.

Zygotene (or *zygonema*). The appearance of the *synaptonemal complex* (SC) and *sex vesicle* (SV), or sex pair, characterizes this stage. The homologous chromosomes become intimately paired and together form the *chiasmata* (Fig. 1-2c).

Pachytene (or *pachynema*). Transitional or early pachytene-stage nuclei contain long stretches of SC associated with densely packed chromatin fibrils. The fully developed SC, visible during the mid-pachytene stage, is a tripartite structure composed of two lateral elements and a central one (8), and is present all along the synapsed homologous chromosomes with its two ends attached to the inner nuclear membrane by basal knobs (Figs. 1-5c and d). Immunohistochemical tests showed that myosin and actin are present along the central element, while the lateral elements seem to be anchored by actin spots to the basal knobs (Figs. 1-5a and b) (9,10). The presence of actin and myosin in SC seems to be related to chromosome pairing, to the contraction of the protein scaffolding of the paired chromosomes, and to the packing of the DNA-histone loops (Fig. 1-2d) during the meiotic prophase. The paired chromosomes of the pachytene stage twist around each other like threads and correspond to the haploid chromosome number (Fig. 1-2d). During the middle and latter parts of the pachytene stages, the chromatin becomes less

←

Figure 1-5a–c. Mouse testis, Meiotic phase. Myosin filaments and actin spots, detected by immunofluorescence and specific antibodies (Figures 1-5a and b, respectively), are visible inside the nuclei of pachytene spermatocytes. Figure 1-5c illustrates the fine structure of a synaptonemal complex with the central area and basal knobs (bk). The white asterisk points to the inner nuclear membrane and the black asterisk to the outer nuclear membrane. Figures 1-5a and b: X 400; Figure 1-5c: X 60,000.

Figure 1-5d–h. Rat testis. Parts of the basal area of the seminiferous epithelium at stage VI. The nucleus of a Sertoli cell (S) is perpendicular to the basement membrane (BM) and the cytoplasm is completely devoid of large lipid bodies, which are apparently replaced by small osmiophilic granules (arrows in Figure 1-5d) positive for Sudan black stain (Figure 1-5g) and DAB reaction (Figure 1-5h). Note in Figure 1-5d the difference in size between a pachytene spermatocyte (P) and a B spermatogonium (B), the latter laying on the basal lamina and beneath the sertolian junctional complex (curved arrow). The pachytene spermatocyte is characterized by synaptonemal complexes (arrowhead) and dispersed chromatin within the nucleus, and by numerous aggregates of partially condensed mitochondria (m) in the cytoplasm. The sertolian mitochondria differ in shape and size. They can be ovoid, elongated or cup-shaped (Figure 1-5e) with tubular or lamellar cristae (Figures 1-5d, e and f). The cup-shaped mitochondria are surrounded by cisternae of rough endoplasmic reticulum (rer) and often contain within the cup smooth endoplasmic reticulum (ser) (Figure 1-5e) and/or lipids. The junctional complex (Figure 1-5e) between Sertoli cells is composed of closely apposed plasma membranes with intercalated tight junctions, of subjacent cortical cytoplasmic filaments (asterisk) and of parallel cisternae of endoplasmic reticulum (arrows). Figure 1-5d: X 6,000; Figure 1-5e: X 38,000; Figure 1-5f: X 15,000; Figure 1-5g: X 600; Figure 1-5h: X 25,000.

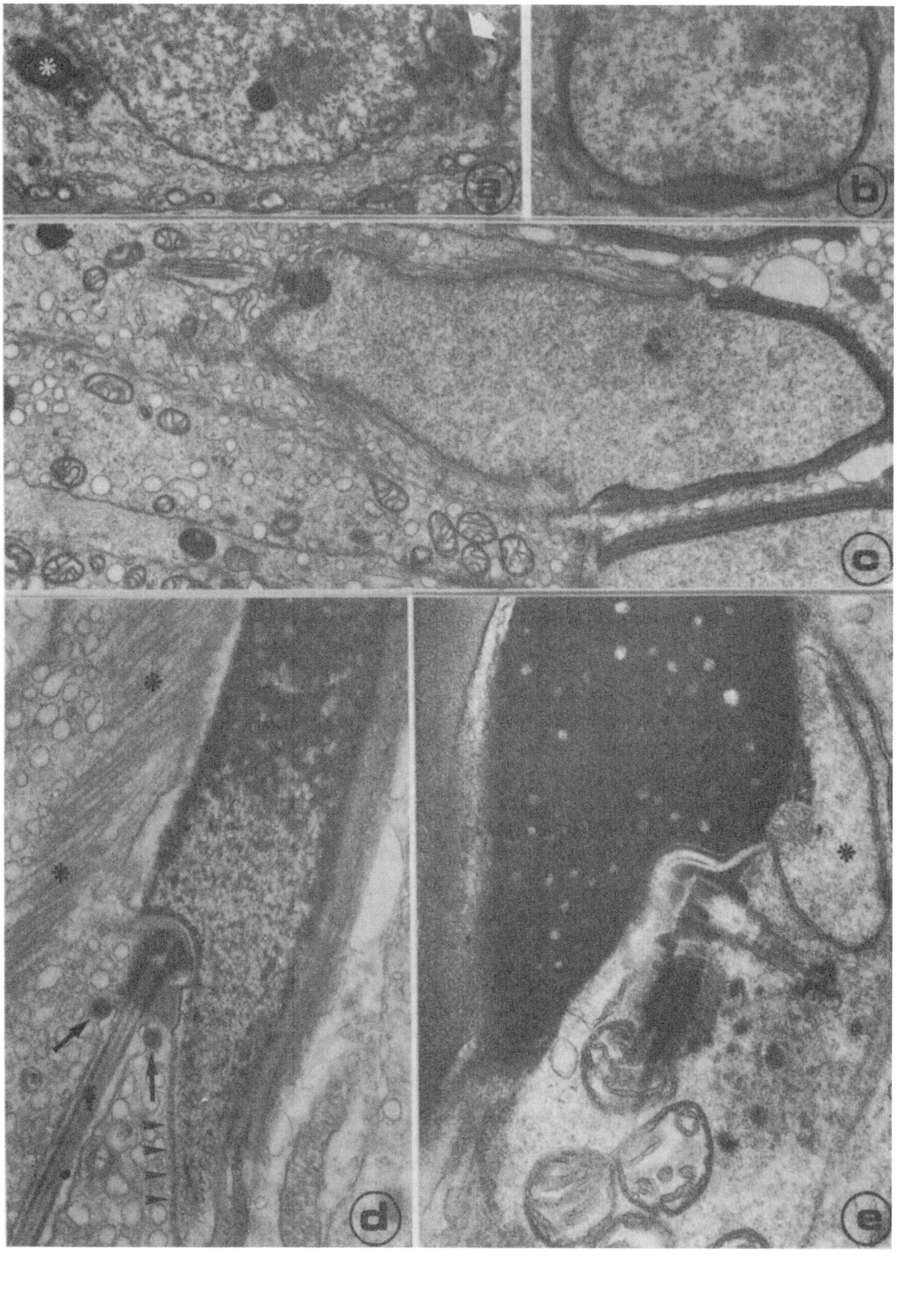
a
b
c
d
e

packed (Figs. 1-2g and 1-5d), while the nucleoli and SV appear more prominent (Fig. 1-2g). The chromosome splits lengthwise to form two chromatids, breakages occur along their length, and portions from different members of the chromosome pairs become joined. This is the process of *crossing over*, by which hereditary material is redistributed among the four chromatids adherent to each other in the chiasmata. The complex unit of four chromatids, representing two homologous chromosomes, is called a *bivalent*, and it forms loops (Fig. 1-2d) at the side of each lateral element of the SC. It is assumed that only the linear segment (chiasmata) of each homologue associated with the SC is involved in crossing over.

Diplotene (or *diplonema*). This stage is marked by the virtual disappearance of the SC and SV and by the development of a round and compact nucleolus. The slips between paired chromosomes and also those between chromatids become visible, while the bivalents still remain adherent because of the chiasmata. As the diplotene stage progresses, the chromatids contract and thicken and are ready for the first meiotic division at the stage of *diakinesis* (Fig. 1-2e), which in turn is followed by metaphase, anaphase and telophase, which are identical to those observed in the mitotic division of somatic cells.

The *second meiotic division* does not differ from a mitotic division, as just mentioned, even though the secondary spermatocytes are haploid cells. In this phase the chromatids of each chromosome separate and, as a result, each of the four cells originating from one primary spermatocyte possesses only one member of each pair of chromosomes present in the spermatogonia (Figs. 1-2a and f). Thus, the haploid condition is achieved.

Besides these main modifications, which allow the development of haploid cells from diploid ones, other important changes occur in primary spermatocytes during the stages of the first meiotic prophase. The most conspicuous are as follows: 1) An increase in volume to many times that of B spermatogonia (Fig. 1-5d). 2) Progressive modifications of mitochondria from the orthodox to the condensed form (Figs. 1-5d and 1-9b) (see as review, 11). The number of these organelles increases and appears to form aggregates which are associated with electron-dense material (Fig. 1-5d). In the rat, these aggregates appear in leptotene spermatocytes, reach maximum development in the pachytene stage (Fig. 1-5d), and disappear in the diplotene stage (Fig. 1-9b) (11); the function of the aggregates is unknown. 3) An increase in RNA synthesis during the pachytene stage with a subsequent, drastic decrease at the diplotene stage, during which phase the nucleolus no longer incorporates ^{3}H uridine (12). 4) By contrast, DNA needed for the meiotic chromosomes is replicated during premeiotic interphase and continues to be synthesized in leptotene and lepto-zygotene spermatocytes. A portion of the nonhistone chromosomal proteins is also synthesized in the zygotene stage (see as review, 8).

3.4. Spermiogenesis or spermatid stage

After meiosis, the last step of spermatogenesis, called *spermiogenesis*, begins. During this period the haploid gamete, or *spermatid*, no longer divides. According to Leblond and Clermont (13), spermiogenesis in the rat is divided into four main phases that are quite similar in all mammals, including man.

Golgi phase. In this phase, the spermatid is characterized by a prominent Golgi apparatus that develops a large spherical vacuole (*idiosome*) in the center, and by a basophilic mass (*chromatoid body*) close to the centrioles (Fig. 1-6a). From one centriole, a thin filament emerges that

←

Figure 1-6a–e. Rat spermatogenesis. Figure 1-6a: spermatid at Golgi phase. The arrow points to the acrosomic granules within the idiosome while the asterisk indicates the chromatoid body. Figure 1-6b: spermatid at cap phase. Figure 1-6c: spermatid at early acrosomic phase. Figure 1-6d: spermatid at late acrosomic phase. Note the anulus (arrows) localized below the centrioles at the beginning of the flagellum (f). Arrowheads point to the basal nuclear pores while asterisks indicate the manchette. Figure 1-6e: spermatids at maturation phase. Two perpendicular centrioles are clearly visible and the mitochondria are condensed and localized close to the flagellum. The asterisk indicates the redundant nuclear membranes. Note the progressive condensation of chromatin and the elongation of nuclei in Figures 1-6c–e. Figures 1-6a and b: X 7,000; Figure 1-6c: X 12,000; Figure 1-6d: X 18,000; Figure 1-6e: X 30,000.

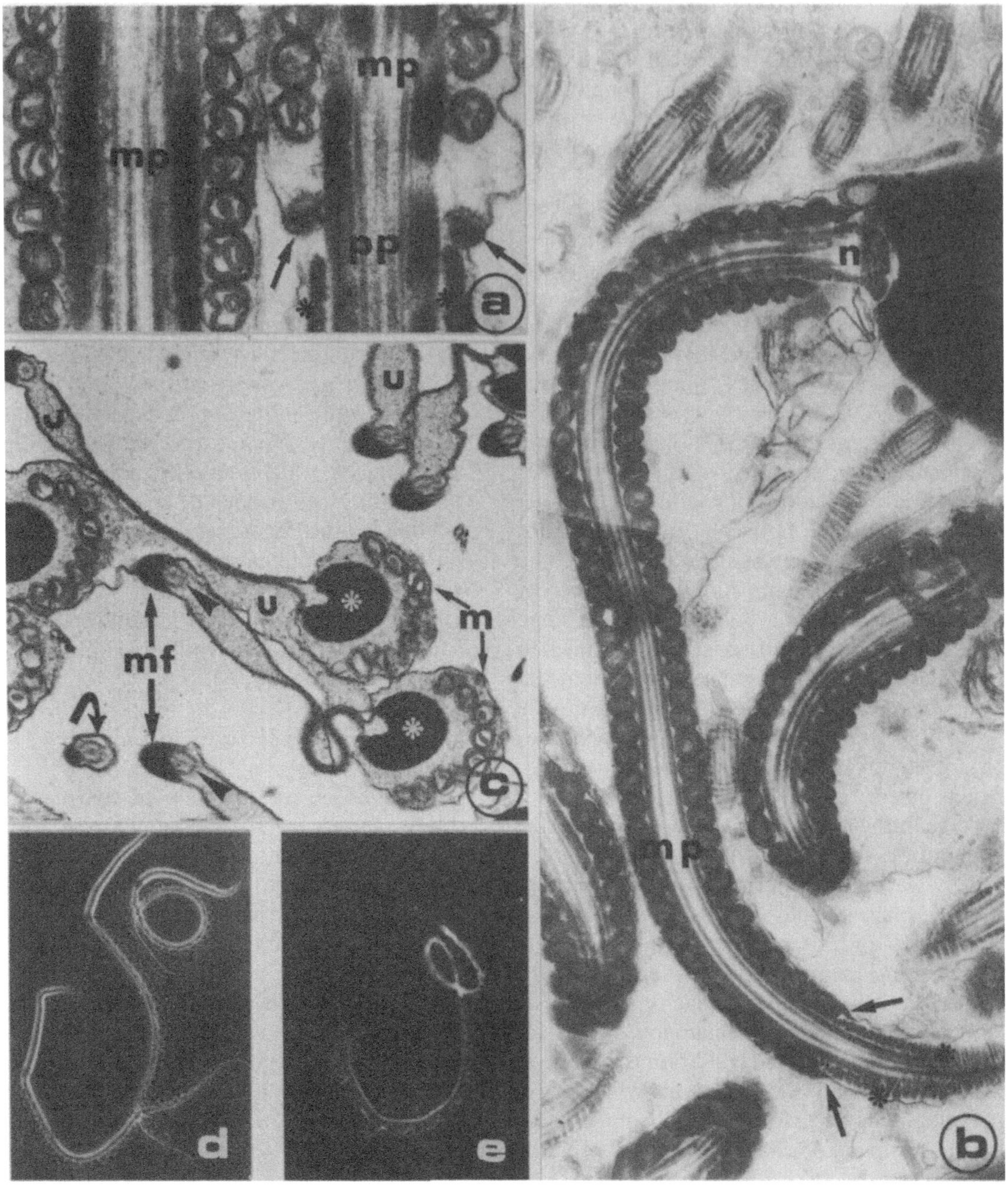

Figure 1-7a–b. Rat testicular spermatozoa (Figure 1-7a) and monkey epididymal spermatozoa (Figure 1-7b). In both mammals these germ cells are characterized by the full development of the middle piece (mp) and its related mitochondrial sheath. Parts of the principal piece (pp) of the tail and its related fibrous sheath (asterisks) are also visible. The anulus (arrows) is localized at the junction between middle and principal pieces. n: neck. Figure 1-7a: X 20,000; Figure 1-7b: X 15,000.

Figure 1-7c–e. Spermatozoa of urodele *Triturus cristatus*. In Figure 1-7c the cross section of the middle piece shows the cup-shaped profile of the axial fiber (asterisks), which is surrounded by a sheath of cytoplasm rich in condensed mitochondria (m). The undulating membrane (u) arises from the cytoplasm in relationship with the cupped zone of the axial fiber and shows the axoneme (arrowheads) and marginal fiber (mf) within its free edge. Curved arrow points to the end piece of a tail, which is composed only by the axoneme and marginal fiber. Figures 1-7d and e illustrate whole spermatozoa by phase-contrast microscopy and after the indirect immunofluorescent test using rabbit antiserum specific for myosin, respectively. It is evident that myosin is in relation with the axial and marginal fibers from the middle piece to the tail. Figure 1-7c: X 18,000; Figures 1-7d and e: X 300.

represents the terminus of the differentiating flagellum. Subsequently, by fusion of granules, a single, large granule, or *acrosomic granule*, appears within the idiosome (Fig. 1-6a). Consequently, according to the gradual development of these structures, the Golgi phase can itself be subdivided into three progressive stages.

Cap phase. This phase consists of four stages during which the acrosomal granule and its related idiosomic membrane flatten to form a randomly oriented cuplike structure over the round nucleus of the spermatid (Figs. 1-2g and 1-6b). The tail filament and its related centriole move toward the opposite side of the nucleus from the acrosomal cap. The spermatids retain the round cytoplasmic shape so that, together with those in the previous phase, they are often termed *round spermatids*.

Acrosomal phase. This spermatid maturation phase can be further divided into seven stages characterized by the gradual orientation of the head cap toward the basal portion of the seminiferous tubule, followed by the progressive differentiation of the manchette, elongation of the nucleus, and condensation of the chromatin. The cytoplasm translocates to the nuclear pole, opposite the acrosome, which now appears horseshoeshaped, giving the spermatid an elongated appearance (*elongated spermatids*) (Figs. 1-6c and d).

Maturation phase. This phase consists of five progressive stages. In the early stages, the elongated nucleus of the spermatid darkens due to maximum chromatin condensation, while clear, randomly distributed holes appear within (Fig. 1-5e). The flagellum gradually reaches full differentiation and different segments can be recognized: the neck (Fig. 1-7b), the middle piece (Figs. 1-7a and b), and the tail, which in turn is divided into principal (Figs. 1-7a and b) and end segments (for ultrastructural details see 14). Simultaneously, some mitochondria move toward the middle piece (Fig. 1-6e), forming a circumferential sheath (Figs. 1-7a and b), while the remaining clustered mitochondria, along with the lipid bodies and ribosomes, are eliminated from the cell in a *residual body*. The *immature* or *testicular spermatozoa* are then released into the tubular lumen, while the residual bodies are engulfed within the apical cytoplasm of the Sertoli cells and are digested by phagocytic activity in the basal cytoplasmic area (Fig. 1-8 VIII–X). However, the testicular spermatozoa are neither morphologically nor physiologically completely mature. Their maturation process continues during the voyage through the male accessory tracts, such as epididymis and ductus deferens, and this process perhaps continues in the female tract after coitus.

In contrast to spermatogonia and spermatocytes, which usually show only minor and often subtle differences throughout phylogenesis, the spermatids usually undergo spectacular development with species-specific features (14,15). The details of spermatid differentiation vary in terms of nuclear shape, type of acrosome, number of mitochondria, and specific flagellar structure that in some instances (urodele amphibians) is provided with undulating membranes (Figs. 1-7c, d, and e).

4. Somatic cells or Sertoli cells

Along the phylogenetic scale, the somatic cells are referred to as supporting, nutritive, sustentacular (Figs. 1-3a and b), nurse, cyst (Figs. 1-3c and e), and Sertoli cells (Figs. 1-2g and 1-5d). These cells seem to be present in all animals except the Porifera (1). However, their morphological characteristics, as described for Amniota, differ markedly in the lower animals (see as review, 1).

The fetal and postnatal development of Sertoli cells, like the development of germ cells, can be divided into three main stages. The first stage corresponds to the extragonadal period of germ cells, while the second one pertains to fetal and postnatal testicular development, during which stage the Sertoli cells are the predominant cell types in the sex cords. Once spermatogenesis enters the third stage, single Sertoli cells become associated with groups or *clones* of differentiating germ cells (Figs. 1-2g and 1-5d). The germ cells are separated by the sertolian *junctional complexes* (Fig. 1-5e) in a *basal compartment*, containing spermatogonia and preleptotene spermatocytes, and an *adluminal compartment*, containing the more advanced spermatocytes and spermatids (see as review, 16, 17). The junctional complexes ap-

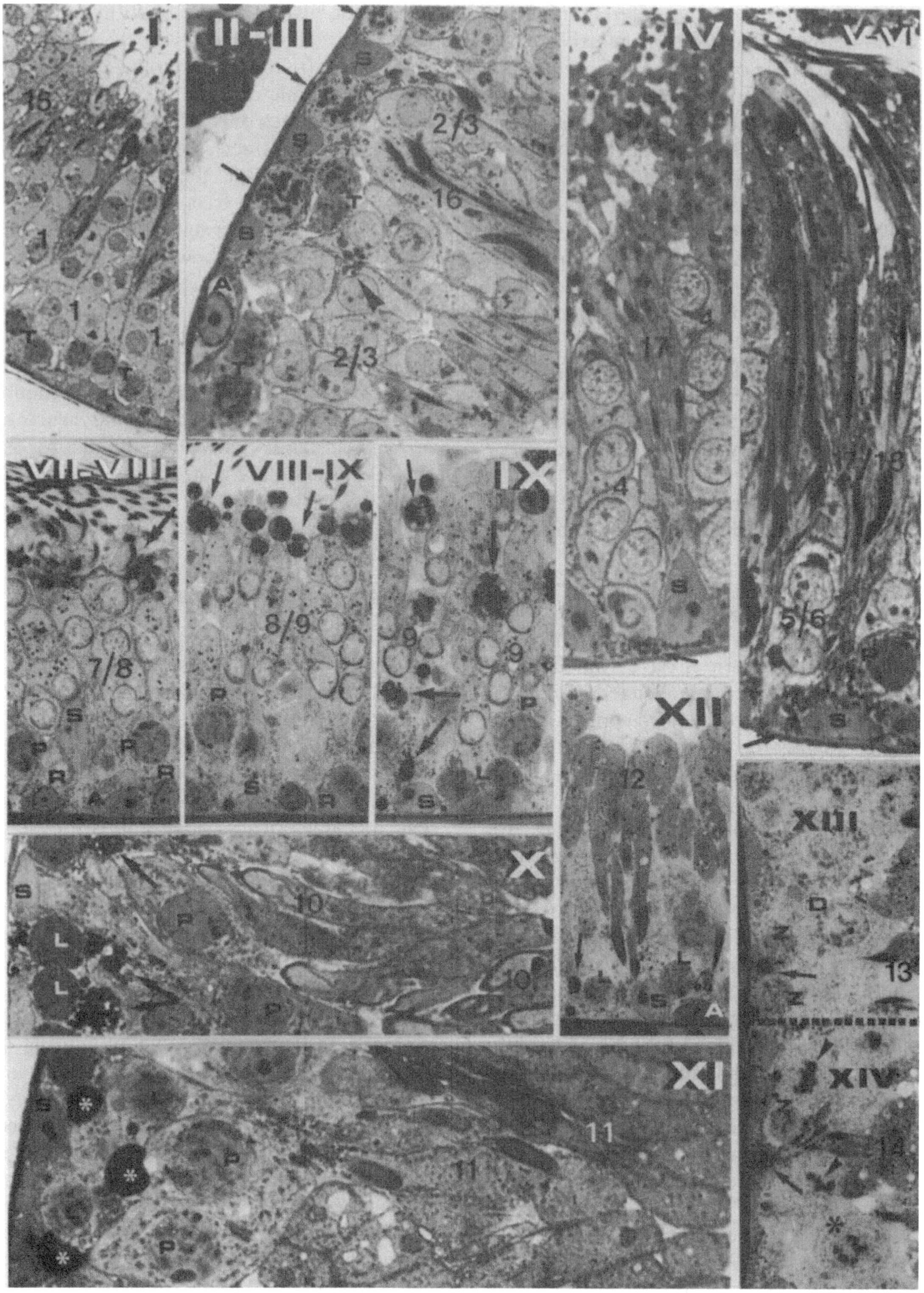
I
II-III
IV
V-VI
VII-VIII
VIII-IX
IX
X
XI
XII
XIII
XIV
2/3
16
17
4
18
5/6
7/8
8/9
9
12
10
11
13

pear to regulate the passage of substances into the adluminal compartment through the basal cytoplasm of Sertoli cells. These mature Sertoli cells are characterized by an invaginated nucleus with a tripartite nucleolus, numerous cytoplasmic processes in the adluminal compartment surrounding the spermatocyte and spermatids (Fig. 1-2g), numerous round and/or elongated cup-shaped mitochondria provided with lamellar and tubular cristae (Figs. 1-5f and e, 1-9a), varying amounts of rough and smooth endoplasmic reticulum, a well-developed Golgi apparatus, numerous lysosomes, peroxisomes (Fig. 1-5h) and cytoplasmic inclusions such as lipid bodies of variable size (Figs. 1-2g,1-5g,1-9a and b), cristalloids (Charcot–Böttcher crystals in man), and lipofuscin granules (Fig. 1-9a). It is also certain that mature Sertoli cells possess phagocytic activity, as demonstrated by resorption and disposal of residual bodies and bacteria (18).

5. Cycle and wave of the seminiferous epithelium (Fig. 1-8 I–XIV)

The *cycle of the seminiferous epithelium*, or simply *cycle*, refers to the morphological changes that occur over time in any given area of the seminiferous epithelium. The cycle can be divided into various stages according to typical cellular associations (13) whose spatial sequence along the tubule gives rise to the *wave of the seminiferous epithelium* (19).

The morphological parameters proposed by Leblond and Clermont (13) for spermatogenesis in the rat are usually followed in order to determine the cycle of the seminiferous epithelium in mammals. In the rat, which shows clear cellular associations between different types of spermatids and spermatocytes, 14 stages in the cycle have been described. Since spermiogenesis in this animal has been divided in 19 steps, some stages of the cycle are characterized by two generations of spermatids. In fact, during the first seven stages of the cycle, the seminiferous epithelium contains young spermatids at Golgi and cap phases (round spermatids), as well as older spermatids at the maturation phase (elongated spermatids). In stage VIII, immature or testicular spermatozoa are released into the tubular lumen (spermiation). Between stage IX and XIV, only one generation of spermatids at the acrosome phase is present. Stage IX is also characterized by phagocytosis of residual bodies, while stage XIV is defined by the first and second meiotic divisions. Thus, a new generation of spermatids appears once again at stage I of the next cycle. Furthermore, different associations between spermatogonia, primary and secondary spermatocytes, and spermatids characterize each stage of the cycle (for details see 13). The entire process of spermatogenesis from A spermatogonia to testicular spermatozoa requires four cycles.

The duration of spermatogenesis, as well as the number of cycles of the seminiferous epithelium, varies according to the mammal in question, and the time lengths are generally not related to the

←

Figure 1-8. Typical cellular associations present in the 14 stages of the spermatogenic cycle of the rat, named from the basal to the apical zone of the tubule, according to the classification of Leblond and Clermont (13). Stage I: transitional spermatocytes (T); spermatids 1 and 15. Stage II–III. A spermatogonia (A), several in mitosis (arrows) intercalated with Sertoli cells (S); transitional spermatocytes (T); spermatids 2/3 and 16; acrosomic granule within an idiosome (arrowhead). Stage IV: perpendicular Sertoli cell (S); spermatids 4 and 17; small lipid droplets (arrow). Stage V–VI: Sertoli cell (S); pachytene spermatocyte (P); spermatids 5/6 and 17/18; small lipid droplets (arrow). Stage VII–VIII: perpendicular Sertoli cell (S); A spermatogonium (A); resting (R) and pachytene (P) spermatocytes; spermatids 7/8; release of spermatids 19 or immature spermatozoa (arrows) into the tubular lumen (spermiation). Stage VIII–IX: Sertoli cell (S); resting (R) and pachytene (P) spermatocytes; spermatids 8/9; apical residual bodies (arrows). Stage IX: Sertoli cell (S); leptotene (L) and pachytene (P) spermatocytes; spermatids 9; translocation of the apical residual bodies toward the mid- and basal regions of Sertoli cells (arrows). Stage X: Sertoli cell (S); leptotene (L) and pachytene (P) spermatocytes; spermatids 10; basal residual bodies (arrows). Stage XI: Sertoli cell (S); leptotene (L) and pachytene (P) spermatocytes; spermatids 11; large lipid droplets (asterisks). Stage XII: Sertoli cell (S); A spermatogonia (A); leptotene spermatocytes (L); spermatids 12; lipid droplet (arrows). Stage XIII: zygotene (Z) and diplotene-diakinesis (D) spermatocytes; spermatids 13; large lipid droplet (arrow). Stage XIV: I meiotic division (arrowhead: metaphase, asterisk: early anaphase); spermatids 14; large lipid droplet (arrow). Note that A and B spermatogonia are not always depicted because of the small tubular sections illustrated in the pictures. However, it should be noted that A spermatogonia are constantly present throughout the cycle, while the presence of B spermatogonia is restricted to stages IV–VI.

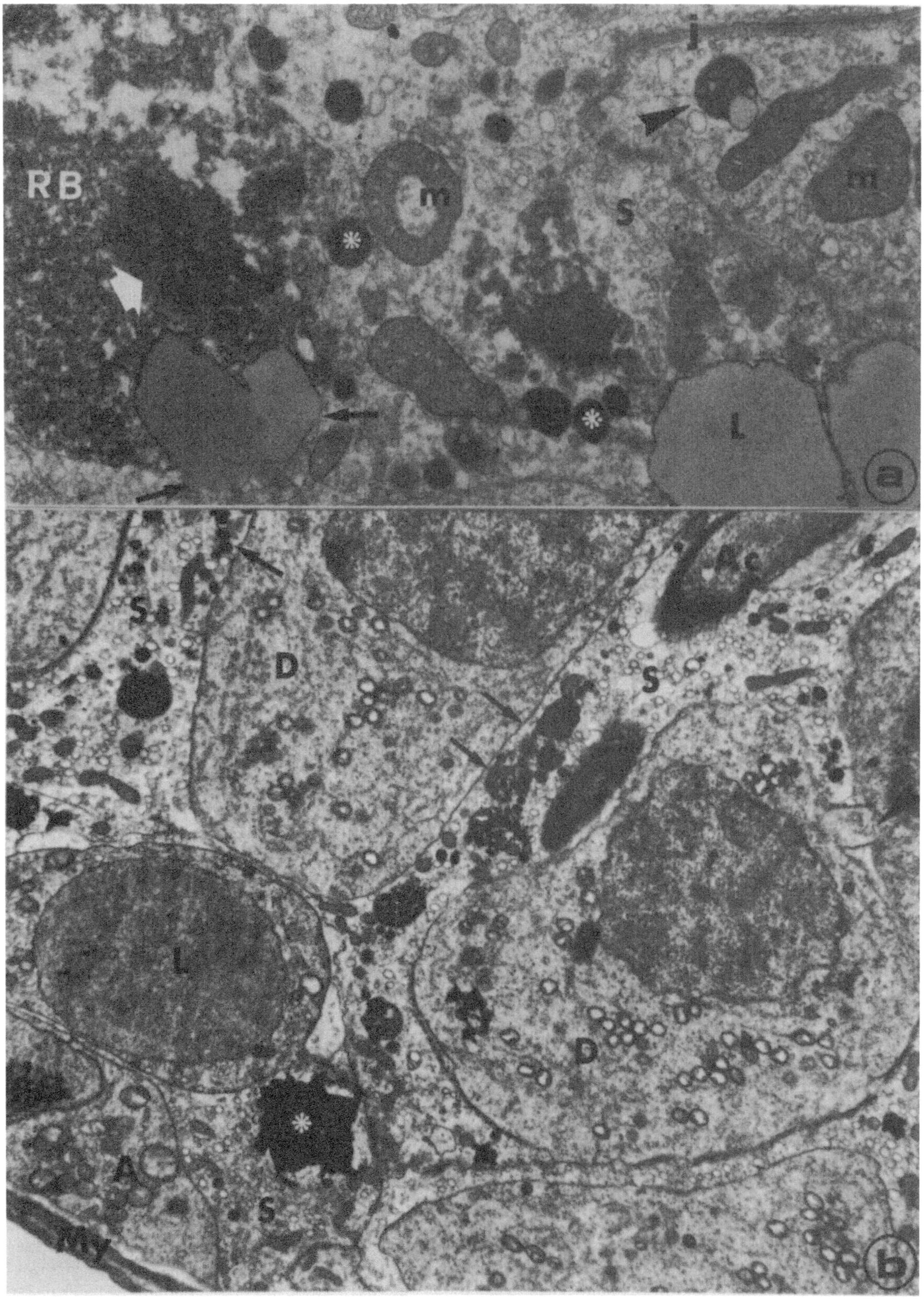
RB
m
S
m
L
a
S
D
Ac
S
L
D
A
S
My
b

Table 1-1. Duration and number of stages of spermatogenesis in different mammals.

	Mouse [20]	Rat [13]	Dog [21]	Monkey [6]	Man [26]
Spermatogenesis in days	34	48	13.6	45	64
Number of stages of the cycle	12	14	8	12	6

stage number identified (Table 1-1).

In the rat, the Sertoli cells show morphological changes during the stages of the cycle. In fact, besides the modifications in their number, shape, and nuclear orientation toward the tubular basement membrane (i.e., in parallel or perpendicular fashion, as reported by Leblond and Clermont, 13), other morphological changes may be observed soon after spermiation in some stages of the cycle (stages VII–VIII) that are related to spermiogenesis. These changes pertain to the position and digestion of the residual bodies, the amount and size of the lipid droplets, the shape of the mitochondria and their cristae, and the development of lipofuscin granules and of dense and/or granular bodies. It is worth noting that some of these bodies seem to contain acid phosphate (22), while others stain positively in the 3, 3′-diaminobenzidine reaction (Fig. 1-5h), thus indicating their lysosomal or peroxisomal nature. These morphological modifications can be grouped as follows: 1) Engulfed residual bodies, containing abundant lipids in the apical cytoplasm and scarce lipid droplets and dense granules in the basal cytoplasm during stage VIII to IX (Fig. 1-8). 2) Migration of the apical residual bodies toward and mid- and basal cytoplasm, and their digestion with subsequent release of lipids into the sertolian cytoplasm during stage IX to X (Figs. 1-8 and 1-9a). This phenomenon is accompanied by an increase in number of lysosomes, peroxisomes, lipofuscin granules and elongated or cup-shaped mitochondria with tubular cristae in the cytoplasm surrounding the residual bodies (Fig. 1-9a). 3) Appearance of large lipid droplets and dense granules in the basal and mid-cytoplasm, respectively, at the acrosomal phase during stage X to XIII (Figs. 1-8 and 1-9b). 4) Fragmentation and decrease in number of large lipid droplets in the sertolian cytoplasm and simultaneous appearance of lipid droplets in the spermatids at the maturation phase during stage II to VI; the spermatid lipid droplets gradually cluster and are then eliminated within the residual bodies during stage VII to VIII (Fig. 1-8). Collectively, these phenomena can be considered as a true lipid cycle (23) in the Sertoli cells. These cells re-elaborate lipids from digested residual bodies and subsequently recycle and transfer them to the spermatids, in which lipids reach the maximum amount during the maturation phase. In other mammals, such as dog and monkey, the lipid-residual body cycle of Sertoli cells is not so clearly evident, suggesting that for these animals, other pathways exist for the metabolism of lipids.

6. Human spermatogenesis

Man and the baboon are apparently the only mammals that do not show a regular and synchronized spermatogenetic process. However, six stages of spermatogenesis have been identified, and these constitute the cycle of the seminiferous epithelium in man (24,25). At variance with what is seen in other mammals, two or three different stages are visible in a single cross section of the human seminiferous tubule. This observation is due to the fact that each stage is formed by a very small number of germ cells. In addition, germ cells at inappropriate stages of development are often observed (Fig. 1-10). Although a spermato-

←

Figure 1-9a–b. Rat testis. Parts of the basal region of the seminiferous epithelium at stage X (Figure 1-9a) and stage XII (Figure 1-9b). Figure 1-9a shows a partially digested residual body (RB), containing disrupted mitochondria (white arrows) and releasing its lipid content (black arrows) into the cytoplasm of the Sertoli cell (S). Numerous dense (white asterisks) and lipofuscin (arrowheads) granules are scattered throughout the sertolian cytoplasm. L: free lipid droplets; J: junctional complex, m: Sertoli cell mitochondria; X 15,000. In Figure 1-9b, A indicates an A spermatogonium, S: the sertolian cytoplasms; L: the leptotene spermatocytes; D: a diplotene spermatocyte with round and condensed mitochondria; Ac: a spermatid at acrosomic phase; asterisk: a large lipid droplet; arrows: the dense granules; arrowhead: an intercellular bridge between diplotene spermatocytes; My the myoid cells. X 12,000.

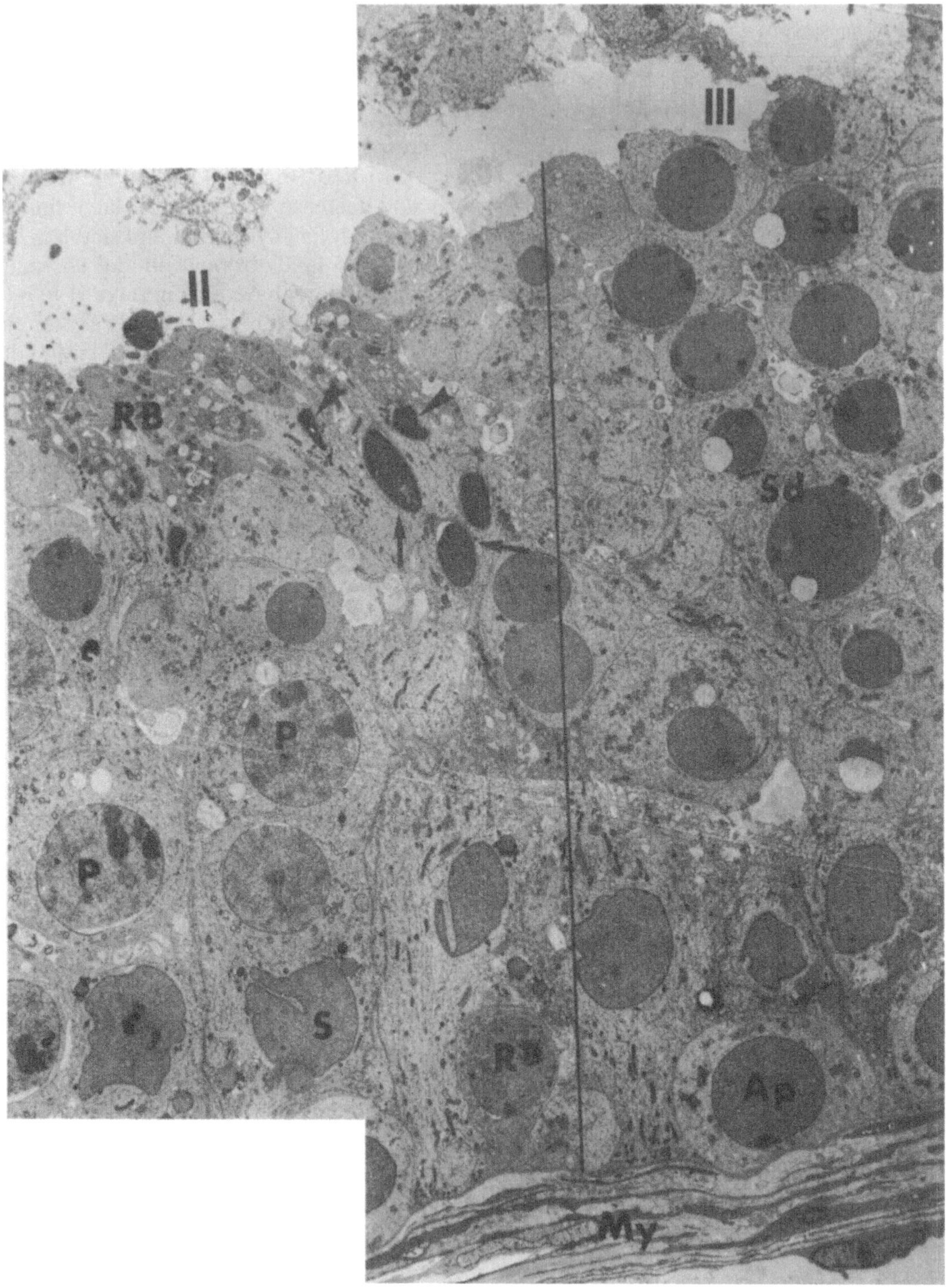

Figure 1-10. Segment of a human seminiferous tubule. On the right, a III stage of the seminiferous epithelium displays its typical organization; on the left, in the context of a II stage, inappropriate spermatids with an incomplete nuclear condensation (arrows) are intermingled with residual bodies (RB) and few mature spermatids (arrowheads). Ap: spermatogonia A pale; R: resting or preleptotene spermatocyte; P: pachytene spermatocytes; Sd: spermatids with the acrosomal vesicle; S: Sertoli cell; My: myoid cell of the tubule wall: X 1,800.

genic wave has not been thought to occur (25), such a process, which seems to be organized in a complex, yet ordered, helical pattern (26), has been identified, even in humans. Therefore, man does not constitute an exception with respect to the regular cyclical progression of spermatogenesis.

In mammals, cyclic changes are not restricted to the germinal eptihelium; in fact, they also involve the production of specific substances and the response to hormonal stimulation by the somatic supporting cells (27), which themselves show cyclic changes in lipids (see previous section) and glycogen content (28).

It has been suggested that the germinal stem cells, along the basal lamina of the seminiferous tubule, are committed to the spermatogenic process when the neighbouring germ cells in the tubule have reached certain defined stages of development: this results in a cycle of the seminiferous epithelium which provides a continuous supply of mature spermatozoa (13). The constant grouping of different germ cells, which form the stages of the cycle, is the result of the interaction between different events of gamete maturation. These interactions seem to be constant throughout the many species which exhibit various durations of spermatogenesis. For example, preleptotene spermatocytes, which constitute the starting point of meiosis, are constantly present at stages VII and VIII in the rat, and at stage II in man (Fig. 1-10), during which stage most mature spermatids are released from the seminiferous epithelium. The constancy of the interactions between different types of germ cells is influenced or orchestrated by interactions between germ cells and all somatic components of the testicular parenchyma rather than by a direct communication between gametes at various stages of development. In a recent review (29), the strict and bi-directional interactions between germ, Sertoli, peritubular myoid, and Leydig cells were discussed comprehensively. While most of the latter information was derived from experiments in rodents, the basic similarities between rodent and human spermatogenesis suggest that, even in man, an interaction of all components of testicular parenchyma that results in continuous sperm production is operative. A disruption of this interaction may cause decreased or absent sperm production, as well as the abnormal development of mature germ cells.

In the following sections, the morphological aspects of testicular physiopathology will be reviewed. For the sake of clarity, the data will deal sequentially with ejaculated sperm, the germinal epithelium, Sertoli cells, the seminiferous tubule wall, and the interstitium.

7. Pathology

7.1. Ejaculated sperm

The quality of human semen is widely variable in fertile men. However, variations of sperm morphology among men are considerably less frequent than differences in other seminal parameters such as sperm motility and density (30). For an individual, sperm morphology appears to be the most predictable parameter of fertility (31). The constancy of sperm shape over time for a given male is clearly demonstrated in men who display specific seminal abnormalities (32). What these observations suggest is that such individual has a characteristic pattern of sperm morphology (32) that is largely defined by the genotype of his spermatogenic cells (33). A careful evaluation of sperm morphology therefore provides some fairly precise information on the efficiency and normality of spermatogenesis.

The occurrence of a large number of immature and malformed ejaculated spermatozoa is certainly the result of an altered process of gamete development. The latter may sometimes involve almost the entire population of spermatozoa (Fig. 1-11a), and can be characterized by a granular aspect of the chromatin, associated with nuclear vacuoles and membranous bodies. In these cases sperm heads show abnormal shapes and are generally classified as amorphous or large (Figs. 1-11a and b), due either to an abnormal shape of the nucleus and/or to cytoplasmic remnants that surround the nucleus and tail (Figs. 1-11a, b and c).

Multinucleate cells often accompany the described malformations (Fig. 1-11a); their origin and meaning will be described in the following section (see Section 7.2). The simultaneous occurrence of immature and malformed, multiple-

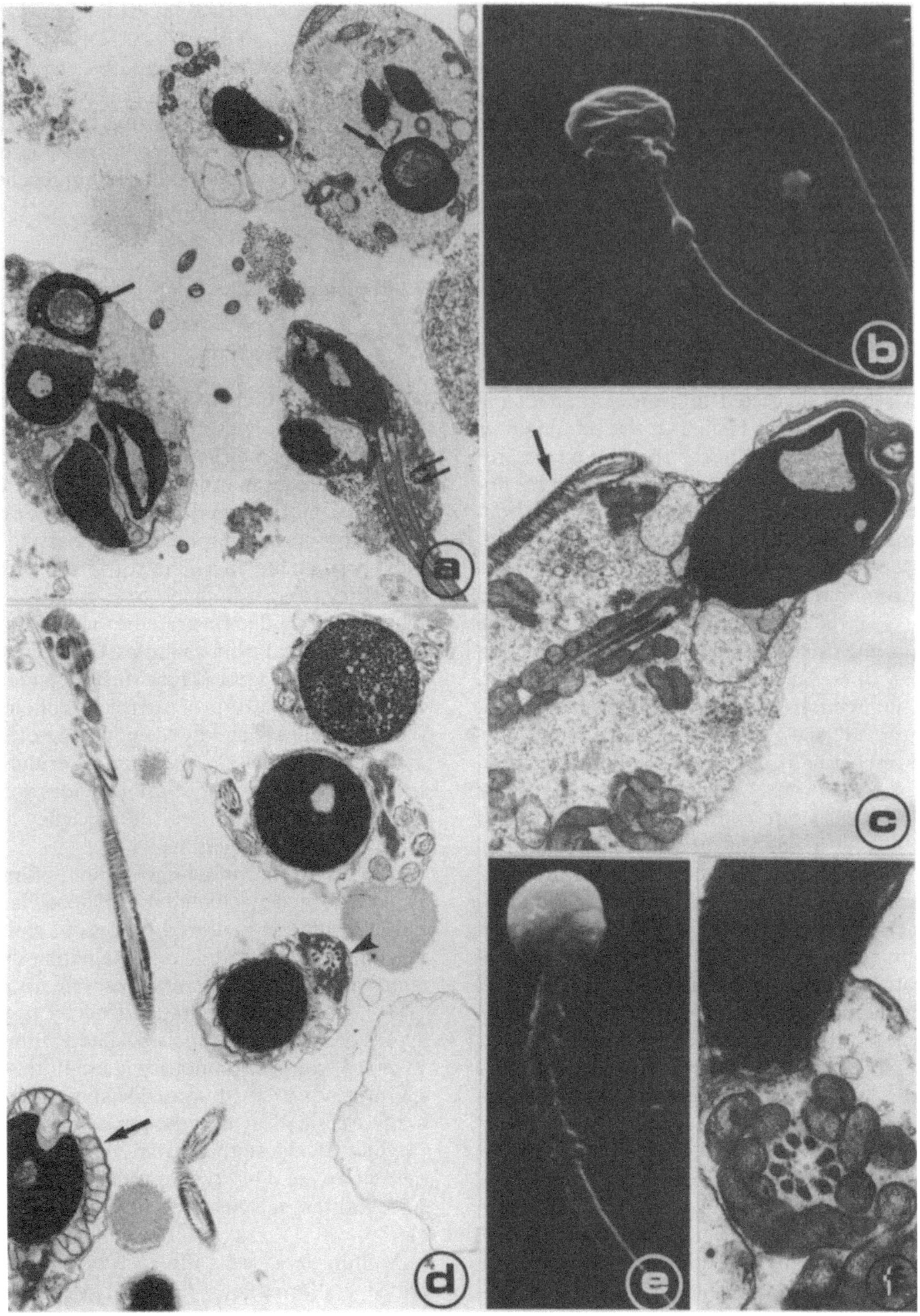
a
b
c
d
e
f

headed spermatozoa in the ejaculate is the main feature of the so-called stress pattern. This is a transient phenomenon in fertile men that occurs after an acute viral infection, and is associated with a depressed sperm count and decreased motility (32). This pattern of abnormal ejaculated sperm is considered to be a specific response to various types of stress (32).

Malformed heads are usually associated with altered acrosomes (Fig. 1-11c), which contain vesicles or membranous inclusions, or which appear extremely thin or invaginated into the nucleus. A condition of special interest concerns the absence of acrosomes in round-headed spermatozoa (Fig. 1-11d and e). This rare genetic disorder, which occasionally occurs in identical twin brothers (34), involves the entire population of sperm (35). Other morphogenetic features of sperm from men affected by this disorder include the presence of immature nuclei (36), large cytoplasmic remnants which contain coiled tails, and irregular, whorled membranes around the nucleus (Fig. 1-11d). It has been suggested that an early loss of the acrosome during spermiogenesis in patients with this defect (36) may be the consequence of missing spermatid proteins that are presumed to maintain the connection between the acrosomal vesicle and the nucleus (37).

It is interesting to note that a fairly constant association exists between the presence of abnormal heads and, as demonstrated by immunohistochemical staining, the absence of acrosomal components such as acrosin. Such an association is observed not only in the above-mentioned genetic disorder (34), but is also consistently detected in a randomized population of infertile men (38) that produce viable spermatozoa with any type of head abnormality. This finding suggests that the factors that control head shape and acrosomal organization and composition are intimately related. Sperm-head defects can occasionally be associated to an abnormal organization of the mid-piece (Fig. 1-11b) that consists of a derangement of the helical organization of the mitochrondria (39).

Tail abnormalities that involve the entire population of spermatozoa have been described infrequently but are associated with a total loss of sperm motility. These situations are apparently due to genetic defects that cause, in some cases, the lack of dynein (40) in both the sperm and the cilia of the respiratory tract. In other men, this condition originates from the lack of the two central tubules of the axoneme (41). While these particular disorders occur rather infrequently, malformations such as partial disorganization of the axoneme (Fig. 1-11f) or the occurrence of double tails (Fig. 11a) are encountered quite often in men presenting with asthenozoospermia (39).

A conclusion that all available data support is that sperm abnormalities are primarily the result of genetic defects of germinal cells that, at least in some murine strains, seem to be linked to autosomal chromosomes. By contrast, Y-linked factors contribute to the control of the overall percentage of abnormal sperm (42). It is worth mentioning that abnormal sperm may also be induced by a number of mutagenic agents, such as chemotherapy drugs, amoebicides, synthetic androgens, heat, mumps virus, and x-radiation (43), suggesting that the prospective evaluation of sperm morphology represents a valuable tool in identifying genetic hazards.

7.2. Germinal epithelium

The pathology of the germinal epithelium is primarily restricted to spermatid maturation of gametes. Malformed spermatids are always visible

←

Figure 1-11a–f. Malformed sperm from human ejaculate. Multinucleate spermatozoa (Figure 1-11a) show an irregular nuclear shape with granular chromatin and nuclear membranous bodies (arrows); a double tail is indicated by two arrows. All cells are embedded in a large cytoplasmic remnant. A giant sperm is depicted in Figure 1-11b and shows an irregular nuclear surface and an abnormally thin middle piece. Figure 1-11c: the sperm shows a malformed acrosome with a large membranous body and a voluminous cytoplasmic remnant containing a coiled tail (arrow) and clusters of mitochondria. Figures 1-11d and e show round-headed spermatozoa without acrosomes; arrow points to an overgrowth of the nuclear membrane. A tail with a disorganized axoneme is embedded in a cytoplasmic remnant along with the nucleus (arrowhead). Figure 1-11f: a sperm with disorganized axonemal tubules. Figures 1-11a and b: X 4,000; Figure 1-11c: X 12,000; Figure 11d: X 8,000; Figure 1-11e: X 7,000; Figure 1-11f: X 43,000.

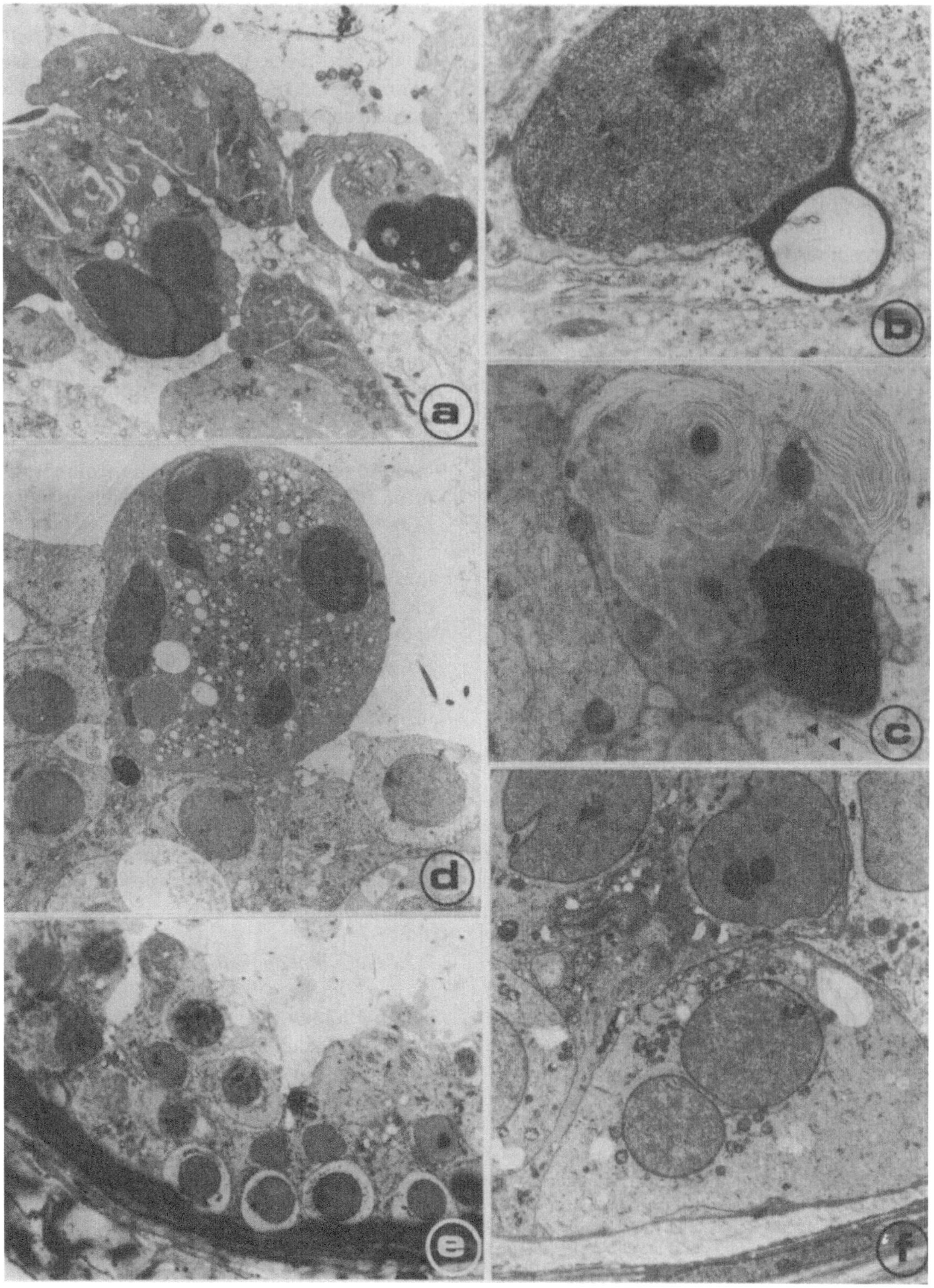
a
b
c
d
e
f

in the seminiferous tubules of normozoospermic men, although in these cases they represent a small percent of maturing haploid elements. By contrast, they are quite numerous in infertile men with teratozoospermia or oligoasthenozoospermia.

As noted in the section dealing with sperm morphology (see Section 7.1), the most common finding in abnormal spermiogenesis consists of the simultaneous occurrence of abnormalities in acrosomal development and nuclear shape. These abnormalities are observed in various conditions such as varicocele (44), unexplained infertility (45), and obstructive azoospermia (46). The most common abnormality of spermatids is the occurrence of two or three nuclei joined by a single acrosome, which results in polynucleate cells (Fig. 1-12a). This abnormality is associated with the appearance of a single or sometimes multiple tail, and with various degrees of nuclear immaturity. Acrosomes of single spermatids of this type show malformations such as large vacuoles and abnormal location around the nucleus (Fig. 1-12b), while an overgrowth of membranes often distorts the cytoplasm (Fig. 1-12c). While most of the abnormal cells undergo degeneration and are resorbed by Sertoli cells, some abnormal spermatids do appear in the ejaculate. In abnormal spermiogenesis, the interactions between germ and Sertoli cells are often altered. In fact, the coat of smooth endoplasmic reticulum and microfilaments in Sertoli cells, which appears around the spermatids during the maturation phase (16), is often missing (Fig. 1-12b) or is displaced from its normal location (Fig. 1-12c). To date, it is not possible to give any significant meaning to this finding with respect to the pathogenesis of altered gamete maturation; it does however demonstrate the existence of some modifications of Sertoli cells in cases of altered spermiogenesis.

Current information suggests that the altered maturation of spermatids may be related to spermatogonial genetic defects which become manifest during the haploid phase of maturation (45). In animal models, in fact, sperm abnormalities induced by mutagenic agents are seen more frequently when the exposed cells happen to be spermatogonia or early primary spermatocytes, while cell exposure during spermiogenesis does not lead to any increase of sperm abnormalities (47).

A finding which has often been observed in severe oligozoospermia or in aging testes (48), as well as in genetic hypogonadism such as Klinefelter's Syndrome (Fig. 12d), is the appearance in the seminiferous epithelium of giant multinucleate germ cells. Abnormal spermatids within the same clone of germ cells form confluent multinucleate cells due to a progressive enlargement of the intercellular bridges. This defect may be experimentally induced by a variety of exogenous noxious agents (49,50) that directly affect germ and Sertoli cells. These giant cells usually undergo degeneration in the testis, although they are occasionally observed in the ejaculate (Fig. 1-11a).

While the defects so far described are consistent with the presence of spermatozoa in the ejaculate, more severe damage to the germinal epithelium results in a maturative arrest of spermatogenesis. This is one of the most common causes of azoospermia, and is characterized by arrested maturation of germ cells at the spermatocytic (Fig. 1-12e) or round spermatid level. A genetic origin for this obscure and relevant cause of sterility has been proposed. A block of meiosis in fact has been reported in man and has been attributed to chromosomal structural defects which result in a failure of sex bivalent pairing (51). An altered response of testicular targets to

←

Figure 1-12a–f. Germinal epithelium in pathologic human testes. Abnormal spermatids feature multiple heads joined by a single acrosome (Figure 1-12a), poorly condensed chromatin (Figures 1-12a and b), large vesicle in a malformed acrosome (Figure 1-12b), and a voluminous membranous body (Figure 1-12c). The subsurface cisternae of the endoplasmic reticulum and microfilaments of Sertoli cells are missing (Figure 1-12b) or displaced (Figure 1-12c) from the typical location around the spermatids (arrowheads). A multinucleated giant cell, formed by abnormal spermatids (Figure 1-12d), is visible in the apical region of the germinal epithelium. A spermatogenic arrest at the level of primary spermatocyte is visible in Figure 1-12e. Figure 1-12f shows a multinucleate spermatogonium A pale. Figure 1-12a: X 3,000; Figure 1-12b: X 14,000; Figure 1-12c: X 12,000; Figure 1-12d: X 1,800; Figure 1-12e: X 1,600; Figure 1-12f: X 2,600.

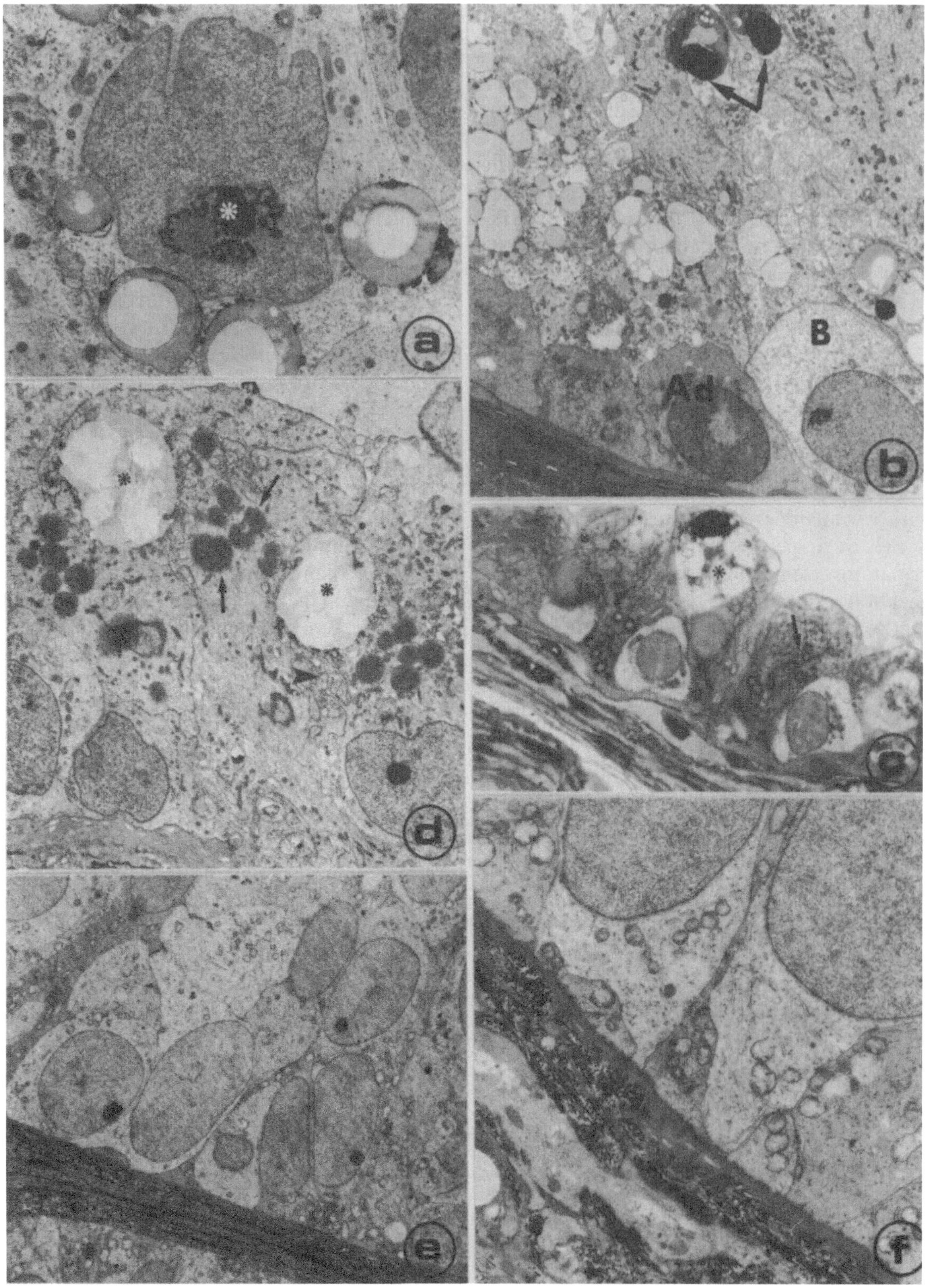
a
B
Ad
b
c
d
e
f

androgens could also explain the occurrence of spermatogenic arrest in cases of testicular feminization (personal observations); the completion of meiosis is in fact dependent on testicular androgen availability (52). The above observation points to the possible occurrence of spermatogenic arrest caused by defects in Sertoli cells, which are the only primary targets of androgens in the seminiferous tubule (53).

In cases of grossly altered spermatogenesis, some abnormal changes are also visible in the spermatogonia. These changes include the appearance of an enlarged nucleus or of multinucleate cells (Fig. 1-12f). DNA content above the normal diploid range has been documented in spermatogonia from cryptorchid testes. These spermatogonia display an increased nuclear size (54). The possibility that these rather immature polyploid spermatogonia evolve into premalignant cells and, later, into testicular tumors, is an important topic to be further evaluated. Carcinoma in situ of the testis, which is followed by the development of malignancies, occurs in cryptorchidism and also in infertile men with deranged spermatogenesis (55).

7.3. *Sertoli cells*

After sexual differentiation, the behavior of germ cells is highly dependent upon the testicular somatic environment (56,57). Sertoli cells mediate all interactions between the body and germ cells, which results in a continuous and regular production of mature gametes. It is therefore expected that altered spermatogenesis could result from a derangement of Sertoli cells. Histopathologic observations, both in experimental and natural models, suggest that Sertoli cells react in the same way under the various conditions associated with deranged spermatogenesis. Testicular biopsies of individuals affected by oligozoospermia and abnormal or immature gametes in their ejaculate generally show a paucity of mature spermatids in the seminiferous tubules, along with an exaggerated exfoliation of germ cells within the lumen. Altered spermiation could be claimed in these cases. Toxic agents that directly affect Sertoli cells do induce in the rat an apical retraction of cytoplasm in the supporting cells, resulting in a premature exfoliation of maturing spermatids (49, 50); indeed, this is the earliest effect seen after introduction of such agents. By contrast, experimental conditons, such as vitamin A deficiency (58) or depletion of luteinizing hormone (LH) and testosterone (59), result in delayed spermiation, while mature spermatids retained in the seminiferous epithelium undergo degeneration and are resorbed by Sertoli cells. It is of interest to note that in experimentally induced androgen depletion (59), a precocious degeneration of primary spermatocytes occurs, and that these cells are abnormally located close to the basal lamina. This suggests a derangement not only in spermiation, but also in the normal displacement of germ cells from the basal compartment carried out by dynamic changes of the Sertoli cell junctional complex (60). All these findings highlight the possible existence of a pathology of the Sertoli cell cytoskeleton, which might result in altered spermatogenesis. In fact, it has been suggested for the rat that actin-rich microfilaments associated with the subsurface endoplasmic reticulum around the spermatids (61) depolymerize before spermiation, and are recycled and repolymerized along the basal junctions of Sertoli cells when a new blood–testis barrier is formed as a consequence of the upward movement of spermatocytes (62). A pathologic behavior of the Sertoli cell cytoskeleton seems to occur also in man, as suggested by the frequent observation of a large increase of microfilaments

←

Figure 1-13a–f. Human Sertoli cells. A typical Sertoli cell with an irregular nucleus and a tripartite nucleolus (asterisk) contains large lipid droplets (Figure 1-13a). The tubule in Figure 1-13b features A dark (Ad) and B spermatogonia (B) along with abnormal spermatids (arrows); Sertoli cells are loaded with lipids. Figures 1-13c and d show a spermatogonial arrest of spermatogenesis. Lipids (arrows) in Sertoli cells are reduced, while large phagosomes (Figure 1-13c) or vacuoles (Figure 1-13d) (asterisks) are still visible. Arrowhead points to interdigitations of the cellular membrane. Figures 1-13e and f show seminiferous cords in a cryptorchid testis, formed by immature Sertoli cells with ovoid nuclei and small nucleoli. Typical junctional complexes are not visible between the somatic cells; the peritubular wall shows an increase in collagen fibers. Figure 1-13a: X 4,000; Figure 1-13b: X 2,000; Figure 1-13c: X 1,600; Figure 1-13d: X 2,200; Figure 1-13e: X 2,500; Figure 1-13f: X 6,000.

in the basal region of Sertoli cells in cases of severely altered or absent germ cells (63,64). This is sometimes associated with the disappearance or dislocation of microfilaments far from the junctional complex (65) where they are thought to participate in maintaining the integrity of the junctions (66).

The most common finding in human Sertoli cells from pathologic testes is a large increase in cytoplasmic lipid droplets and phagosomes (Fig. 1-13a–d). This is mainly observed when spermatogenesis is intact up to spermatid stages, and seems to be related to phagocytosis and disposal of abnormal gametes. In these cases, Sertoli cells do not show any changes in the infolded nucleus, tripartite nucleolus, or appearance of the cytoplasmic organelles (Fig. 13a). An increase in lipids and phagosomes, due to the disposal of degenerated germ cells, may also be demonstrated experimentally by inducing a derangement of spermatogenesis with different agents such as antifertility drugs, x-radiation, hypophysectomy, etc. (59). This finding suggests that the phagocytic activity of Sertoli cells is not affected under these conditions.

The rather normal appearance of Sertoli cells in numerous pathologic conditions associated with altered spermatogenesis does not imply that this cell is unaffected by, or resistant to, various insults. Dysfunctions of supporting cells are in fact documented in conditions of altered spermatogenesis and include a reduced secretion of androgen-binding protein (ABP) (67), inhibin (68,69), and seminal transferrin (70). To date, however, it is not possible to say whether or not the biochemical dysfunctions of Sertoli cells are primitive or the result of damage to the germ cells. Normal secretion by Sertoli cells of specific proteins such as ABP or the plasminogen activator seems to require cooperation by the germ cells (71,72).

The most conclusive example of pathologic Sertoli cells in testicular biopsies is reported for the cryptorchid testes and is characterized by the presence of immature supporting cells of the so-called hypoplastic seminiferous tubules (64). In this condition the presence of cords formed by packed somatic cells resembling prepubertal Sertoli cells is often observed close to the seminiferous tubules that show various degrees of damage (Figs. 1-13e and f). These cells display an ovoid nucleus that lacks the typical invaginations and the large tripartite nucleolus. The cytoplasm does not show any development of the smooth endoplasmic reticulum, annulate lamellae or Charcot–Böttcher crystals, and typical junctional complexes are also missing. The lumen is usually absent in these cords, suggesting a lack of fluid secretion that, in the normal testis, appears at puberty along with the signs of differentiation of Sertoli cells (73). These features are typical of the seminiferous cords seen in hypogonadotropic hypogonadism in man (74) and are experimentally induced in adult rats by depletion of androgens and LH (59). They are also occasionally observed in the aging human testes (75), and are visible in testes of androgen-insensitive individuals (personal observations). The rather common occurrence in cryptorchid testes of these immature areas of testicular structures, along with tubules of different degrees of development for both Sertoli and germ cells, suggests that a partial resistance to androgen may underlie these cases and explain the frequently inevitable sterility of these patients.

7.4. Tubule wall

The seminiferous epithelium is bound by a basal lamina and a tubule wall formed by three to four circular layers of elongated myoid cells and fibroblastlike cells (Fig. 1-10). Collagen and fibrillar material are intermingled between the cells, whose contraction allows the progression of the luminal content toward the excurrent pathways. The location of this structure between the tubule and interstitium certainly influences communication between the two testicular compartments. A growing interest has been recently focused on the cellular components of the tubule wall because a deep funtional interaction between peritubular cells and Sertoli or Leydig cells has been demonstrated. Briefly, myoid cells seem to modulate the synthesis of ABP (76), transferrin (77), and plasminogen activator (78) by Sertoli cell cultures through the production of still-unknown macromolecules, some of which seem to be dependent upon androgens (76). A cooperative relationship between Sertoli and myoid cells is also required in order for each element to participate in vitro in the synthesis of different components of the extracellular matrix, which forms the basal lamina

(79). By contrast, peritubular cells seem to constitute a reservoir from which Leydig cells could differentiate (80,81). This does not imply that myoid cells transform into Leydig cells. Observations of the development of mature Leydig cells in the rat (82) suggest that fibroblastlike cells along the periphery of the tubule wall could be the precursors of Leydig cells. From the above studies, it appears clear that the tubule wall may influence the spermatogenic process both through a regulation of the passage of signals between the interstitium and the tubule, and through a direct interaction with Sertoli cells.

The tubule wall undergoes relevant changes in some pathologic conditions. In general, the changes are observed both in man and in other mammals with germinal aplasia, or whenever spermatogenesis is greatly perturbed. These changes include an increase in collagen fibers, which results in a widening of the space interposed between the tubule basal lamina and myoid cells (Fig. 1-14b) (85,86). This widening is also associated with a large increase in microfibrillar material which is interposed between the outer layers of myoid cells (Fig. 1-14b). Similar changes have been observed in rodents after induction of cryptorchidism (85,86) and hypophysectomy (87). Myoid cells do not usually show relevant modifications. However, in experimental cryptorchidism in the rat, a decrease of actin and contractile filaments in the cytoplasm of these cells (88) has been documented, suggesting that in this condition, myoid cells suffer a partial loss of the contractile element specializations. In the cryptorchid human testis, a peculiar change is characterized by the stratification of the basal lamina (Fig. 14a) (84). A slight thickening due to a folding of the basal lamina is also observed in other pathological conditions. These situations are characterized by perturbed spermatogenesis that is related to the collapse of the tubules.

Recent findings on the cooperation between Sertoli and myoid cells (76–79) suggest that the pathology of the tubule wall may result from an altered interaction between these two cell types. The subsequent fibrosis could then preclude the reversibility of the damage to the seminiferous epithelium and myoid cells, resulting in tubular hyalinosis and the complete disappearance of germ and Sertoli cells. This is a finding which is frequently encountered in testicular failures due to genetic (Klinefelter's Syndrome), infectious (mumps orchitis), physical (x-irradiation), and pharmacologic (estrogen excess, antineoplastic drugs) factors.

7.5. *Interstitium*

Spermatogenesis is dependent upon a high concentration of testosterone produced by Leydig cells (88). The cytological features of these cells have been extensively investigated, as has the role of their cytoplasmic organelles, namely the endoplasmic reticulum and mitochondria (Fig. 1-14c), in the different steps of testosterone synthesis (80). Normal development of Leydig cells depends upon the availability of LH (59) but can be induced by human chorionic gonadotropins in individuals affected by hypogonadotropic hypogonadism (89). In the rat, the rate of testosterone production by each Leydig cell is strictly dependent upon the amount of smooth endoplasmic reticulum (90), while cellular size is largely influenced in a cyclic way by the seminiferous epithelium (91): size, in fact, increases in proximity to the tubules at stages VII and VIII of the cycle, where there is maximum concentration of endogenous testosterone (27). This paracrine control of size and function of Leydig cells is claimed to be due to some factors that are released by Sertoli cells into the interstitium (29), and that contribute to the optimization of Leydig cell function with respect to the efficiency of spermatogenesis. In man, due to the small size of each stage along the tubules, it is much more difficult to evaluate a cyclic behavior of Leydig cells with respect to the stages of the seminiferous epithelium. However, the frequent observation in human testicular biopsies of Leydig cells with different cytologic appearances, side by side in the interstitial area (Fig. 14e), suggests that, also in this species, interstitial cells undergo morphofunctional changes that are perhaps related to different messages from neighboring seminiferous tubules. The physiology of Leydig cells is also influenced by macrophages (Fig. 1-14b), which in fact seem to stimulate testosterone synthesis by cultured Leydig cells through secretory factors (92).

Leydig cells undergo a number of changes in pathologic conditions that are associated with altered spermatogenesis. These abnormal changes include lipid and lipofuscin deposits (Fig. 1-14d),

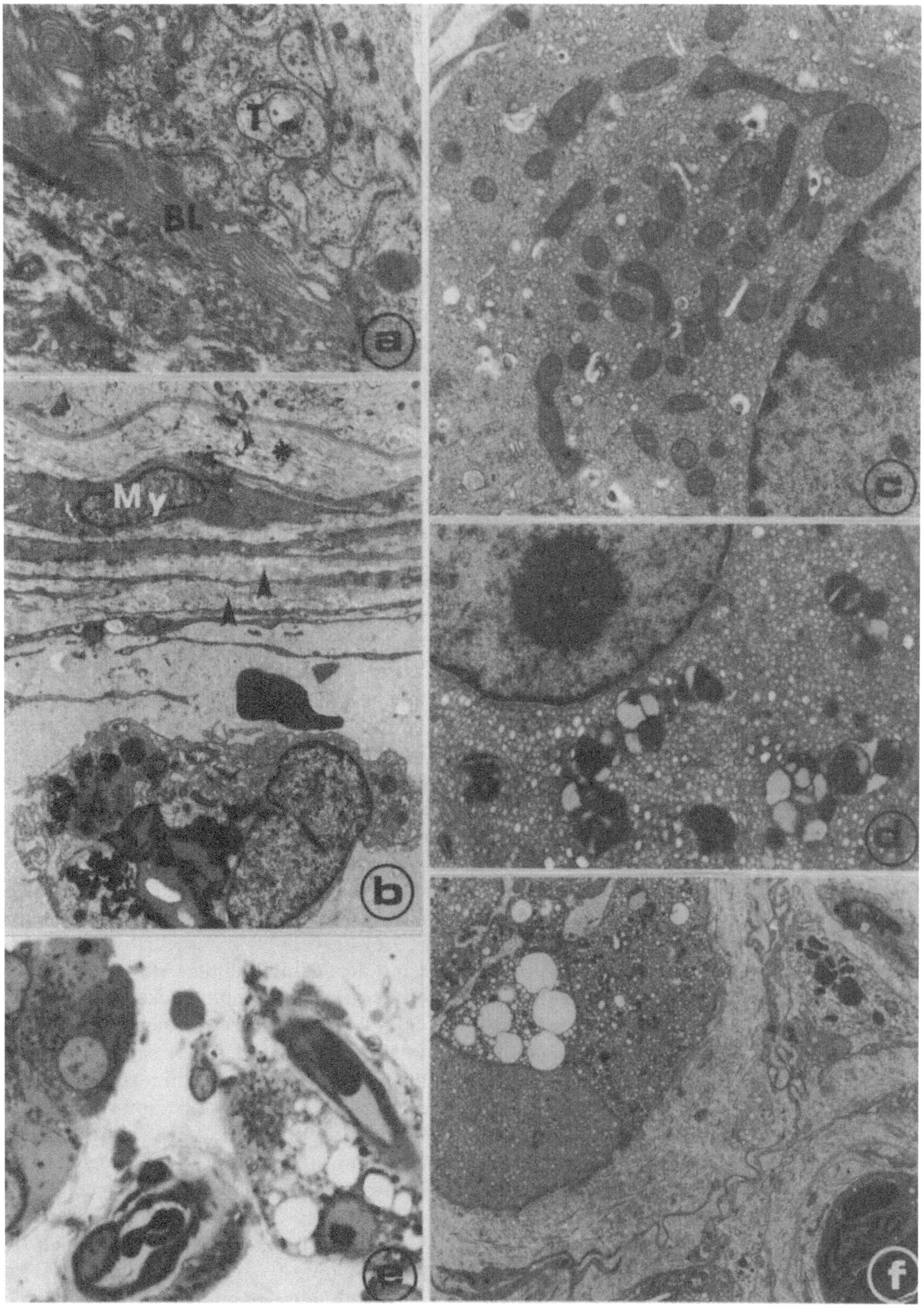
T
BL
a
My
b
c
d
e
f

cytoplasm vacuolization due to dilatation of the endoplasmic reticulum (Fig. 1-14e, f), and sclerosis due to collagen deposits around the cell (Fig. 1-14f). Leydig cell changes are usually observed in testes that, besides perturbed spermatogenesis, also exhibit thickened blood vessel walls (Fig. 1-14f), such as cryptorchid testes (84), testes of oligozoospermic patients with left-sided varicocele (93), and aging testes (94). The vascular changes may be relevant to the induction of deranged spermatogenesis and interstitial damage through a reduced supply of blood-borne substances.

8. Conclusions

In this chapter we have attempted to provide an overview of mammalian spermatogenesis and its morphological changes in human diseases. Space considerations prevent a long discussion of the most important features. However, it is worth remarking that a short survey on testicular pathophysiology is presented in order to emphasize the deep relationship between germinal and somatic components of the testis. In particular, we have stressed the concept that deranged spermatogenesis is always associated with some morphofunctional change in the somatic cells, both in the tubular and extratubular compartments, that could result either from a direct action of some noxious agent(s) and/or from an altered communication between the different components of the testicular parenchyma.

References

1. Roosen-Runge EC: The Process of Spermatogenesis in animals. London, New York, Melbourne: Cambridge University Press, 1977.
2. Clermont Y, Leblond CP: Renewal of spermatogonia in the rat. *Am J Anat* 93: 475–502, 1953.
3. Clermont Y, Bustos-Obregon E: Reexamination of spermatogonial renewal in the rat by means of seminiferous tubules mounted "in toto". *Am J Anat* 122: 237–247, 1968.
4. Huckins C: The spermatogonial stem cell population in adult rats. I. Their morphology, proliferation and maturation. *Anat Record* 169: 533–548, 1971.
5. Oakberg EF: Spermatogonial stem cell renewal in the mouse. *Anat Record* 169: 515–531, 1971.
6. Clermont Y: Two classes of spermatogonial stem cells in the monkey (*Cercopithecus aethiops*). *Am J Anat* 126: 57–72, 1969.
7. Clermont Y: Dynamics of human spermatogenesis. In: The Human Testis. E Rosemberg, CA Paulsen (eds), New York, London: Plenum Press, p 47–61, 1970.
8. Solari AJ, Tres LL: Ultrastructure and histochemistry of the nucleus during male meiotic prophase. In: The Human Testis. E Rosemberg, CA Paulsen (eds), New York, London: Plenum Press, p 127–138, 1970.
9. De Martino C, Capanna E, Nicotra MR, Natali PG: Immunochemical localization of contractile proteins in mammalian meiotic chromosomes. *Cell Tiss Res* 213: 159–278, 1980.
10. Capanna E, De Martino C, Natali PG: Presence and role of contractile proteins in the vertebrate synaptonemal complex. *Boll Zool* 47: 231–235, 1980.
11. De Martino C, Floridi A, Marcante ML, Malorni W, Scorza Barcellona P, Bellocci M, Silvestrini B: Morphological, histochemical and biochemical studies on germ cell mitochondria of normal rats. *Cell Tiss Res* 196: 1–22, 1979.
12. Stefanini M, De Martino C, D'Agostino A, Agrestini A, Monesi V: Nucleolar activity of rat primary spermatocytes. *Exptl Cell Res* 86: 166–170, 1974.
13. Leblond CP, Clermont Y: Definition of the stages of the cycle of the seminiferous epithelium in the rat. *Ann NY Acad Sci* 55: 548–573, 1952.
14. Fawcett DW: A comparative view of sperm ultrastructure. *Biol Reprod* 2, suppl. 2: 90–127, 1970.
15. Baccetti B: Comparative Spermatology. New York, London: Academic Press, 1970.
16. Fawcett DW: Ultrastructure and function of Sertoli cell. In: Handbook of Physiology, vol. V. DW Hamilton, RO Greep (eds), Washington DC: American Physiology

←

Figure 1-14a–f. Human tubule wall and interstitium. In Figure 1-14a, a multilayered basal lamina (BL) is visible in a tubule (T) of a cryptorchid testis. A typical thickening of the seminiferous tubule wall (Figure 1-14b) features an increase of collagen fibers (asterisk) and a microfibrillar deposit (arrowheads). A large macrophage is visible in the interstitial area. My: myoid cell. Figures 1-14c–f show adult Leydig cells. A normal cell in Figure 1-14c shows large mitochondria with tubular cristae and packed tubular endoplasmic reticulum. Lipofuscins and a dilated endoplasmic reticulum are visible in Figure 1-14d. Figure 1-14e shows an interstitial area containing side-by-side normal cells (left) along with cells with large cytoplasmic vacuoles (right). In Figure 1-14f, Leydig cells with various degrees of dilatation of endoplasmic reticulum are surrounded by large amounts of collagen fibers which are also responsible of the thickening of the blood vessel wall (lower right corner). Figure 1-14a: X 11,500; Figure 1-14b: X 4,000; Figure 1-14c: X 9,000; Figure 1-14d: X 7,500; Figure 1-14e: X 2,500; Figure 1-14f: X 3,000.

Society, p. 21–55, 1975.
17. Setchell BP, Waites GMH: The blood–testis barrier. In: Handbook of Physiology, vol V. DW Hamilton, RO Greep (eds), Washington D.C.: American Physiology Society, p. 143–172, 1975.
18. De Martino C, Marcante ML, Floridi A, Citro G, Bellocci M, Cantafora A, Natali PG: Sertoli cells of adult rats in vitro. *Cell Tiss Res* 176: 69–90, 1977.
19. Perey B, Clermont Y, Leblond CP: The wave of the seminiferous epithelium in the rat. *Am J Anat* 108: 47–77, 1961.
20. Oakberg AF: A description of spermiogenesis in the mouse and its use in analysis of the cycle of the seminiferous epithelium and germ cell renewal. *Am J Anat* 99: 391–413, 1956.
21. Foote RH, Swierstra EE, Hunt WL: Spermatogenesis in the dog. *Anat Rec* 173: 341–352, 1972.
22. Tice LW, Barnett RJ: The fine structural localization of some testicular phosphatases. *Anat Rec* 147: 43–56, 1963.
23. Kerr JB, De Kretser DM: Cyclic variations in Sertoli cell lipid content throughout the spermatogenic cycle in the rat. *J Reprod Fert* 43: 1–8, 1975.
24. Chowhury AK, Marshall G: Irregular pattern of spermatogenesis in the baboon (papio anubis) and its possible mechanism. In: Testicular Development, Structure and Function. A Steinberger, E Steinberger (eds), New York: Raven Press, p 129–137, 1980.
25. Heller CG, Clermont Y: Kinetics of germinal epithelium in man. *Rec Progr Horm Res* 20: 545–575, 1964.
26. Schulze W, Rehder U: Organization and morphogenesis of the human seminiferous epithelium. *Cell Tiss Res* 237: 395–407, 1984.
27. Parvinen M: Regulation of the seminiferous epithelium. *Endocr Rev* 3: 404–417, 1982.
28. Fabbrini A, Re M, Conti C: Glycogen in the normal human testis. *J Endocrinol* 43: 499–506, 1969.
29. Sharpe RM: Paracrine control of the testis. *Clinics Endocrinol Metab* 15: 185–207, 1986.
30. MacLeod J: Semen quality in one thousand men of known fertility and in eight hundred cases of infertile marriage. *Fertil Steril* 2: 115–139, 1951.
31. Sherins RJ, Brightwell D, Sternthal PM: Longitudinal analysis of semen of fertile and infertile men. In: The Testis in Normal and Infertile Men. P Troen, NR Nankin (eds), New York: Raven Press, p 473–488, 1977.
32. MacLeod J: The significance of deviations in human sperm morphology. *Adv Exp Med Biol* 10: 481–494, 1970.
33. Beatty RA: The genetics of size and shape of spermatozoan organelles. In: The Genetics of the Spermatozoon. Proceedings of the International Symposium, Edinburgh 1971. RA Beatty, S Glyeckschn-Waelsch (eds), Edinburgh: Univ. Edinburgh, p 97–115, 1972.
34. Flörke-Gerloff S, Töpfer-Petersen E, Müller-Esterl W, Mansouri A, Schatz R., Schirren C, Schill W, Engel W: Biochemical and genetic investigation of round-headed spermatozoa in infertile men including two brothers and their father. *Andrologia* 16: 187–202, 1984.
35. Schirren CG, Holstein AF, Schirren C: Über die morphogenese rundköpfiger spermatozoen des menschen. *Andrologie* 3: 117–125, 1971.
36. Baccetti B, Renieri T, Rosati F, Selmi MG, Casanova S: Further observations on the morphogenesis of the round-headed human spermatozoa. *Andrologia* 9: 255–264, 1977.
37. Flörke-Gerloff S, Krause W, Töpfer-Petersen E, Tschesche H, Müller-Esterl W, Engel W: On the teratogenesis of round-headed spermatozoa: investigations with antibodies against acrosin, an intraacrosomally located acrosin-inhibitor, and the outer acrosomal membrane. *Andrologia* 17: 126–138, 1985.
38. S. Francavilla, B. Bruno, G. Poccia F. Francavilla, G. Properzi, and R. Santucci: Fluorescence microscopic detection of Acrosin in different morphologic types of human spermatozoa. *Andrologia*, 1988 (in press).
39. Renieri T: Submicroscopical observations on abnormal human spermatozoa. *J Submicr Cytol* 6: 421–432, 1974.
40. Afzelius BA: A human syndrome caused by immotile cilia. *Science* 193: 317–319, 1976.
41. Baccetti B, Burrini AG, Maver A, Pallini V, Renieri T: "9 + 0" immotile spermatozoa in an infertile man. *Andrologia* 11: 437–443, 1979.
42. Krzanowska H: Inheritance of sperm head abnormality types in mice: the role of the Y chromosome. *Genet Res* 28: 189–198, 1976.
43. Wyrobeck AJ, Bruce WR: The induction of sperm-shape abnormalities in mice and humans. In: Chemical Mutagens: Principles and Methods for Their Detection, vol. 5. A Hollaender, FJ de Serres (eds), New York: Plenum Press, p 257–285, 1978.
44. Cameron DF, Snydle FE, Ross MH, Dryle DM: Ultrastructural alterations in the adluminal testicular compartment in man with varicocele. *Fertil Steril* 33: 526–533, 1980.
45. Vegni-Telluri M, Menchini-Fabris F, Renieri T: A possible haploid effect in acrosome malformations of human spermatozoa. *Andrologia* 9: 315–322, 1977.
46. Camatini M., Franchi E, Faleri M: Ultrastructure of acrosomal malformations in men with obstructive azoospermia. *Arch Androl* 1: 203–208, 1978.
47. Wyrobeck AJ, Bruce WR: Chemical induction of sperm abnormalities in mice. *Proc Natl Acad Sci USA* 72: 4425–4429, 1975.
48. Holstein AF, Eckmann C: Multinucleated spermatocytes and spermatids in human seminiferous tubules. *Andrologia* 18: 5–16, 1986.
49. De Martino C, Stefanini M, Agrestini A, Cocchia D, Morelli M, Scorza-Barcellona P: Antispermatogenic activity of 1-p chlorobenzyl-1H-indazol-3-carboxylic acid (AF312/TS) in rats. *Exp Mol Pathol* 23: 321–356, 1975.
50. Francavilla S, De Martino C, Cordeschi G, Properzi G, Martini M, Campana A, Scorza-Barcellona P, Fabbrini A: Antispermatogenic effect of indazolecarboxylic acids in the rat. In: IV European Workshop on Molecular and Cellular Endocrinology of the Testis. Amsterdam: Excerpta Medica, 289–295, 1986.
51. Chandley AC: Male infertility and meiosis in man. In: Oligospermia: Recent Progress in Andrology. G Frajese, ESE Hafez, C Conti, A Fabbrini (eds), New York: Raven Press, p. 247–265, 1981.
52. Chowdhury AK: Dependence of testicular germ cells on

hormones: a quantitative study in hypophysectomized testosterone-treated rats. *J Endocrinol* 83: 331–340, 1979.
53. Purvis K, Hansson V: Hormonal regulation of spermatogenesis. Regulation of target cell response. *Int J Androl* Suppl 3: 88–95, 1981.
54. Muller J, Skakkebaeck NS: Abnormal germ cells in maldescended testes: a study of cell density, nuclear size and deoxyribonucleic acid content in testicular biopsies from 50 boys. *J Urol* 131: 730–733, 1984.
55. Skakkebaeck NE: Carcinoma in situ of the testis: frequency and relationship to invasive germ cell tumors in infertile men. *Histopathology* 2: 157–170, 1978.
56. Upadhyay S, Zamboni L: Ectopic germ cells: natural model for the study of germ cell sexual differentiation. *Proc Natl Acad Sci USA* 79: 6584–6588, 1982.
57. Francavilla S, Zamboni L: Differentiation of mouse ectopic germinal cells in intra- and perigonadal locations. *J Exp Zool* 233: 101–109, 1985.
58. Huang HFS, Marshall GR: Failure of spermatid release under various vitamin A states. An indication of delayed spermiation. *Biol Reprod* 28: 1163–1172, 1983.
59. Dym M, Madhwa Raj HG: Response of adult rat Sertoli cells and Leydig cells to depletion of luteinizing hormone and testosterone. *Biol Reprod* 17: 676–696, 1977.
60. Russell LD: Movement of spermatocytes from the basal to the adluminal compartment of the rat testis. *Am J Anat* 148: 313–328, 1977.
61. Toyama Y: Actin-like filaments in the Sertoli cell junctional specializations in the swine and mouse testis. *Anat Rec 186*: 477–492, 1976.
62. Russell L, Myers P, Ostenburg J, Malone J: Sertoli ectoplasmic specializations during spermatogenesis. In: Testicular Development, Structure and Functions. A Steinberger, E Steinberger (eds), New York: Raven Press, p 55–63, 1980.
63. Chemes HE, Dym M, Fawcett D, Javadpour N, Sherins R: Patho-physiological observations of Sertoli cells in patients with germinal aplasia or severe germ cell depletion. Ultrastructural findings and hormone levels. *Biol Reprod* 17: 108–123, 1977.
64. Schulze C, Holstein AF, Schirren C, Körner F: On the morphology of human Sertoli cells under normal conditions and in patients with impaired fertility. *Andrologia* 8: 167–178, 1976.
65. Schleiermacher E: Ultrastructural changes of the intercellular relationship in impaired human spermatogenesis. *Hum Genet* 54: 391–404, 1980.
66. Connell CJ: A freeze-fracture and lanthanum tracer study of the complex junctions between Sertoli cells of the canine testis. *J Cell Biol* 76: 57–75, 1978.
67. Kerr JB, Rich KA, De Kretser DM: Effects of experimental cryptorchidism on the ultrastructure and function of the Sertoli cell and peritubular tissue of rat testis. *Biol Reprod* 21: 823–838, 1979.
68. Au CL, Robertson DM, De Kretser DM: In vitro bioassay of inhibin in testes of normal and cryptorchid rats. *Endocrinology* 112: 239–244, 1983.
69. Scott RS, Burger HG: Inhibin is absent from azoospermic semen of infertile men. *Nature* 285: 246–247, 1980.
70. Serio M, Orlando C, Caldini A, Barni T, Forti G, Casano R, Giannotti P, Natali A, Maver A, Chelo E, Vannelli B, Balboni G: Sertoli cell proteins in human seminal plasma. In: Reproductive Medicine. E Steinberger, G Frajese, A Steinberger (eds), New York: Raven Press, p 465–472, 1986.
71. Vihko KK, Suominen JJO, Parvinen M: Cellular regulation of plasminogen activator secretion during spermatogenesis. *Biol Reprod* 31: 383–389, 1984.
72. Galdieri M, Monaco L, Stefanini M: Secretion of androgen binding protein by Sertoli cells is influenced by contact with germ cells. *J Androl* 5: 409–415, 1984.
73. Tindall DJ, Vitale R, Means A: Androgen binding protein as a biochemical marker of formation of the blood-testis barrier. *Endocrinology*. 97: 636–648, 1975.
74. De Kretser DM, Burger HG: Ultrastructural studies of the human Sertoli cell in normal men and males with hypogonadotropic hypogonadism before and after gonadotropin treatment. In: Gonadotropins. BB Saxena, CG Beling, HM Gandy (eds), New York: Wiley Interscience, p 640–656, 1972.
75. Paniagua R, Amat P, Nistal M, Martin A: Ultrastructural changes in Sertoli cells in aging humans. *Int J Androl* 8: 295–312, 1985.
76. Skinner MK, Fritz IB: Androgen stimulation of Sertoli cell function is enhanced by peritubular cells. *Mol Cell Endocrinol* 40: 115–122, 1985.
77. Holmes SD, Lipschultz LI, Smith RG: Regulation of transferrin secretion by human Sertoli cells cultured in the presence or absence of human peritubular cells. *J Clin Endocrinol Metab* 59: 1058–1062, 1984.
78. Hettle JA, Waller EK, Balekjian E, Fritz IB: Controls of plasminogen activator activity in rat seminiferous tubules and Sertoli cells. IV European Workshop on Molecular and Cellular Endocrinology of the Testis. Miniposter book, Capri, Italy: p 43, 1986.
79. Tung PS, Fritz IB: Interactions of Sertoli cells with myoid cells in vitro. *Biol Reprod* 23: 207–217, 1980.
80. Christensen AK: Leydig cells. In: Handbook of Physiology, Male Reproductive Tract, vol. V. DW Hamilton, Ro Greep (eds), Washington: American Physiological Society, p 57–94, 1975.
81. Schulze C, Holstein AF: Leydig cells within the lamina propria of seminiferous tubules in four patients with azoospermia. *Andrologia* 10: 444–452, 1978.
82. Kerr JB, Sharpe RM: Stimulatory effect of follicle-stimulating hormone on rat Leydig cells. *Cell Tiss Res* 239: 405–415, 1985.
83. De Kretser DM, Kerr JB, Paulsen CA: The peritubular tissue in the normal and pathological human testis. An ultrastructural study. *Biol Reprod* 12: 317–324, 1975.
84. Francavilla S, Santiemma V, Francavilla F, De Martino C, Santucci R, Fabbrini A: Ultrastructural changes in the seminiferous tubule wall and intertubular blood vessels in human cryptorchidism. *Arch Androl* 2: 21–30, 1979.
85. Kerr JB, Rich KA, De Kretser DM: Effects of experimental cryptorchidism on the ultrastructure and function of Sertoli cell and peritubular tissue in the rat testis. *Biol Reprod* 21: 823–838, 1979.
86. Bellocci M, De Martino C, Malorni W, Francavilla S, Santiemma V, Natali PG: Experimental cryptorchidism in

the rat. *Ultramicroscopy* 5: 417A, 1980.
87. Ross MH, Grant L: On the structural integrity of basement membrane. *Exp Cell Res* 50: 277–285, 1968.
88. Steinberger E, Smith KD, Tcholakian RK, Chowdhury M, Steinberger A, Fisher M, Paulsen CA: Steroidogenesis in human testes. In: Male Fertility and Sterility. RE Mancini, L Martini (eds), New York: Academic Press, p 149–174, 1973.
89. De Kretser DM: Changes in the fine structure of the human testicular interstitial cells after treatment with human gonadotropins. Z. Zellforsch. *Mikrosk Anat* 83: 344–358, 1967.
90. Mori I, Christensen K: Morphometric analysis of Leydig cells in the normal testis. *J Cell Biol* 84: 340–354, 1980.
91. Bergh A: Local differences in Leydig cell morphology in the adult rat testis: evidence for a local control of Leydig cells by adjacent seminiferous tubules. *Int J Androl* 5: 325–330, 1982.
92. Yee JB, Hutson JC: Effects of testicular macrophage-conditioned medium on Leydig cells in cultures. *Endocrinology* 116: 2682–2684, 1985.
93. Fabbrini A, Santiemma V, Francavilla S, Moscardelli S, Francavilla F, Incorvati N., De Martino C, Bellocci M: Leydig cell morphology and function in varicocele. In: Oligozoospermia: Recent Progress in Andrology. G Frajese, ESE Hafez, C Conti, A Fabbrini (eds), New York: Raven Press, p 77–94, 1981.
94. Vermeulen A, Rubens R, Verdonck AL: Testosterone secretion and metabolism in male senescence. *J Clin Endocrinol Metab* 34: 730–735, 1972.

Author's address:
Professor C. De Martino
Regina Elena Institute for Cancer Research
Viale R. Elena
Rome, Italy

CHAPTER 2

Life history of the human female germ cell: Ultrastructural aspects

SAYOKO MAKABE, STEFANIA A. NOTTOLA & PIETRO M. MOTTA

1. Introduction

The haploid germ cells are produced in the adult gonad after completion of the first meiotic division, even though the diploid germ-cell line is established during the early stages of embryogenesis, when those cells are known as primordial germ cells (PGCs) (1).

Over the last century, several scientists have investigated the complex phenomena of germ cell line development and gonadal differentiation, which occur during organogenesis and fetal growth (2–4). More recent studies of humans and of various vertebrates have dealt with morphological and experimental aspects related to a) the foreign origin of the PGCs with respect to the gonadal tissue, b) the difficult path of their migration to the genital ridges and c) the settlement and the differentiation of the gonadal anlage at the time of their colonization (5–17) (for review, see 18, 19).

Recent technical progress and the large amount of data obtained have allowed investigators to precisely define some difficult aspects that were previously encountered; these aspects principally concern the study of the complex relationship between germ and somatic cells during migration and after the settlement stage. Furthermore, such interrelationships should be considered in the light of the close association occurring between germ line development and sexual differentiation of the gonad.

The purpose of this chapter is to analyze the fine structure of human female germ cells in comparison with results obtained from other vertebrates, especially in mammals, at comparable developmental stages, in order to more clearly define the events occurring during the early life history of these cells. Particularly, this chapter addresses 1) the origin and development of the germ cells, with special attention to their developmental stages preceding oogenesis, 2) the relationship between female gametes and ovarian somatic cells, and 3) the possible mechanisms responsible for the dramatic reduction in the pool of oocytes at birth.

2. Origin and general characteristics of primordial germ cells

During the second half of the last century, Waldeyer's theory (2) that the germ cell line had an intragonadal origin and differentiated from mesothelial elements covering the genital ridges (the so-called *germinal epithelium*) enjoyed considerable support (20–29). Afterwards, a different hypothesis was suggested. According to this hypothesis, the total pool of germ cells at birth might arise not only from germ cells that proliferated from the surface germinal epithelium of the ovary, but also by a number of other PGCs formed early in extragonadal sites and which migrated to the ovary (for review see 30).

Since the beginning of this century, other authors (5–8,10,12,13,16,31–33), using histochemical and experimental techniques, in addition to the classic morphological approach, have observed that PGCs do not originate in the gonads. This finding has also been confirmed by recent data obtained in both humans and a large number of animals. PGCs in fact originate before gonadal differentiation and even before the development of the gonadal anlage (18,27,30,34,35). Furthermore, PGCs are the sole precursors of all the germ cells present in the gonads of adults (germ plasm continuity) (30,36,37). This current concept mainly derives from observations during germ cell line development of the fate of particular histochemical and radioactive markers present in PGCs (32,38,39). It seems that the first evidence of germ plasm continuity can be recognizable morphologically in a certain cytoplasmic region of the zygote, where the determinants of the germ cell line are thought to be segregated from the remaining somatic cytoplasm; such determinants will give rise to the next generation of germ cells (3,4).

In mammals, PGCs are first detected in the endoderm of the dorsal wall of the yolk sac, near the developing allantois (Diagram A); the earliest stage at which they are recognized in humans is the third week postfertilization (p.f.) (1,12,13, 18,40–42). Thus, the origin of mammalian PGCs is similar to that of other amniotes, being not only extragonadal but also extraembryonic (1,37).

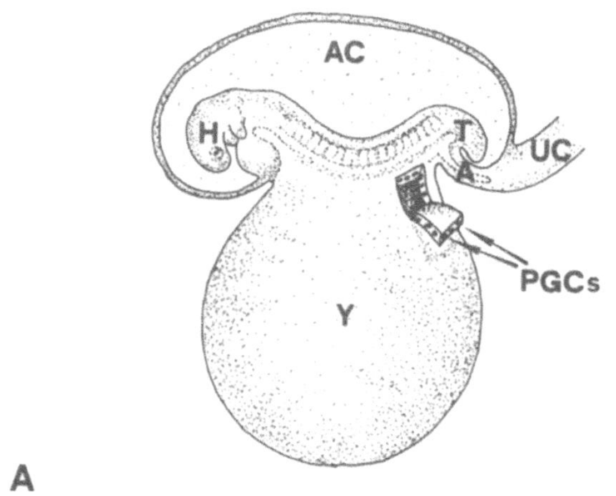

Diagram A. Drawing of a three-week-old human embryo, showing the position of the PGCs in the wall of the yolk sac (Y), near the developing allantois (A). H: head end of the embryo; T: tail end of the embryo; UC: future umbelical cord; AC: amniotic cavity.
Redrawn from: Langman's Medical Embryology, 1985 (42).

PGCs of vertebrates, especially those of amniotes, show common features (43).

With light microscopy (LM), PGCs appear as large and round elements situated among the smaller endodermal cells of the yolk sac; their size and shape (in human embryos PGCs are rounded cells of about 15–20 μm in diameter), together with intense staining for alkaline phosphatase, make these cells easily distinguishable from other somatic cells (1,35,37,41,44,45). However, whether and where the non-staining precursors of PGCs appear earlier in development is still a mystery (1). PGCs can also be detected by their cytoplasmic content of esterases (1,32,37,46) and for their pronounced affinity for toluidine blue (19); the latter feature was first revealed by Zamboni and Merchant for the mouse (27).

By transmission electron microscopy (TEM), PGCs can be easily identified as voluminous and rounded elements. In humans and in most other mammals, the PGC nucleus is extremely large and vesicular and many pores are present in the well-developed nuclear membrane. The chromatin is finely granular and uniformly dispersed; one or two prominent reticular nucleoli are present in the nucleoplasm (Figs. 2-1,2-2,2-3,2-6) (1,27, 41,44,47,48). Relatively few organelles are distributed in the cytoplasm, which appears pale and allows this cell type to be clearly distinguished from adjacent somatic cells (Figs. 2-1,2-2,2-6) (27,35,41). Mitochondria are usually disposed in a perinuclear arrangement and are round or oval in shape. Even in the same cell, cristae vary in shape from lamellar to tubulo-vesicular. In humans, round mitochondria with vesicular cristae seem to be more numerous than the other forms (27,44). The cytoplasm is characterized by Golgi complexes (generally a single one per cell) and cisternae of rough endoplasmic reticulum (RER), which are occasionally continuous with the perinuclear space. Numerous free ribosomes and polysomes are randomly distributed in the ground substance with vesicles, lipid droplets, and membrane-bound dense bodies (Fig. 2-3) (27,41,44). Microfilaments (41,44) and microtubules (27,41) are also observed. In addition, abundant intracytoplasmic glycogen particles are common in human PGCs, particularly at the beginning of their migration (41,44). Lipids and glycogen may be stored as an energy reserve

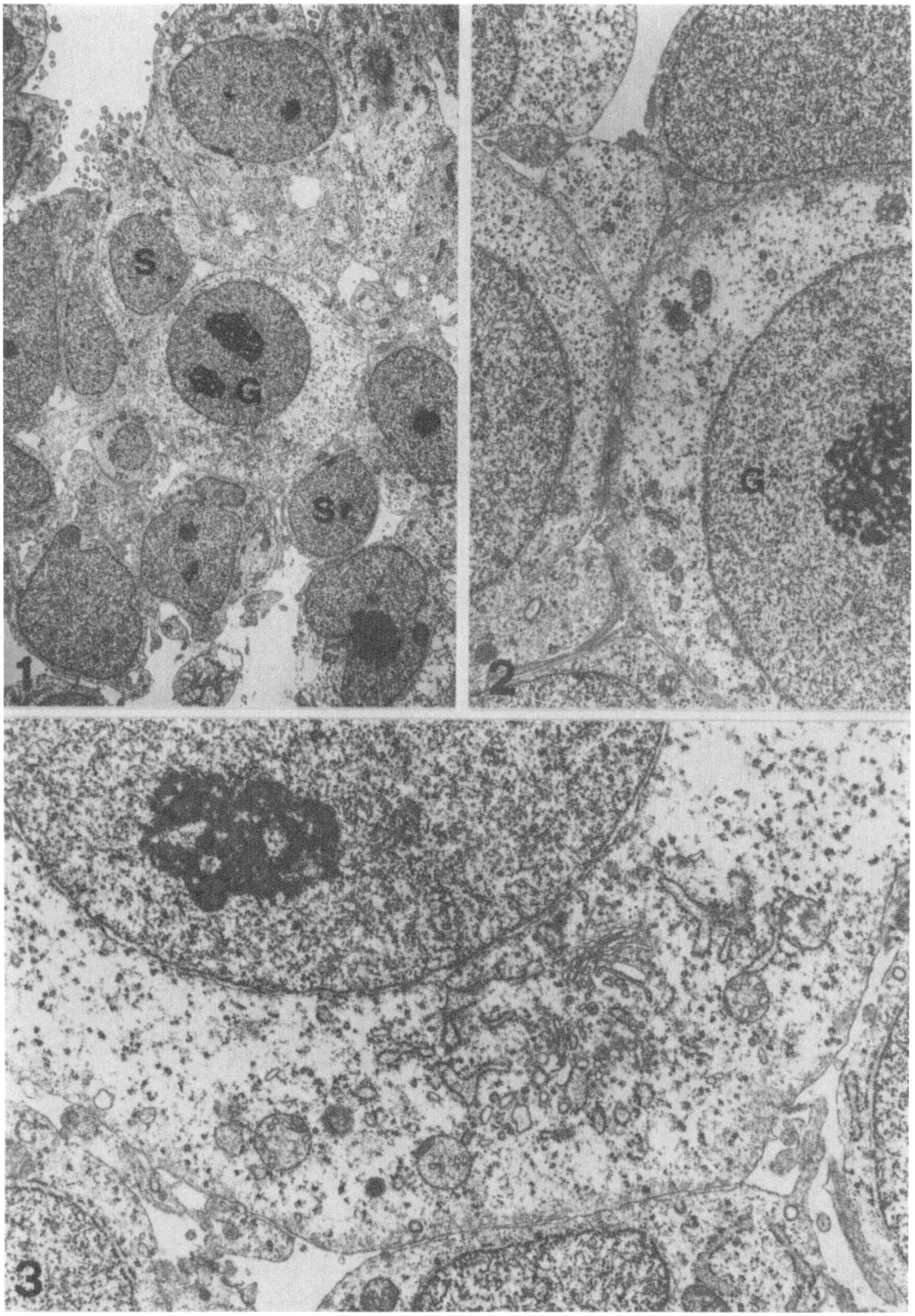

Figure 2-1. Five-week-old human embryo. A migrating PGC (G) is seen among the somatic cells (S) of the dorsal mesentery. The cytoplasm appears pale; the rounded nucleus contains a finely granular and uniformly dispersed chromatin, and two distinct nucleoli. TEM X 3,400.

Figure 2-2. Five-week-old human embryo. A PGC (G) showing a few organelles in the cytoplasm is seen in an area adjacent to that shown in Figure 2-1. TEM X 10,300.

Figure 2-3. Five-week-old human embryo. Close view of the PGC seen in Figure 2-1. Round mitochondria with tubular cristae, a Golgi complex, RER cysternae, free ribosomes and polysomes are seen in the cytoplasm. TEM X 16,750.

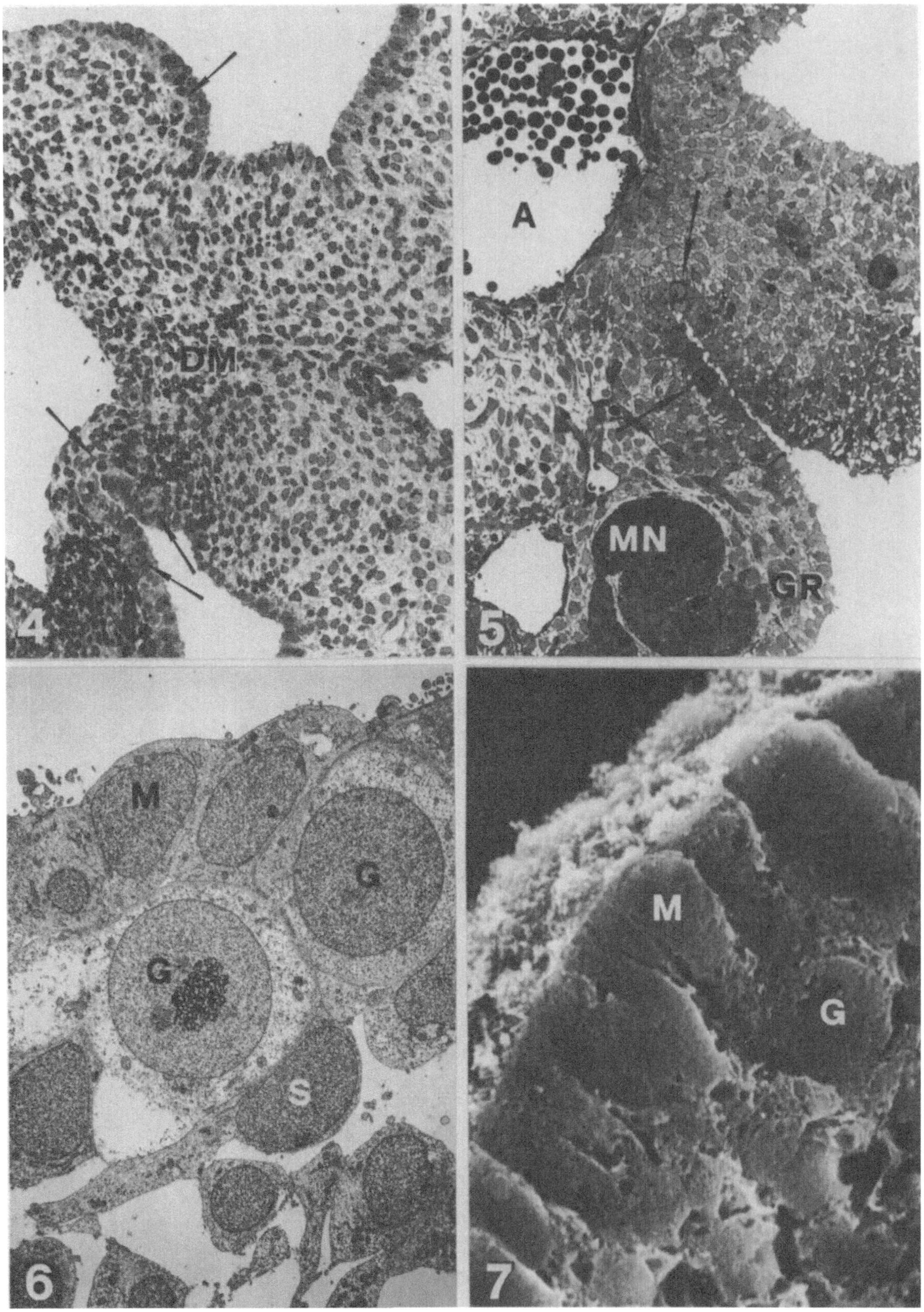

Figures 2-4,2-5. Five-week-old human embryos. Numerous migrating PGCs (arrows) are present in the dorsal mesentery (DM) and near the genital ridge (GR). MN: mesonephros; A: dorsal aorta. Figure 2-4: LM X 300. Figure 2-5: LM X 340.

Figure 2-6. Five-week-old human embryo. Two PGCs (G) are seen in the dorsal mesentery. M: mesothelial cells; S: stromal cell. TEM X 4,500.

Figure 2-7. Five-week-old human embryo. A voluminous, irregularly rounded cell (G), probably a PGC, is also evident by SEM in the dorsal mesentery. M: mesothelial cells. SEM X 3,900.

for use during the migration of the PGCs to the genital ridges (1,41,48). Intracytoplasmic, highly electron-dense yolk granules, as well as large amounts of lipid droplets are present in PGCs of the chick embryo. These inclusions likely correspond to the small rounded elevations of the cell surface that can be detected by scanning electron microscopy (SEM) (49). The yolk material is specific for PGCs of the oviparous vertebrates and, together with lipid inclusions, presumably serves to store energy (43,49), as appears to be the situation for human PGCs. By contrast, in some mammals, particularly rodents, glycogen and lipids are poorly represented in the cytoplasm of the PGCs. However a consistent cytoplasmic feature in rodents is the presence of multivesicular bodies (27).

PGCs appear to incorporate exogenous substances (gases and nutrients) necessary for metabolism mainly by their close association, during migration to the gonads, with neighboring somatic cells (cells of the yolk-sac endoderm, epithelial cells of the gut, mesothelial cells covering the mesentery and the genital ridges, mesenchymal cells, and cells of mesonephric origin). Such associations can be identified as focal areas of close contact between the cell membranes (1,27, 41,49,50). In these areas, the cortical cytoplasm of opposed cells may elaborate a number of finger-like projections (27). According to some investigators, intercellular junctions between PGCs and opposing somatic cells include desmosomes (41, 44), intermediate junctions, and tight junctions (41). Another important function may be ascribed to such intercellular connections; i.e., the exchange of molecular information by which migrating PGCs can find their way to the gonads (41).

In the cytoplasm of PGCs of many mammals (30,51–53), as well as in human oogonia and oocytes (29,54,55), a peculiar material can be observed. This substance, morphologically comparable to the germ plasm of amphibia (56) and to the polar granules of insects (57) has been termed *nuage* (58). The nuage contains RNA and appears, by TEM, as a large fibro-granular mass associated with ribosomes (see Fig. 2-18) or as an electron-dense substance located between clusters of mitochondria (1,29,30,54,59). A similar structure was also observed in certain cells of the early rabbit embryoblast (60). Because this material seems to be closely related to the earliest differentiation of the germ-cell line, the discovery of the nuage in germ cells is considered by some authors to be a structural marker that supports the germ plasm continuity theory (1,30).

Finally, rudimentary cilia have recently been described for the PGCs of the fetal pig. However, their role is still uncertain (19).

3. Migration of primordial germ cells to the gonads

Proliferating PGCs migrate from the yolk-sac endoderm to the gonadal anlage (1,17,18,41,42). In the four-week-old human embryo, many PGCs are found in the hindgut (Diagram B); subsequently, they penetrate the mesenchyme surrounding the epithelium through breaks in the basal lamina (*separation stage*). A few PGCs are seen in the mesothelium, which forms the outermost layer of the wall of the gut (27,37,41). In the five-week-old human embryo, PGCs arrive in the dorsal mesentery, which is formed by stromal cells and by a single layer of cylindrical mesothelial cells (*migration stage*) (Diagram B; Figs. 2-4, 2-6,2-7). Finally, after passage through the dorsal mesentery and the mesenchyme near the genital

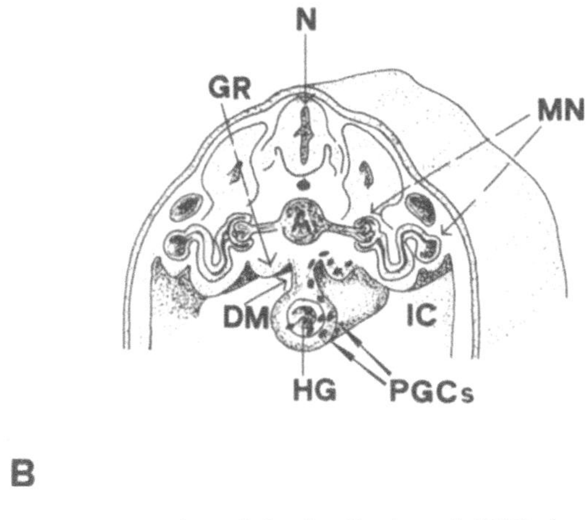

Diagram B. Drawing of the localization of PGCs in early human embryos: hindgut (HG) = 4th week; dorsal mesentery (DM) = 5th week; genital ridge (GR) = 5th–6th week, MN: mesonephric tubuli and glomeruli; A: dorsal aorta; N: neural tube; IC: intraembryonic coeloma.
Redrawn from: Netter, FH, Embriologia Umana ed Anomalie Congenite, CIBA GEIGY, 1983.

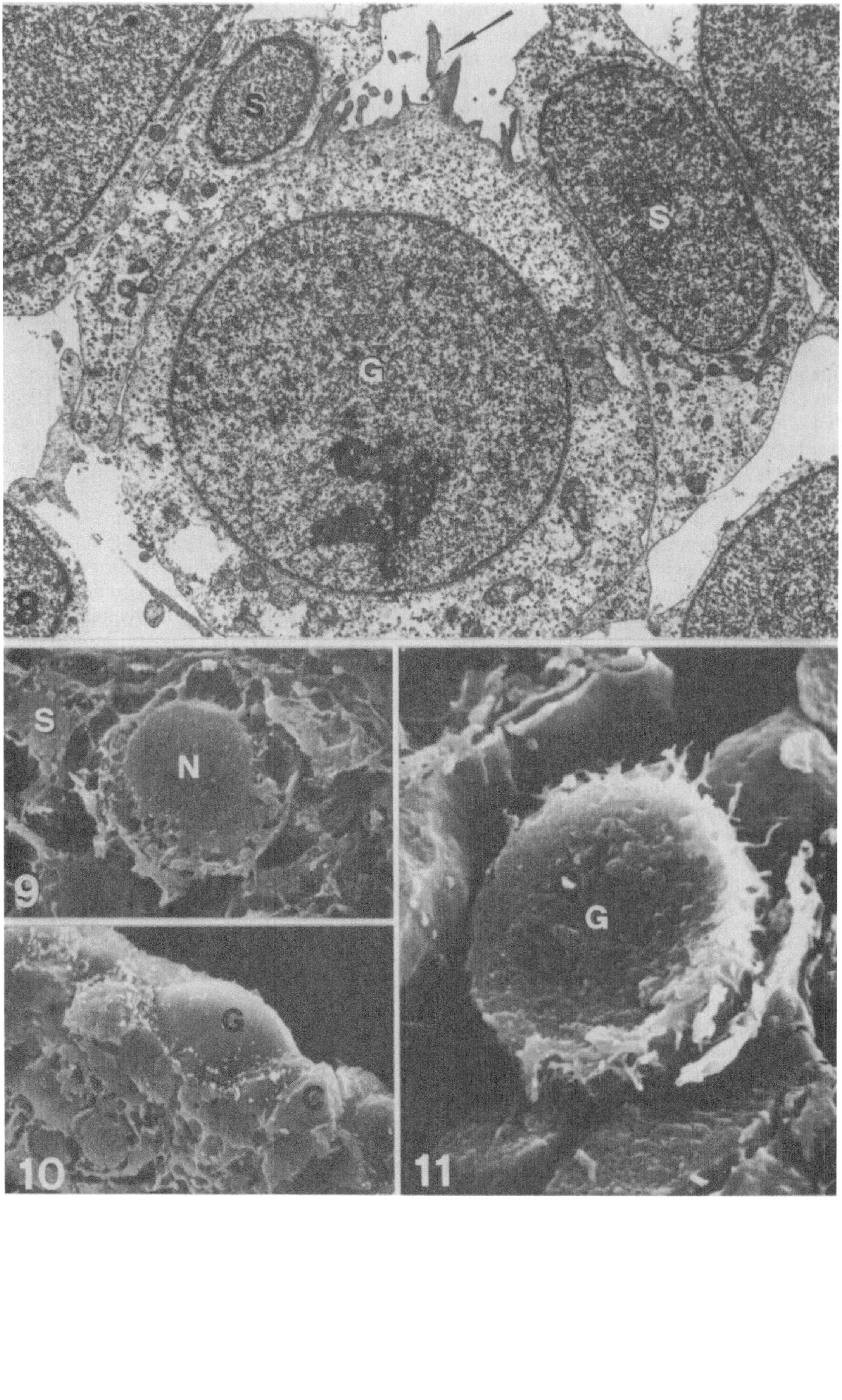
S
S
G
8
S
N
9
G
10
G
11

ridges, PGCs reach the gonadal anlage (*settlement stage*) (Diagram B; Figs. 2-5,2-10) (18,27,37,41). Although it is possible that PGCs are carried to the gonads by the bloodstream, the occurrence of this phenomenon in mammals is assumed to be very rare (18,41). In birds, however, the vascular supply seems to be the principal means by which the PGCs reach their final position (14,18,43,49, 61,62).

Regardless of the manner in which PGCs migrate, some of these cells miss their target and reach unusual sites, often to become trapped in ectopic positions (19,63). Although the majority of these germ cells degenerate (18), others, depending upon the genetic sex of the embryo and the particular ectopic location, may differentiate with the PGCs that have normally colonized the gonads (64,65). Indeed, some extragonadal PGCs may occasionally go on proliferating and give rise to germ cell tumors (1,19,30,35,44,66).

At the separation stage, the shape of the cell and sometimes the nuclear profile are somewhat irregular. The cells may also show a spindle shape (with the major axis varying from 12 to 30 μm in diameter) (41,44).

During the migration stage, the PGC nucleus is extremely large and, as a rule, has a smooth, rounded outline (Figs. 2-1,2-2,2-3,2-6,2-8). This aspect is also evident under SEM (Fig. 2-9). The nucleolus is conspicuous, but slightly altered in organization if compared to earlier stages of germ cell development. The amount of cytoplasmic glycogen and lipids is decreased. By contrast, mitochondria, Golgi complexes and RER cisternae are more developed than previously. The plasma membrane of migrating PGCs exhibits high levels of alkaline phosphatase activity (41).

During the migration stage, the PGCs often retain the irregular shape observed at the separation stage. In addition, they elaborate cytoplasmic protrusions (Fig. 2-8) and pseudopodia containing microfilaments and microtubules (41,44,51). Membrane extensions (Fig. 2-11) and particularly pseudopodia are also detected by SEM (67). The presence of membrane extensions has been observed in migrating human PGCs, as well as in other vertebrates. In the chick embryo, such extensions are present mainly during later stages of migration (43); in rodents, cytoplasmic prolongations are associated with the polarized distribution of cytoplasmic organelles and with deformation of the nuclear profile (27,30,53).

These ultrastructural data confirm that in humans, and in many mammals, the PGCs can actively migrate by their own ameboid movements (10,12,18,41,68). Even the PGCs of birds, after passage through the wall of the vessels, migrate towards the gonadal anlage by ameboid movements (43). Further aspects supporting this view include demonstrations of such ameboid motion by time-lapse photography (1,69,70) and by other experimental means involving cultured samples (17,30,71,72).

At least in some vertebrates, PGC migration seems to be partially dependent on the emission of chemotactic substances from the gonad-forming areas (12,18,30,37,70,73,74). However, such inductive influences are probably responsible for directing the PGCs at later stages of migration, when they are already in proximity to the gonadal anlage (18). These chemotactic factors, which seem to exert a non-species-specific attraction for the germ cells, have not yet been chemically identified (1,19,37). Furthermore, some authors emphasize the possibility that, both in vivo and in vitro, the oriented migration of PGCs of some vertebrates toward the genital ridge can be partially guided by the substrate (69,75–78).

Finally, according to some authors (79), the migration of mammalian PGCs to the gonads

←

Figure 2-8. Five-week-old human embryo. A migrating PGC (G) provided with cytoplasmic protrusions (arrow) is observed among somatic cells (S) in an area similar to that shown in Figure 2-5. The nucleus shows a smooth, rounded profile, and contains a centrally located nucleolus. TEM X 10,500.

Figure 2-9. Five-week-old human embryo. A migrating PGC is seen in proximity of the gonadal ridge. The rounded outline of the PGC nucleus (N) during the migration stage is also evident by SEM. S: somatic cell. SEM X 6,000.

Figure 2-10. Five-week-old human embryo. A rounded cell (G) with a smooth surface, probably a PGC, is observed among the cells of the coelomic epithelium (C) covering the gonadal ridge. SEM X 3,800.

Figure 2-11. Five-week-old human embryo. A PGC (G) provided with membrane extensions is seen in the interstitial tissue underlying the mesodermal cells of the dorsal mesentery. SEM X 8,500.

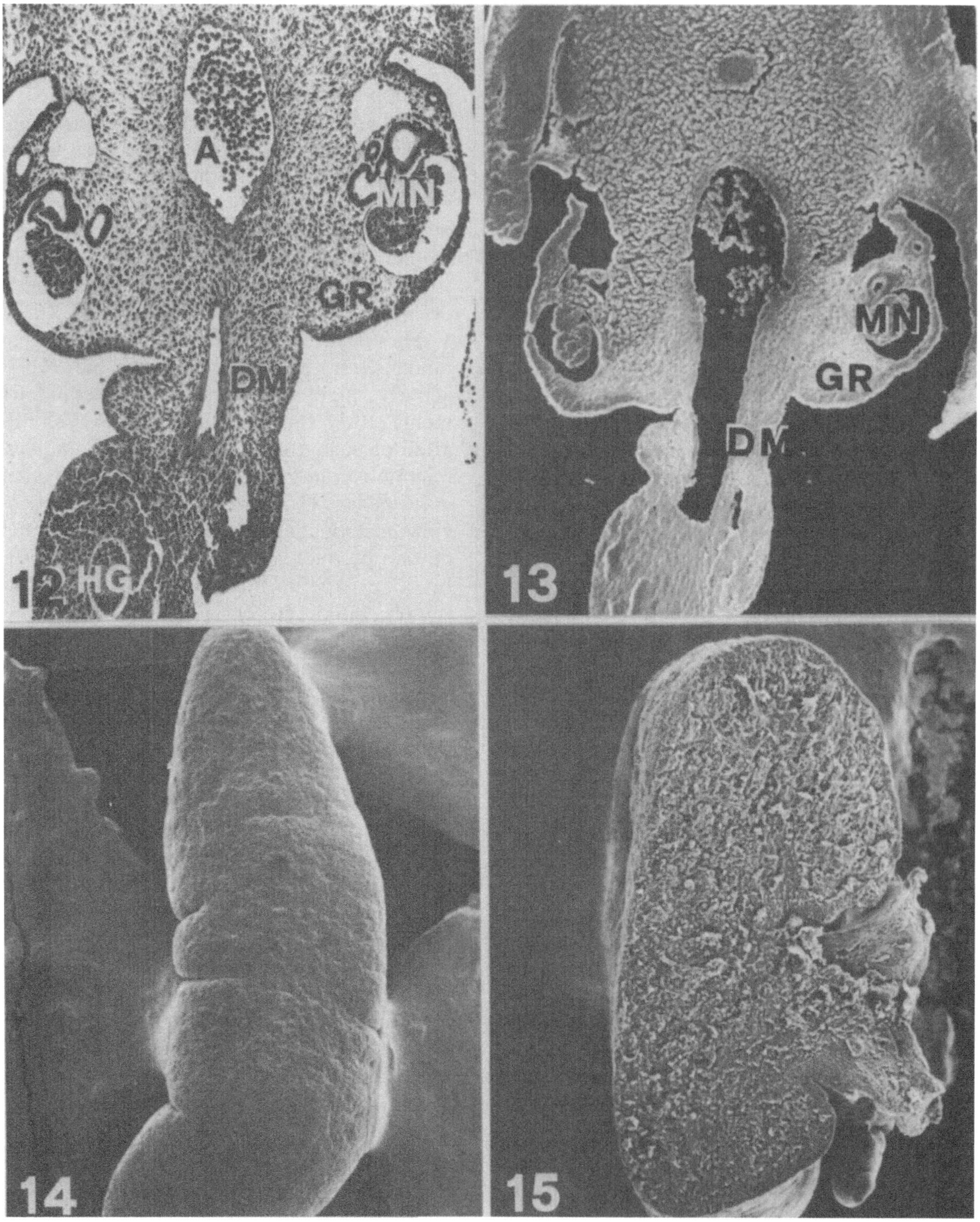

Figures 2-12,2-13. Transverse sections of a 4/5-week-old human embryo. The genital ridges (GR) are forming between the developing mesonephric kidney (MN) and the dorsal mesentery root (DM). A: dorsal aorta; HG: hindgut. Figure 2-12: LM X 150. Figure 2-13: SEM X 140.

Figure 2-14. Nine- to 10-week-old human embryo. General view of the ovary. SEM X 60.

Figure 2-15. Nine- to 10-week-old human embryo. A section of an ovary similar to that shown in Figure 2-14; cortical and medullary areas become apparent. SEM X 85.

could be caused by morphogenetic movements of surrounding cells. Such movements, thought to occur in amphibia (18,80), would be caused by the differential growth of the germ layers during organogenesis; thus, the ultimate location of PGCs would be the result of other active processes (18,19,30).

However, it seems that in mammals, the same events occur only during the early separation stage of PGC migration, when the yolk-sac endoderm makes an invagination into the embryo, forming part of the developing gut (30,41). In this way, PGCs could be transferred passively from their original extraembryonic sites within the developing embryo (30).

4. Colonization of the developing gonads

In humans, PGCs colonize the gonadal primordia during the latter part of the fifth week or at the beginning of the sixth week p.f. (11,42,81).

The first sign of gonadal development is recognizable in the four-week-old human embryo, in which two longitudinal prominences, the so-called genital ridges, appear bilaterally (Figs. 2-12,2-13). These ridges, lying between the developing mesonephric kidney and the dorsal mesentery root, subsequently shorten and protrude into the coelomic cavity as hemispherical bulges. They are formed mainly by the proliferating coelomic epithelium and by the underlying mesenchymal tissue, which contains blood vessels (18,26,37, 42,82). At least in humans and in some other mammalian species, the genital ridges also contain cells of mesonephric origin that derive via the *rete system* (83,84) from the mesonephric capsules of the glomeruli and/or the mesonephric tubules (1,19,27,83–87). It seems that the cells of the coelomic epithelium may attract the PGCs, whereas the mesonephric cells induce both migratory and proliferative activities of the germ cells (87–89). After the PGCs reach the gonadal tissue, they rapidly increase in number by means of frequent mitoses; these PGC divisions are attended by a corresponding hyperplasia of the surrounding somatic components. Both of these elements result in increased size of the gonads (90).

This system is morphologically indifferent as to sex but, at the time of the PGCs' arrival, it possesses all the cellular elements necessary for differentiation (37). The length of the indifferent phase varies according to species (91). In humans, the gonads acquire their female or male morphological characteristics only during the seventh week p.f.; PGCs induce such changes, which, in the female embryo, essentially consist of the development of the cortex and the involution of the medulla (Figs. 2-14,2-15) (18,37,42,82,87,92) (see Chapter 3, this volume).

Upon their arrival at the gonad, proliferating PGCs become non-uniformly distributed among the somatic elements; they tend to occupy the most superficial areas of the organ (44,93), but a few PGCs can also be observed in the developing medulla. By TEM, cortical and subcortical human PGCs usually appear round or elliptical and elaborate small cytoplasmic processes that are randomly distributed over their entire surface (41,44). These features are also detectable in rodents (27,50,53) and in the chick embryo (43, 49). On the other hand some PGCs, especially those found in the primitive medulla, have a relatively irregular shape (41). Gonadal PGCs have a large nucleus, often eccentrically located, and an abundant cytoplasm (94). Within the cytoplasm, glycogen particles and lipid droplets are relatively reduced in number, especially in comparison to those present in migrating PGCs (39,41,68). Small mitochondria, polyribosomes, elements of the endoplasmic reticulum and Golgi cisternae are well-represented (44,93). Alkaline phosphatase activity is demonstrable (41). These ultrastructural features are also recognizable in gonadal PGCs that are still present during the third month of ovarian development (41).

By SEM, numerous PGCs can be detected in the cortical areas of the developing ovary in 7–12-week-old human embryos. As observed by TEM, these primordial cells are relatively smooth-surfaced, voluminous, rounded elements that often display short projections (see Figs. 2-39,2-40). It is also possible to detect a number of PGCs possessing one or two ameboid projections; the presence of such cytoplasmic projections at this stage suggests that some of the gonadal PGCs retain their motile capacity and, possibly, wander through the developing ovarian tissue (90,94,95).

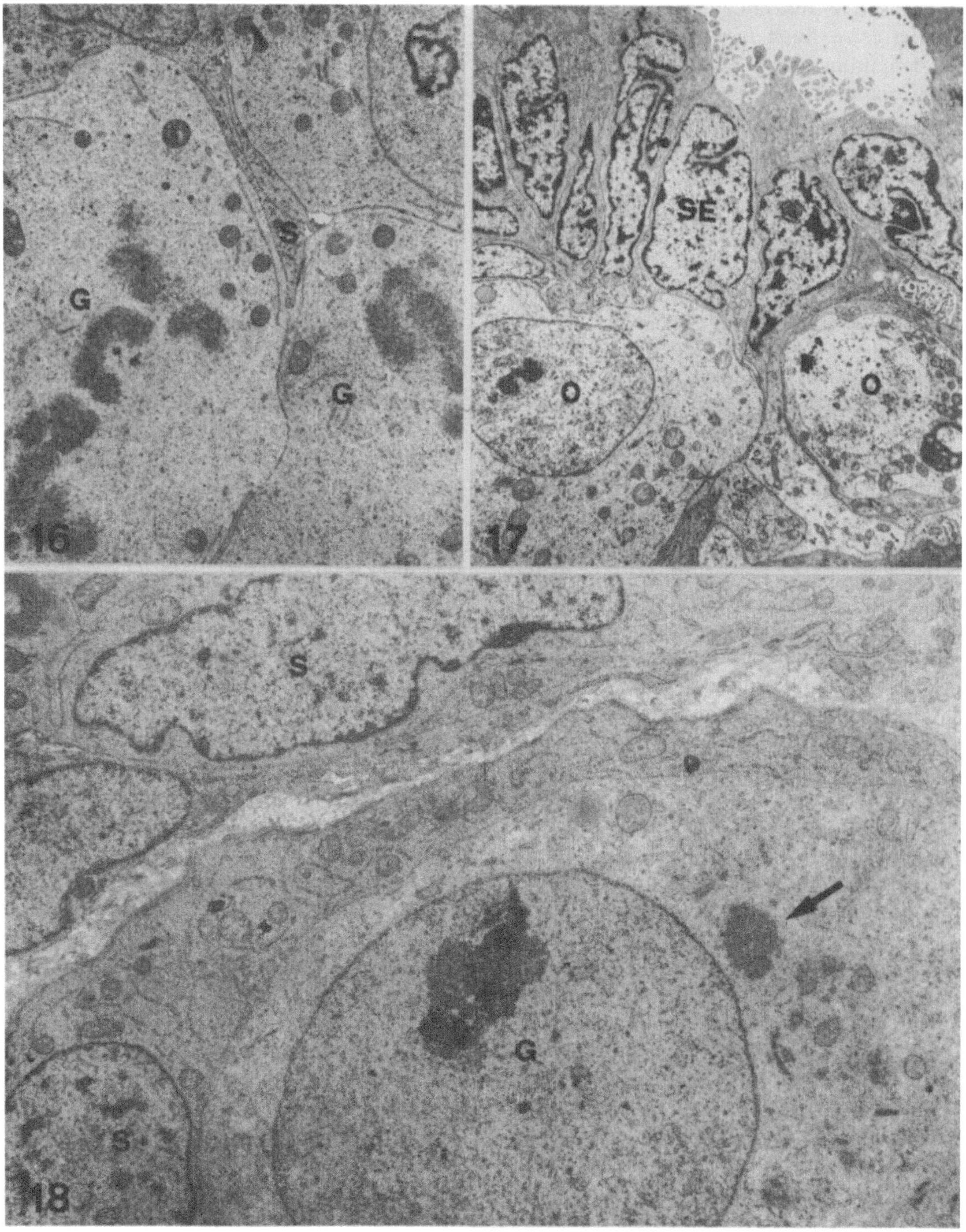

Figure 2-16. Human ovary at 15/16 weeks. Nests of dividing oogonia (G) are observed in the cortical areas of the ovary; a few organelles are seen in their cytoplasm. S: somatic cell. TEM X 5,300.
Figure 2-17. Human ovary at term. Two oocytes (O) are seen just beneath the surface epithelium (SE) of the ovary; numerous organelles are present in their abundant cytoplasm. TEM X 5,900.
Figure 2-18. Human ovary at 14 weeks. A conspicuous fibro-granular body (*nuage*) (arrow) is identified in the cytoplasm of an oocyte (G). S: somatic cells. TEM X 8,700.

5. Germ-cell differentiation

Germ-cell differentiation begins in the cortical areas of the developing ovary (in humans it occurs around the ninth week p.f.) with the formation of oogonia from proliferating PGCs (1,18,93,96) (see Chapter 3, this volume). By TEM, human oogonia appear rather similar to PGCs, but the cell body of the former displays an even more regular and smooth outline (29,97). In addition, oogonia possess a large, round nucleus located at the center of the cell; the nucleoplasm contains one to three distinct reticular nucleoli (84). Within the scanty cytoplasm, a reduced number of organelles are seen. They include Golgi membranes, free ribosomes, round or oval mitochondria usually disposed in a perinuclear arrangement, and poorly developed elements of the endoplasmic reticulum (Fig. 2-16). Lipid inclusions and glycogen granules seem to gradually disappear, whereas alkaline phosphatase reactivity is retained (1,29,84,93,96–103). Desmosome-like structures and small gap junctions can be observed among oogonia and adjacent somatic cells (93,96).

After numerous mitotic divisions (Fig. 2-16) oogonia differentiate into oocytes (see Chapter 3, this volume). Meiosis then begins, usually in the innermost areas of the cortex; in humans, it occurs during the 12th–13th week p.f. (1,18,84, 93,104). These oocytes are in the first meiotic prophase and may be quiescent for as long as 40 years or more, after which the resumption of meiosis occurs (42,93).

The nucleus of the oocyte undergoes morphological changes that lead to the occurrence of synaptonemal complexes during the pachytene stage. By contrast, the organization of the cytoplasm does not change markedly during oogenesis. Nevertheless, cytoplasmic organelles become more numerous, often disposed in perinuclear arrangement. The internal cytoplasmic structure is more complex than that observed in oogonia (Fig. 2-17), and cell dimensions increase as oocyte development progresses towards the diplotene stage (42,93,97,100). In addition, germ-cell alkaline phosphatase activity is lost after commencement of meiosis (1).

Oogonia and oocytes are frequently arranged to form nests within the cortical areas of the developing ovary. In humans, this occurs mainly during the third to fourth month of gestation. These nests, easily detectable by TEM and SEM, are surrounded by a layer of small, irregular and/or flattened somatic cells (Figs. 2-16,2-19, 2-20). It is believed that such structures are formed by germ cells derived from a single PGC, rarely from more than one (18,29,42,100,105).

The germ cells that constitute a nest, most clearly in 10–22-week-old human embryos and fetuses (106), are often joined by intercellular bridges (Fig. 2-21). These syncytia are thought to derive from the incomplete division of the cell body during the last mitotic divisions (93,100, 101,107–113). The cytoplasmic bridges interconnecting sister germ cells are ultrastructurally similar in all species studied. When observed by TEM, they appear cylindrical, sometimes provided with slight convexities and large blebs (Fig. 2-21); moreover, the bridges are bounded laterally by a membrane that is continuous with the plasma membrane of the conjoined cells. A peculiar aspect of these channels is a band of electron-dense material located just beneath the membrane. In addition, cellular organelles such as mitochondria, ribosomes, membranes of the endoplasmic reticulum, vesicles, microfilaments, and microtubules are commonly seen within the cytoplasm of the bridges (93,100,106,113). SEM studies of analogous structures present during testicular germ cell differentiation have confirmed the cylindrical nature of the bridges. The bridges appear to vary in length and diameter, and scattered microvilli are often noticeable on their surface (93,114,115).

The most widely accepted theory on the function of intercellular bridges is that they coordinate and synchronize germ-cell differentiation (84,93,100,110–113,116,117). However, the existence of bridges between developing cells in humans, whose ovarian development seems to be asynchronous (118), makes this an inadequate explanation for their presence (100). On the other hand, the bridges could be involved in degenerative processes affecting the germ cells, particularly during oogenesis. According to the *nurse-cell* concept, modified by Ruby et al. (112) for mammals, some of the germ cells that constitute a nest (nurse cells) might transfer part of their cytoplasmic organelles to the conjoined germ cells

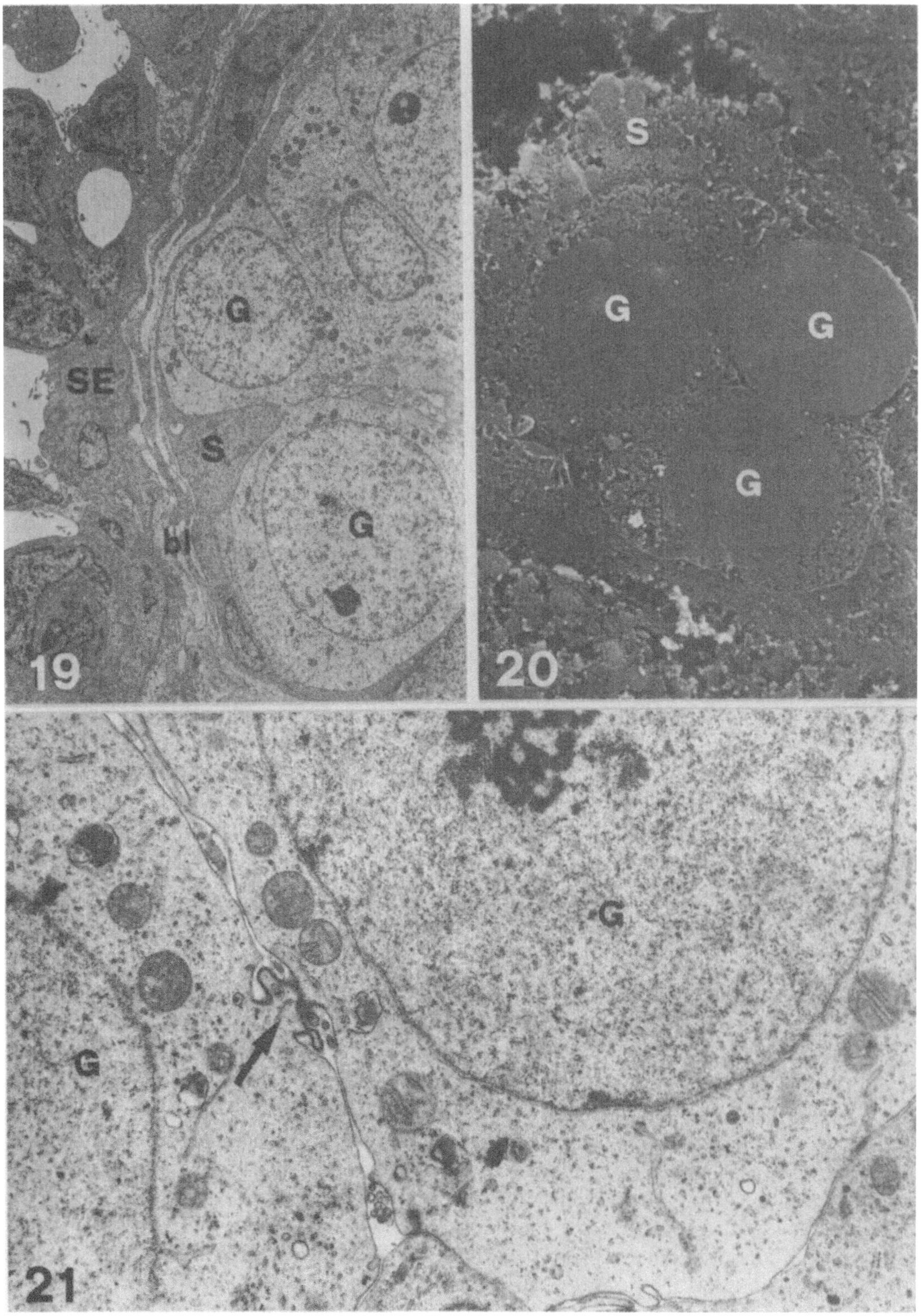

Figure 2-19. Human ovary at 20 weeks. Nests of germ cells (G), associated to irregular and/or flattened somatic cells (S) are located in the cortical areas of the ovary and intermingle with the cells of the surface epithelium (SE). bl: basal lamina. TEM X 4,700.
Figure 2-20. Human ovary at 14/15 weeks. Nests of germ cells (G), associated with somatic cells (S) are also detected by SEM. SEM X 5,800.
Figure 2-21. Human ovary at 15 weeks. Two germ cells (G) are joined by an intercellular bridge (arrow), provided with convexities and blebs. TEM X 12,000.

through the intercellular bridges. As a result of this passage of organelles, the former eventually degenerate, whereas the latter might develop into maturing oocytes (100). Nevertheless, according to other results, such preferential distribution of organelles does not seem to occur among joined germ cells (93). Furthermore, because germ cells showing regressive changes can be linked by intercellular bridges, an alternative implication of the bridges in degenerative processes might be the synchronization of such events (93,106,113).

It has also been suggested that the bridges may have other functions, such as the restriction of the number of mature germ cells derived from a single PGC and the limitation of germ cell motility. Finally, it should not be excluded that such bridges may have a genetic regulatory function, or that they play some role in the pathogenesis of germ-cell tumors (93).

6. Differentiation of follicular cells

The origin of the somatic elements that surround the germ-cell nests and give rise to the follicular cells is still debated.

According to some authors, follicular cells originate from cords of proliferating coelomic epithelial cells (*sex cords*) that intermingle with germ cells and subsequently incorporate them within the follicles (11,42,118,119). Even opinions about the role of the superficial epithelium in ovarian differentiation, and especially in the formation of the sex cords, differ (18,19,26,28,82,84, 89,91,96,120–132).

Mesenchymal stromal cells have been considered to directly contribute cells to the follicles (19,82,125). In addition, several LM and TEM studies have renewed interest in the hypothesis, originally proposed by Waldeyer (2), that the sex cords arise from the mesonephros, and that the walls of the primitive follicles are formed by the proliferation of mesonephric cells derived from the rete system (19,83–86,134–136). In mammals, these cells might also play a role in the onset of meiosis and in the differentiation of the follicle by the secretion of a meiosis-inducing substance (MIS) (1,19,84,137).

Nevertheless, it seems possible that both the coelomic epithelium and the mesonephric cells of the underlying proliferating cords contribute towards the formation of early follicles, at least in humans (29,88,90,95,138,139) and in some vertebrates (1,83,87,88) (see Chapter 3, this volume).

In ovaries of 13–21-week-old human fetuses, such cords appear, by TEM and SEM, as irregular masses composed of somatic cells. These cells assume a typical cordlike arrangement in the cortico-medullary region, whereas at the periphery of the cortex, some of these masses are confluent with ingrowths derived from the intermingling surface epithelium (Fig. 2-22). In such cases, the basal lamina covering the cords is continuous with that underlying the surface epithelium (90,94). Rare PGCs and numerous oogonia and oocytes are scattered among the somatic cells of the cords and of the surface epithelium (82,101, 118,139). As previously mentioned, zones of attachment between germ and supporting cells can be detected (Fig. 2-23) (140), although such attachments are more numerous in later stages (90). By TEM, the superficial cells show a columnar arrangement (Fig. 2-25), and occasionally, it is even possible to see irregularly shaped cells, whose features are often distorted because of their relationship with the germ cells. The nuclei of the superficial cells are oval, occasionally indented, and contain uniformly distributed chromatin and large nucleoli. The cytoplasm contains free ribosomes, elongated mitochondria, RER elements, Golgi membranes and, occasionally, lipid droplets. The lateral aspect of the cells is provided with desmosomes and irregular, often interdigitating, fingerlike projections. Mainly in the areas where confluence between the underlying cords and the surface epithelium can be observed, the latter appears to be composed of a number of irregularly invaginated layers. Within these layers, the superficial cells may show somewhat irregular features and form large intercellular lacunae bordered by microvilli (Fig. 2-26) (90). By SEM, microvilli, isolated cilia and blebs frequently occur on the exposed surface of the coelomic epithelium (Figs. 2-30,2-31,2-32,2-33) (90,95). The somatic cells that form the sex cords are similar to those of the surface epithelium, not only where the two are in continuity but also where they are apparently separated by the interposition of a nascent tunica albuginea. The albuginea consists of a basal lamina, fibroblasts

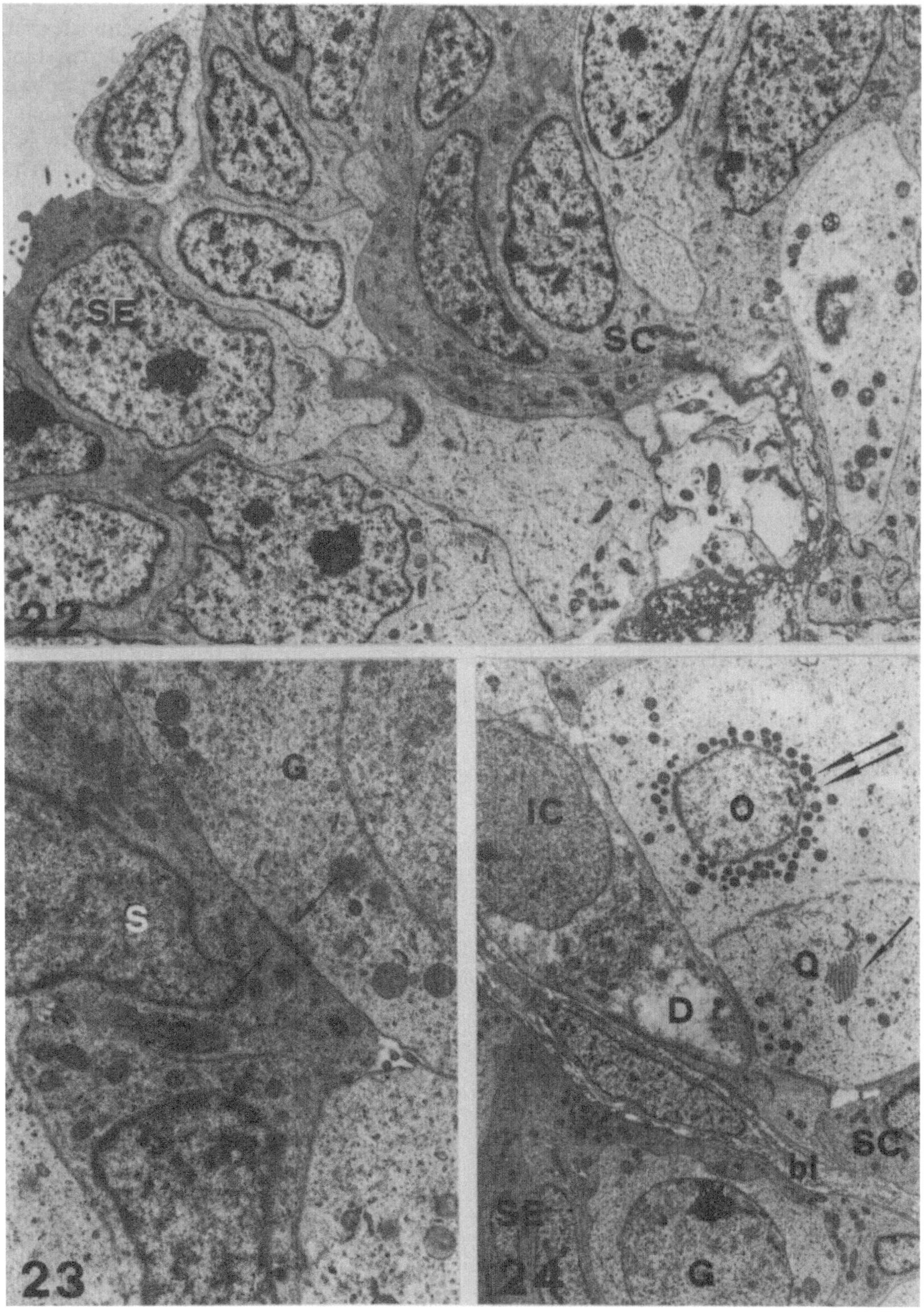
SE
SC
G
22
G
S
23
IC
O
Q
D
SC
bl
SE
G
24

derived from the differentiation of the primitive mesenchymal cells, and collagen fibers (90).

During these developmental stages, both the surface epithelium and the sex cords actively proliferate. However, this aspect is even more pronounced in ovaries of 17–21-week-old human fetuses. Such morphodynamic activity lends an irregular shape to the ovarian surface, due to the formation of papillary projections and deep infoldings (Figs. 2-25,2-28) (90,94,118,121,141). Also during these stages, the tunica albuginea becomes more evident (90).

In ovaries obtained from human fetuses of 22 weeks until term of gestation, the surface epithelium generally appears to be one cell thick, although a few areas in which the epithelium is multilayered and forms infoldings and papillary protrusions can be observed (Figs. 2-27,2-29) (90,94). During these developmental stages, the formation of follicles occurs (94). Single oocytes, which are the most common germ cells detectable during these later stages of differentiation, are invested by a layer of flattened and/or polyhedral cells, which correspond to the future follicular cells (Figs. 2-34,2-36,2-37,2-38). Thus, intercellular bridges are eliminated, possibly by follicular cells or by degeneration of adjacent germ cells (82). Numerous gap junctions are present between the oocyte and the follicular cells (140). In the mouse, such contacts can be detected between oocytes and follicular cells as early as fetal day 17, and their number increases with fetal age (136). The increase in junctional complexes seems to occur in response to the increasing metabolic activity by the oocyte, which becomes excluded from the vascular compartment during the formation of the follicle (136). These junctions are important in the maintenance of the oocyte, allowing it to incorporate nutrients and favouring the synchronous differentiation of the oocyte-follicular cell complex (136,140,142). When this direct intercellular communication does not occur, the oocyte degenerates (136, 143). In addition, a basal lamina lies between the follicular cells and the surrounding somatic cells. The segregation of such primitive follicles is apparently dependent upon both discontinuities of the basal lamina covering the cords and extensive proliferation of somatic elements. These events are observed mostly in abundantly vascularized areas, probably because of the greater transport of nutrients and hormonal factors in such zones (90). Although follicles are formed normally even in anencephalic fetuses, their growth is severely retarded (144).

As recently observed by electron microscopy, both light and dark somatic cells can always be seen both in the surface epithelium and in the sex cords during all phases of embryonic and fetal development (Figs. 2-22,2-24,2-25). These two types of somatic cells, also present in differentiating ovaries of other vertebrates, generally do not show any significant differences in paraffin-embedded specimens, whereas they are easily distinguished from one another in plastic-embedded and glutaraldehyde-osmium tetroxide-fixed specimens (19,87). Light cells possess nuclear and cytoplasmic features similar to dark cells, although the former stain less intensely than the latter. Moreover, light cells often show an epithelioid shape and short cellular projections, whereas dark cells are usually characterized by a fibroblastlike appearance which resembles the cells of the rete system (83,84). This contrasting morphology is also recognizable in the follicle-forming cells. Finally, light cells seem to originate directly from the coelomic epithelium; on the contrary, dark cells probably arise from the

←

Figure 2-22. Human ovary at 14 weeks. At the periphery of the ovarian cortex, the sex cords (SC) appear confluent with ingrowths deriving from the intermingling surface epithelium (SE). The somatic cells of the sex cords are similar to those of the surface epithelium, and both of them are composed by *light* and *dark* cells. TEM X 7,000.

Figure 2-23. Human ovary at 16 weeks. Note a wide zone of attachment (arrows) between a germ cell (G) and a supporting somatic cell (S). TEM X 11,000.

Figure 2-24. Human ovary at 14 weeks. A spindle-shaped cell (IC) is present in the cortical area of the ovary; lipid droplets (D) are seen in the cytoplasm. This cell likely corresponds to a primitive (interstitial) steroidogenic cell. Two adjacent oocytes (O) show a stack of annulate lamellae (arrow) and numerous mitochondria around the nucleus (double arrow). G: germ cell; bl: basal lamina; SE: surface cells; SC: sex cords. TEM X 4,830.

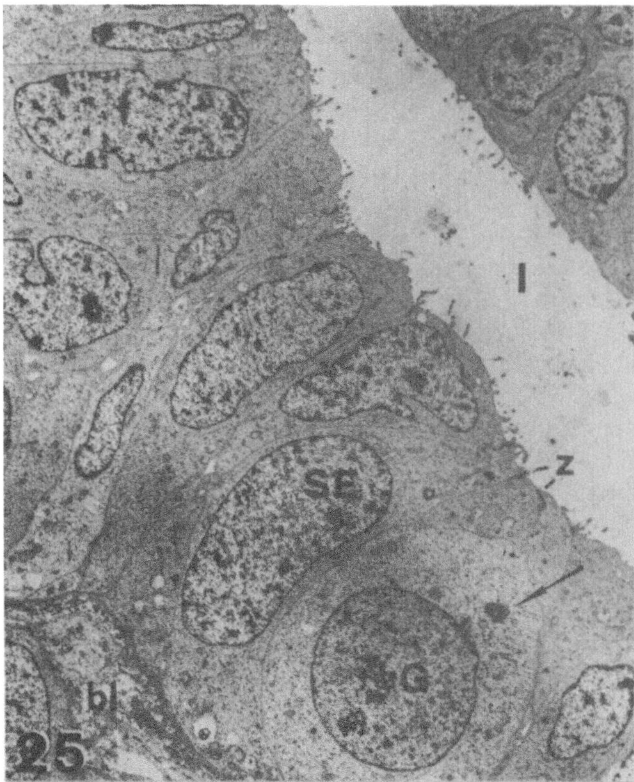

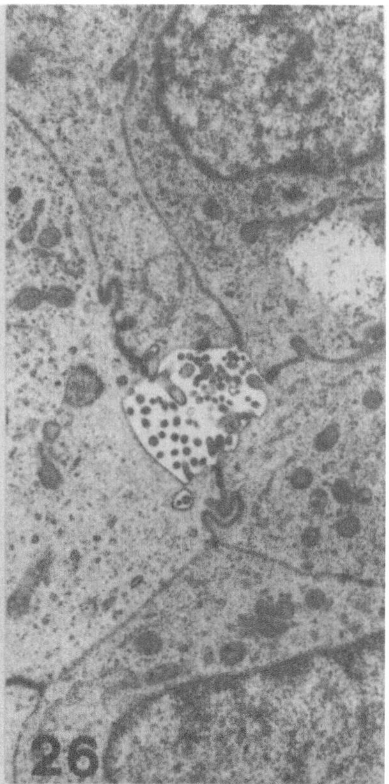

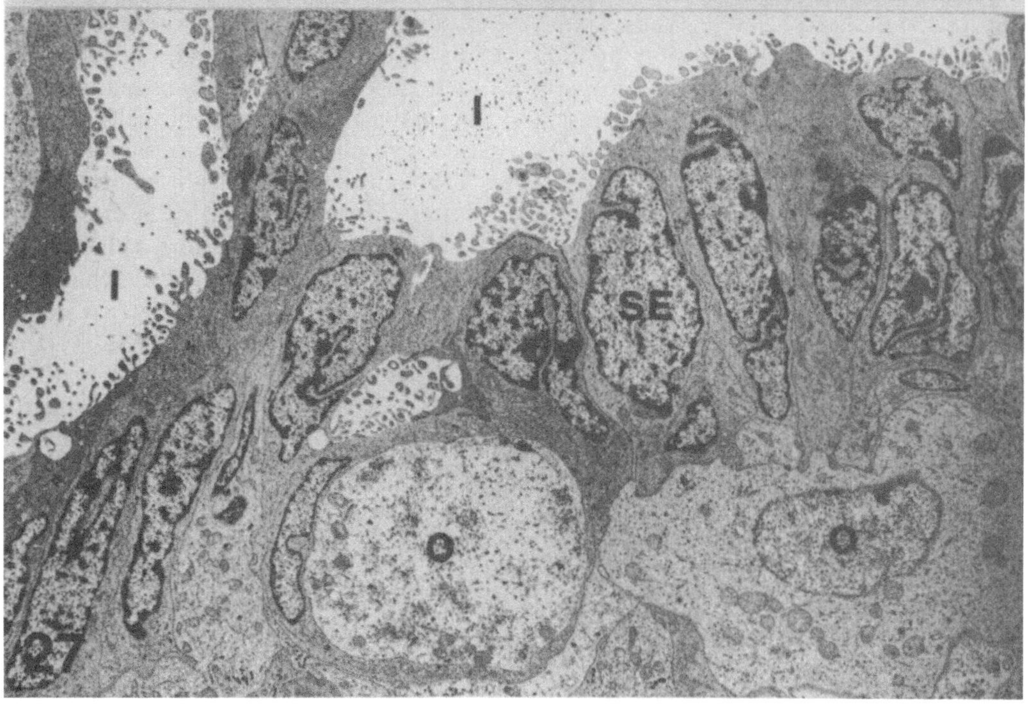

Figure 2-25. Human ovary at 18 weeks. The superficial cells (SE) have a columnar arrangement. Attachment zones (z) are evident between adjacent superficial cells. Germ cells (G) are contained within the surface epithelium. Note the presence of nuage material (arrow) in the germ cell's cytoplasm. bl: basal lamina; G: germ cell; I: infolding of the surface epithelium. TEM X 6,000.
Figure 2-26. Human ovary at 15 weeks. Stratified surface epithelial cells form intercellular lacunae bordered by microvilli. TEM X 11,000.
Figure 2-27. Human ovary at term. A few areas in which the epithelium of the ovarian surface (SE) forms infoldings (I) can be still observed in these later stage of development. Oocytes (O) are closely associated to it. TEM X 6,140.

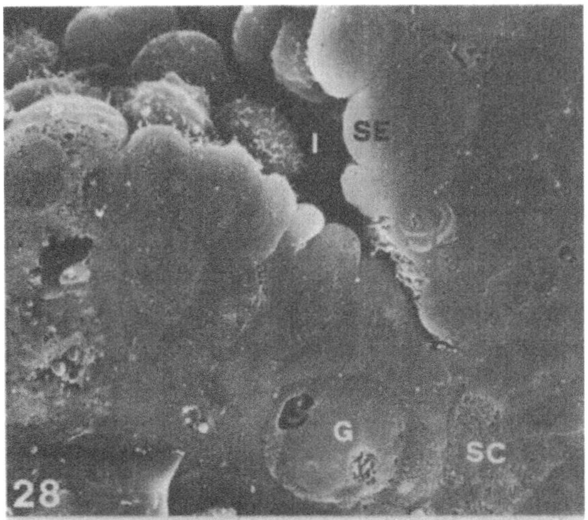

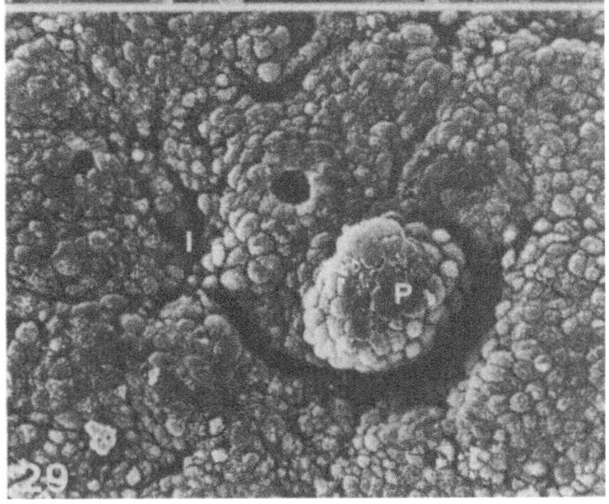

Figure 2-28. Human ovary at 17/18 weeks. Deep infoldings, similar to those observed by TEM (Figure 2-25), are more easily detectable by SEM (I). SE: surface epithelium; G: germ cell; SC: somatic cells of the cords. SEM X 3,000.

Figure 2-29. Human ovary at term. Infoldings (I) and papillary protrusions (P) are easily detected by SEM in ovaries at term. SEM X 1,000.

mesonephros, and intermingle with the light cells during gonadal development up to the period of follicle formation in the ovary (87) (see Chapter 3, this volume). A similar double origin has also been proposed for the primitive interstitial cells and for the cells of the theca interna of the ovary (Fig. 2-24) (90,145). On the contrary, other authors suggest that the primitive interstitial cells are derived mainly from mesonephric cells (19, 84,146), or from medullary fibroblastlike cells (147,148).

7. Reduction in number of the germ cells

7.1. Germ cell degeneration

In the human embryo, only a few hundred PGCs are present at the beginning of their migration to the gonadal anlage. As mentioned previously, the number of PGCs rapidly increases by mitosis during their migration to, and after reaching, the gonads. In addition, oogonia, which originate from PGCs in the developing ovary, not only differentiate into oocytes but also show a remarkable mitotic activity. Thus, during the fifth month p.f., the number of germ cells in the ovary reaches a peak of 6–7 million (1,42,149). At about this time, the degeneration of germ cells begins. This process affects oogonia and, above all, primary oocytes in zygotene and pachytene stages (1,42, 104,149).

There is some evidence that oogonia degenerate through defective formation of the meiotic spindle (98). Moreover, during the early stages of prophase of the first meiotic division, the oocytes are subjected to crossing-over, which seems to be a high-risk procedure, resulting in chromosomal breaks and other genetic errors, so that few oocytes survive to develop further. It may also be possible that metabolic alterations and/or deficiencies resulting from both inadequate vascularization and defective connections with follicular cells are involved (1,37,97,136,149,150). However, the primum movens of the degenerative process that drastically reduces the pool of oocytes at birth still remains obscure (37,94,97, 106,151).

Irregular mitotic figures in some oogonia, and pyknotic changes in oocyte nuclei, often associated with a retracted, eosinophilic cytoplasm, are the histologic features that herald germ-cell degeneration (37). By TEM, the degenerating germ cells show swollen nuclei, with alterations in the nuclear profile and various degrees of chromatin condensation (Fig. 2-35). The chromosomes may also be condensed in homogeneous strands. The cytoplasm contains both normal and abnormal mitochondria; the latter are swollen and show disrupted cristae. Dilatation of the endoplasmic reticulum can also be seen (97,106). Intercellular bridges are observed between early degenerating cells, but they are rare, distorted or disrupted

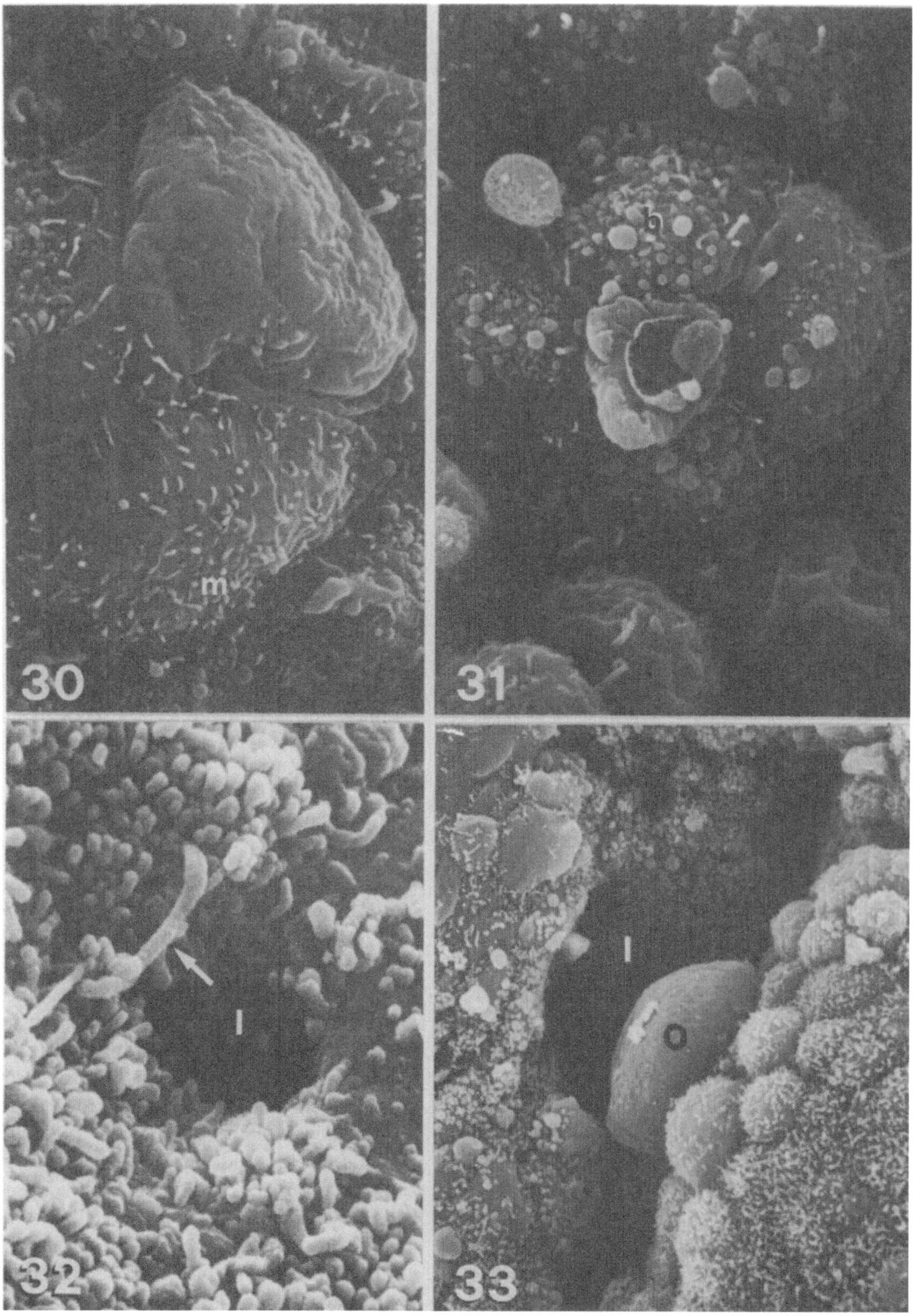

Figures 2-30,2-31,2-32. Human ovaries at 13/14 weeks. Microvilli (m), blebs (b), and isolated cilia (arrow) are seen on the exposed surface of the coelomic epithelium. I: infolding of the surface epithelium. Figure 2-30: SEM X 8,000. Figure 2-31: SEM X 8,000. Figure 2-32: SEM X 13,200.

Figure 2-33. Human ovary at term. The surface expansions of the coelomic epithelium are also evident in this stage of development. I: infolding of the superficial epithelium; O: oocyte emerging on the surface of the ovary. SEM X 4,500.

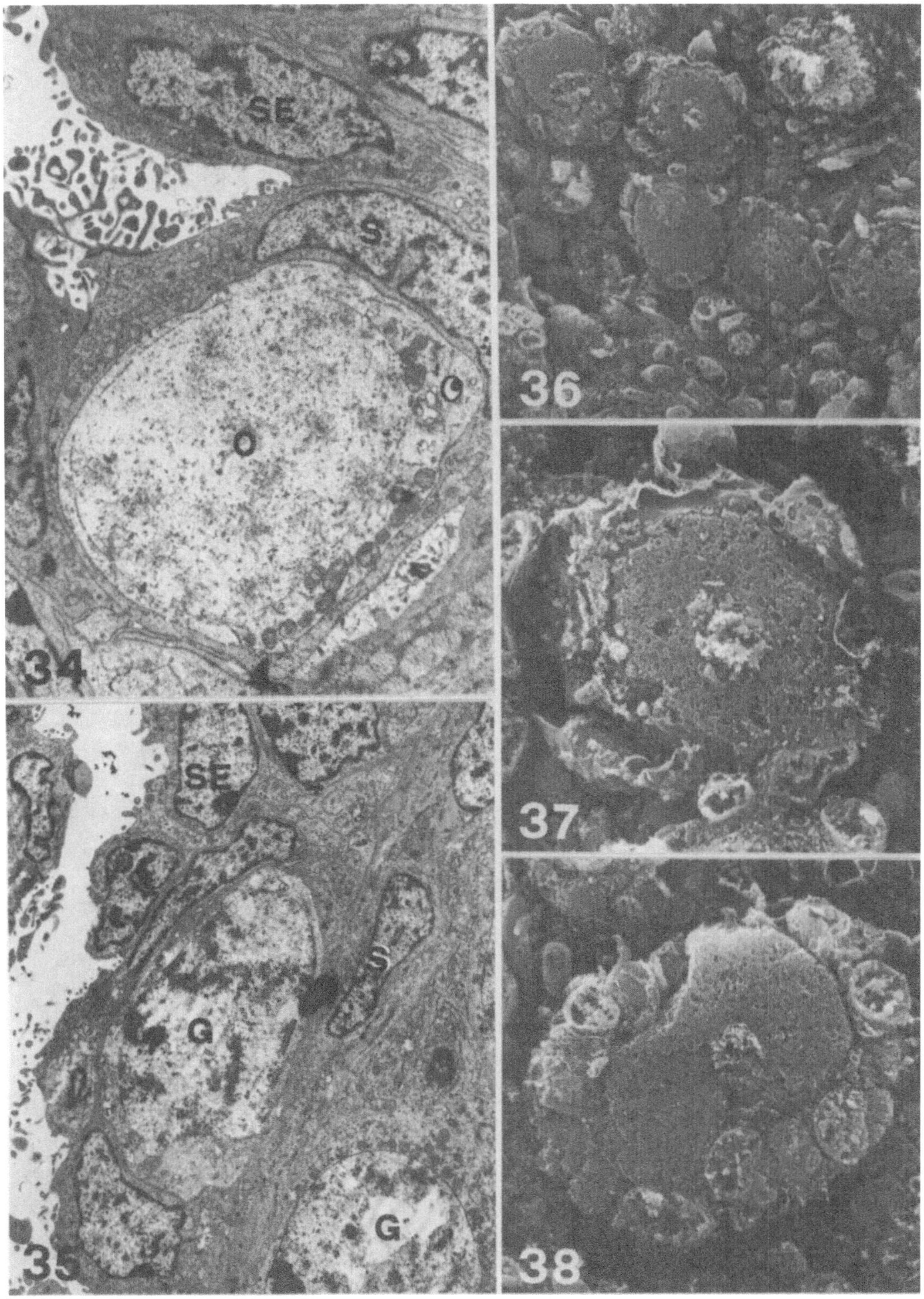

Figure 2-34. Human ovary at term. An oocyte (O), invested by a layer of flattened somatic cells (S) that correspond to the future follicular cells, is contained within the surface epithelium (SE) of the ovary. TEM X 7,100.

Figure 2-35. Human ovary at 19/20 weeks. These oocytes (G) show swollen nuclei with peripheral chromatin condensation and appear to undergo degeneration. S: somatic cells; SE: surface epithelium. TEM X 6,000.

Figures 2-36,2-37,2-38. Human ovary at term. Individual ooctyes, contained in primitive follicles, are scattered into the ovarian cortex. Some of these appear atretic: note the altered shape of the follicle and the vacuolized cytoplasm of the oocyte in Figure 2-38. Figure 2-36: SEM X 1,800. Figure 2-37: SEM X 4,600. Figure 2-38: SEM X 4,600.

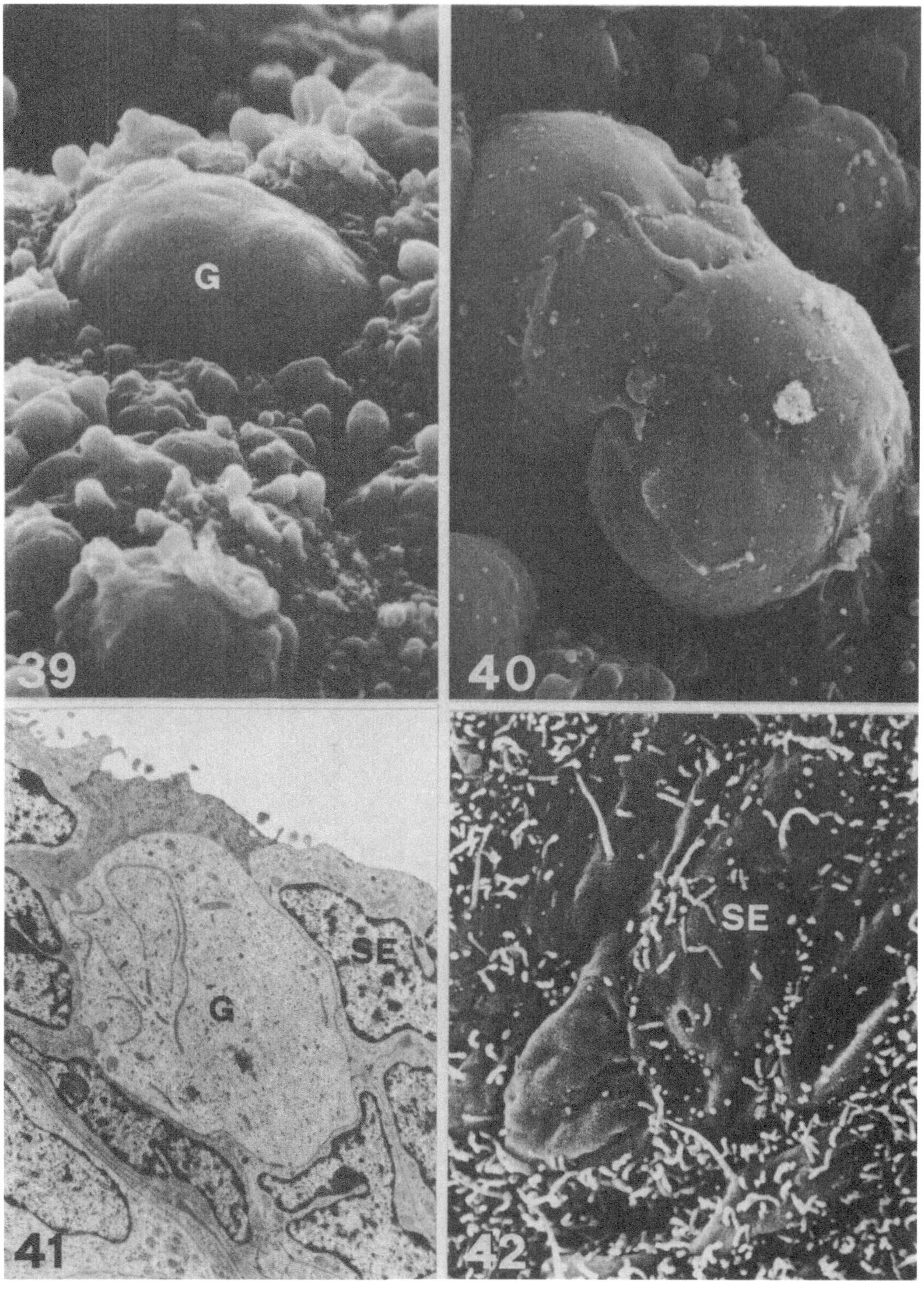
G
39
40
SE
G
41
SE
42

between cells in advanced stages of degeneration. As already noted, it has been suggested that these bridges are somehow involved in these regressive processes (29,93,100,106).

Degeneration can affect not only the germ cells, but also the adjacent follicular cells; this aspect is called *atresia* (Figs. 2-36,2-38). On the contrary, in some cases, the latter survive and are even capable of phagocytizing the degenerating germ cells (37,97,98,101,106,150).

Thus, atretic events begin in the ovary during fetal development, continue throughout the span of reproductive life, and affect primordial, primary, and more developed follicles. This fact, together with the inability of the oocytes to increase their number by mitotic divisions, leads us to observe that ovulated oocytes represent only a small proportion of the total number of gametes produced during oogenesis (1,37).

7.2. Elimination of germ cells from the ovarian surface

The extrusion of the germ cells from the surface of the gonad into the coelomic cavity is another mechanism that seems to promote a reduction in the number of germ cells present within the ovary. This morphodynamic aspect has been studied by LM, TEM and SEM in the mouse (152–155) and human (29,90,94,95,139,156).

According to Bonilla-Musoles et al. (156), who demonstrated this phenomenon in humans, and to Byskov and Rasmussen (154), who noted it in newborn mice, this process of oocyte depletion is the most important one, especially if compared to degenerative events such as germ cell necrosis, autolysis, and follicular atresia. Nevertheless, at least in humans, the latter degenerative processes appear most responsible for the drastic reduction in the pool of oocytes at birth, even though the elimination of some oocytes from the surface of the ovary has been observed (29,94).

During the seventh to eighth week of development, some PGCs are located in rather superficial areas of the gonad. These cells are mixed with coelomic epithelial cells or are observed in the process of emerging from the ovarian surface (Figs. 2-39,2-40) (29). As mentioned earlier, these PGCs may have an irregularly rounded body and occasionally elaborate one or two ameboid prolongations (90,94). The presence of residual ameboid aspects, mainly during these early stages, seems to suggest the possibility of an active local migration of PGCs towards the epithelial covering of the gonad. Subsequently, in 12–24-week-old human fetuses, the most frequently observed germ cells are rounded oogonia and oocytes, often found in clusters and associated with small, proliferating somatic elements (Figs. 2-19,2-20). These germ cells may be detected in superficial areas of the ovary, where the somatic cells may appear to be flattened (Figs. 2-41,2-42) and/or desquamating (Fig. 2-43), probably because of the pressure exerted by the subjacent proliferating clusters. In this extremely proliferative stage of ovarian development, germ cells, as a result of morphogenetic events, could be passively pushed into the superficial epithelium (Figs. 2-41,2-43) and, once free on the ovarian surface, could be eliminated into the peritoneal cavity, sometimes accompanied by satellite somatic cells. In fact, germ cells can frequently be observed on certain areas of the ovarian surface (Fig. 2-44).

In ovaries at term, occasional oocytes, either isolated or associated in small groups of two to four elements, can be seen within the surface epithelium or free on the ovarian surface (Fig. 2-45). These observations suggest that the mechanism of oocyte extrusion might continue to operate even

←

Figure 2-39. Human ovary at seven to eight weeks. A large, irregularly rounded PGC (G) is observed among the superficial cells (SE), and appears to emerge from the ovarian surface. SEM X 7,600.

Figure 2-40. Human ovary at seven to eight weeks. Two PGCs probably just after mitosis and still in contact by means of cytoplasmic projections are seen on the surface epithelium. SEM X 11,300.

Figures 2-41,2-42. Human ovaries at 24 weeks. These pictures show a pleomorphic surface epithelium with very flattened superficial cells (SE). In Figure 2-41, a mitotic germ cell (G) is seen within the surface epithelium. Figure 2-41: TEM X 7,510. Figure 2-42: SEM X 7,700.

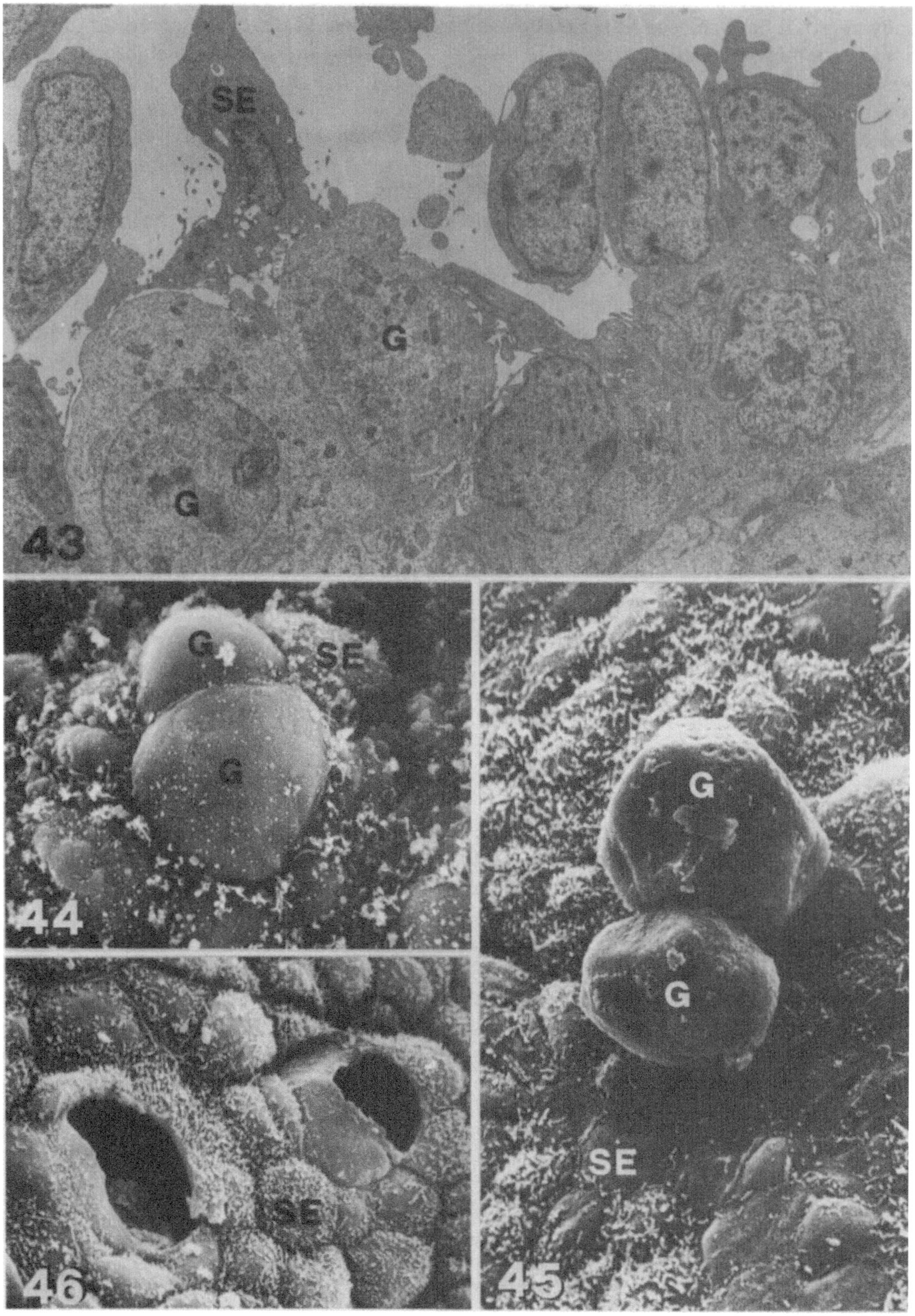
43
SE
G
G
44
G
SE
G
45
G
G
SE
46
SE

during the period between birth and puberty (29). In other areas, rounded pits are evident and likely correspond to temporary breaks in the surface epithelium through which germ cells may be extruded onto the ovarian surface (Fig. 2-46) (90).

In general, these observations seem to indicate that the germ cells eliminated from the ovarian surface into the coelomic cavity could be the same cells that earlier become isolated within the stroma and that had temporarily been loosely associated with the somatic cells. In later stages, during which the sexual differentiation of the ovary into cortex and medulla takes place, and during which the superficial epithelium is separated from the underlying tissues by a basal lamina, germ cells that had previously been associated with the most superficial areas of the ovary become incorporated into the superficial epithelium. From this location they can ultimately be eliminated in the peritoneal cavity. On the other hand, some of these residual germ cells can even give rise to some abnormal cell growths, such as teratomas (29,90,94,95). By contrast, germ cells that in earlier developmental stages had been located deep within the ovarian tissues and were closely associated with the somatic cords maintain this relationship until the time of formation of the primordial follicles. It should be recalled that these germ–somatic cell complexes (future follicles) are always surrounded by an intact basal lamina. The germ cells contained in these early follicles are destined for further development, or degeneration (follicular growth and/or atresia). It does not seem likely that oocytes contained within the follicles could be extruded into the peritoneal cavity. Nevertheless, the fusion and subsequent disappearance of both basal laminae (i.e., of that beneath the superficial epithelium and that surrounding the follicle) may occasionally be detected. This observation suggests that the loss of the basal lamina may provide an ancillary mechanism for the elimination of oocytes (155). Therefore, the formation of a continuous basal lamina delimiting both follicles and surface epithelium seems to be of great importance in the determination of the destiny of the germ cells (29,94).

8. Conclusions

TEM and SEM, associated with classic LM and other experimental techiques, offer a more accurate description than previously available of the origin and fate of the human female germ cells, as well as a more comprehensive view of the three-dimensional details of the dynamic morphological events involved in human female gonadogenesis.

Unfortunately, some aspects in the life history of the oocyte remain unresolved. For example, how and where germ cells differentiate from somatic elements are still open questions. Moreover, the origin of the follicular somatic components, which are strictly associated with the germ cells and which frequently experience the same destiny, remains obscure. Other questions, such as those relevant to function(s) of the intercellular bridges, and the process that initiates the severe reduction in the number of intragonadal germ cells, are also still unresolved. Finally, the apparent significance of aberrant germ cell mitotic activity in the induction of neoplastic processes, and the obvious necessity of better analyzing the etiopathogenesis and associated subcellular aspects of such processes, must be emphasized for future studies.

←

Figure 2-43. Human ovary at 20 weeks. Cells of the surface epithelium (SE) appear desquamating into the peritoneal cavity. Some germ cells (G) are emerging on the surface of the ovary. TEM X 7,400.

Figure 2-44. Human ovary at 17/18 weeks. Two germ cells (G) appear to leave the ovary, emerging among surface cells (SE). SEM X 5,900.

Figure 2-45. Human ovary at term. Two smooth-surfaced, large and rounded cells (G), probably corresponding to germ cells, are just extruded on the ovarian surface (SE). SEM X 5,750.

Figure 2-46. Human ovary at term. These rounded pits probably correspond to temporary openings in the surface epithelium (SE), through which germ cells may be extruded onto the ovarian surface. SEM X 6,000.

References

1. Byskov AG: Primordial germ cells and regulation of meiosis. In: Reproduction in Mammals. 2nd edition, Book 1, Germ Cells and Fertilization. CR Austin, RV Short (eds), Cambridge: Cambridge University Press, p 1–16, 1982.
2. Waldeyer W: Eierstock und Ei. Leipzig: Engelmann, 1870.
3. Nussbaum M: Zur differenzierung des Geschlechts im theirreich. *Arch Mikrosk Anat* 18: 1–120, 1880.
4. Weissmann A: Die Continuitat des Keimplasmas als Grundlage einer Theorie der Verebung. Quoted by Zuckerman and Baker, 1977 (see 18).
5. Fuss A: Uber extraregionare Geschlechtzellen bei einem menschlichen Embryo von 4 Wochen. *Anat Anz* 39: 407–409, 1911.
6. Fuss A: Uber die Geschlechtzellen des Menschen under der Saugetiere. *Arch Mikr Anat* 81: 1–23, 1913.
7. Politzer G: Uber Zahl, Lage und Befaschaffenheit der "Urkeinzellen" eines menschlichen Embryo mit 26–27 Ursegmentenpaaren. *Anat Entwickl* 87: 766–780, 1928.
8. Politzer G: Die Keimbahn des Menschen. *Anat Entwickl* 100: 331–361, 1933.
9. Muratori G: Embryonal germ cells of the chick in hanging-drop cultures. *Contrib Embryol Carnegie Inst* 26: 59–69, 1937.
10. Everett NB: Observational and experimental evidences relating to the origin and differentiation of the definitive germ cells in mice. *J Exp Zool* 92:49–91, 1943.
11. Gillman J: The development of the gonads in man, with a consideration of the role of fetal endocrines and the histogenesis of ovarian tumors. *Contrib Embryol Carnegie Inst* 32: 81–131, 1948.
12. Witschi E: Migration of the germ cells of human embryos from the yolk sac to the primitive gonadal folds. *Contrib Embryol Carnegie Inst* 32: 67–80, 1948.
13. Chiquoine AD: The identification, origina dn migration of the primordial germ cells in the mouse embryo. *Anat Rec* 118: 135–146, 1954.
14. Simon D: La migration des celules germinales de l'embryon de poulet vers les ébauches gonadiques; preuves expérimentales. *CR Soc Biol* (Paris), 151: 1576–1580, 1957.
15. Mintz B: Germ cell origin and history in the mouse: genetic and histochemical evidence. *Anat Rec* 127: 335–336, 1957.
16. Mintz B: Continuity of the female germ cells line from embryo to adult. *Arch Anat Microsc Morphol Exp* 48: 155–172, 1959.
17. Blandau RJ, White BJ, Rumery RE: Observations on the movements of the living primordial germ cells in the mouse. *Fertil Steril* 14: 482–489, 1963.
18. Zuckerman S, Baker TG: The development of the ovary and the process of oogenesis. In: The Ovary. 2nd edition. Vol. 1. S Zuckerman, BJ Weir (eds), New York, London: Academic Press, p 41–67, 1977.
19. Byskov AG: Differentiation of mammalian embryonic gonad. *Physiol Rev* 66: 71–117, 1986.
20. Kingery HM: Oogenesis in the white mouse. *J Morph Physiol* 30: 261–315, 1917.
21. Brambell FWR: The development and morphology of the gonads of the mouse. I. The morphology of the indifferent gonad and of the ovary. *Proc Roy Soc London (B)* 101: 391–409, 1927.
22. Stiere H: Die Entwicklung der Keimzellen under der Zwischenzellen in der Hoden anlage des Menschen. *Z Mikr Anat Forsch* 10: 225–285, 1927.
23. Neumann HO: Was wissen Wir uber die Keimbahn des Menschen? *Ark Gyn* 136: 197–244, 1929.
24. Watzka AM; Weibliche Genitalorgane. Das ovarium. In: Handbuch der Mikroskopischen Anatomie des Menschen. Vol 7. F Mollendorf, W Bargmann (eds), Berlin: Springer-Verlag, p 1–178, 1957.
25. Baca M, Zamboni L: The fine structure of human follicular oocytes. *J Ultrastruct Res* 19: 354–381, 1967.
26. Mossman HW, Duke KL: Comparative Morphology of the Mammalian Ovary. Madison, WI: University of Wisconsin Press, 1973.
27. Zamboni L, Merchant H: The fine morphology of mouse primordial germ cells in extragonadal locations. *Am J Anat* 137: 299–336, 1973.
28. Van Blerkom J, Motta PM: The Cellular Basis of Mammalian Reproduction. Munich, Baltimore: Urban & Schwarzenberg, 1979.
29. Motta PM, Makabe S: Germ cells in the ovarian surface during fetal development in humans. A three-dimensional microanatomical study by scanning and transmission electron microscopy. *J Submicrosc Cytol* 18: 271–290, 1986.
30. Eddy EM, Clark JM, Gong D, Fenderson BA: Review article: origin and migration of primordial germ cells in mammals. *Gamete Res* 4: 333–362, 1981.
31. Rubaschkin W: Zur Frage von der Entstehung der Keimzellen bei Saugetierembryonen. *Anat Anz* 32: 222–224, 1908.
32. McKay DG, Hertig AT, Adams EC, Danziger S: Histochemical observations on the germ cells of human embryos. *Anat Rec* 117: 201–219, 1953.
33. Bennett D: Developmental analysis of a mutation with pleiotrophic effects in the mouse. *J Morphol* 98: 199–233, 1956.
34. Nieuwkoop PD, Satasurya LA: Primordial germ cells in the chordates. In: Developmental and Cell Biology Series. Vol 7. M Abercrombie, DR Newth, JA Torrey (eds), Cambridge: Cambridge University Press, 1979.
35. Eddy EM: Origin of the germ cell line. In: Ultrastructure of Reproduction. J Van Blerkom, PM Motta (eds), Boston, The Hague: Martinus Nijhoff, p 1–11, 1984.
36. Baker TG: Oogenesis and ovarian development. In: Reproductive Biology. H Balin, SR Glasser (eds), New York: Excerpta Medica Foundation, 1972.
37. Franchi LL, Baker TG: Oogenesis and follicular growth. In: Human Reproduction. ESE Hafez, TN Evans (eds), New York, London: Harper and Row, p 53–83, 1973.
38. Rudkin, GT, Griech HA: On the persistence of oocyte nuclei from fetus to maturity in the laboratory mouse. *J Cell Biol* 12: 169–175, 1962.

39. Falin LI: The development of the genital glands and the origin of germ cells in human embryogenesis. *Acta Anat* 72: 195–232, 1969.
40. Witschi E: Embryology of the ovary. In: The Ovary. Vol. 3. HG Grady, DE Smith (eds), Baltimore: Williams and Wilkins. p 1–10, 1962.
41. Fujimoto T, Miyayama Y, Fuyuta M: The origin, migration and fine morphology of human primordial germ cells. *Anat Rec* 188: 315–330, 1977.
42. Sadler TW; Langman's Medical Embryology. Fifth Edition. Baltimore, London: Williams & Wilkins, 1985.
43. Ukeshima A, Fujimoto T: Ultrastructure of primordial germ cells in early chick embryo. In: Ultrastructure of Reproduction. J Van Blerkom, PM Motta (eds), Boston, The Hague: Martinus Nijhoff, p 12–18, 1984.
44. Fukuda T: Ultrastructure of primordial germ cells in human embryo. *Virchows Arch B Cell Path* 20: 85–89, 1976.
45. Merchant-Larios H, Mendlovic F, Alvarez-Buylla A: Characterization of alkaline phosphatase from primordial germ cells and ontogenesis of this enzyme in the mouse. *Differentiation* 29: 145–151, 1985.
46. McKay DG, Hertig AT, Adams EC, Danziger S: Histochemical horizons in human embryos. *Anat Rec* 122: 125–151, 1955.
47. Fuyuta M, Miyayama Y, Fujimoto T: Identification and fine morphology of the primordial germ cells in mouse and man. *J Kumamoto Med Assoc* 50: 106–114, 1976.
48. Miyayama Y, Fuyuta M, Fujimoto T: Ultrastructural observations on the origin, ameboid features and associations of human primordial germ cells. *Acta Anat Nippon* 52: 255–268, 1977.
49. Ukeshima A, Fujimoto T: Scanning electron microscopy of primordial germ cells in early chick embryos. *J Electron Microsc* 27: 19–24, 1978.
50. Clark JM, Eddy EM: Fine structural observations on the origin and association of primordial germ cells of the mouse. *Dev Biol* 47: 136–155, 1975.
51. Spiegelmann M, Bennett D: A light- and electron-microscopic study of primordial germ cells in the early mouse embryo. *J Embryol Exp Morphol* 30: 97–118, 1973.
52. Eddy EM: Fine structural observations on the form and distribution of nuage in germ cells of the rat. *Anat Rec* 178: 731–758, 1974.
53. Eddy EM, Clark JM: Electron microscopic study of migrating primordial germ cells in the rat. In: Electron Microscopic Concepts of Secretion. M Hess (ed), New York: Wiley & Sons, p 154–167, 1975.
54. Eddy EM: Germ plasm and the differentiation of the germ cell line. *Int Rev Cytol* 43: 229–280, 1975.
55. Kellokumpu-Lehtinen PL, Soderstrom KO: Occurrence of nuage in fetal human germ cells. *Cell Tissue Res* 194: 171–177, 1978.
56. Smith LD, Williams MA: Germinal plasm and determination of the primordial germ cells. In: The Developmental Biology of Reproduction. CL Marbert, J Papaconstantinou (eds), New York: Academic Press. p 3–24, 1975.
57. Mahowald AP, Allis CD, Karrer KM, Underwood EM, Waring GL: Germ plasm and pole cells in Drosophila. In: Determinants of Spatial Organization, 37th Symposium of the Society for Developmental Biology. S Subtelny, IR Konigsberg, (eds), New York: Academic Press, p 127–146, 1979.
58. André J, Rouiller C: L'ultrastructure de la membrane nucléaire des ovocytes de l'araignée (Tegenaria Domestica). In: Proceedings of the European Conference on Electron Microscopy, Stockholm, 1956. FS Sjostrand, J Rhodin (eds), New York: Academic Press. p 162–164, 1957.
59. Takeuchi IK, Sonta S, Takeuchi YK: Basic proteins in the nuage of growing Chinese hamster oocytes. *Acta Histochem Cytochem* 19: 1–10, 1986.
60. Motta PM, Van Blerkom J: Presence d'un material caracteristique granulaire dans le cytoplasma de l'ovocyte et dans les primiers stades de la differenciation des cellules embryonnaires. *Bull Ass Anat* 58: 947–953, 1974.
61. Fujimoto T, Ukeshima A, Kyiofuji R: Light- and electron-microscopic studies on the origin and migration of the primordial germ cells in the chick. *Acta Anat Nippon* 50: 22–40, 1975.
62. Fujimoto T, Ukeshima A, Kyiofuji R. The origin, migration and morphology of the primordial germ cells in the chick embryo. *Anat Rec* 185: 139–154, 1976.
63. Jirasek JE: Principles of reproductive embryology. In: Disorders of Sexual Differentiation. Etiology and Clinical Delineation. JL Simpson (ed), New York: Academic Press, p 51–72, 1976.
64. Upadhyay S, Zamboni L: Ectopic germ cells: Natural model for the study of germ cell sexual differentiation. *Proc Natl Acad Sci USA* 79: 6584–6588, 1982.
65. Francavilla S, Zamboni L: Differentiation of mouse ectopic germinal cells in intra- and perigonadal localizations. *J Exp Zool* 233: 101–109, 1985.
66. McIlhinney RAJ: The biology of human germ cells tumours: experimental approaches. In: Current Problems in Germ Cell Differentiation. A McLaren, CC Wylie (eds), Cambridge: Cambridge University Press, p 175–197, 1983.
67. Lin TH, Ukeshima A, Maeyama M, Fujimoto T: Scanning electron microscopic study on human primordial germ cells during the migration period. *Okajimas Folia Anat Jpn* 59: 305–320, 1982.
68. Fuyuta M, Miyayama Y, Fujimoto T: Histochemical identification of primordial germ cells in human embryos by PAS reaction. *Okajimas Folia Anat Jpn* 51: 251–262, 1974.
69. Kuwana T, Fujimoto T: Active locomotion of human primordial germ cells in vitro. *Anat Rec* 205: 21–26, 1983.
70. Kuwana T, Maeda-Suga H, Fujimoto T: Attraction of chick primordial germ cells by gonadal anlage in vitro. *Anat Rec* 215: 403–406, 1986.
71. Tarkowski AK: Germ cells in natural and experimental chimeras in mammals. *Philos Trans R Soc London, Ser B* 259: 107–112, 1970.
72. Donovan PJ, Stott D, Cairns LA, Heasman J, Wylie CC: Migratory and postmigratory mouse primordial germ

cells behave differently in culture. *Cell* 44: 831–838, 1986.
73. Dubois R, Cumminge D: Chimiotactisme et organization biologique. Etude de l'installation de le lignée germinale dans le ébauches gonadiques chez l'embryon de poulet. *Ann Biol* 13:241–248, 1974.
74. Swartz WJ: Effect of steroids on definitive localization of primordial germ cells in the chick embryo. *Am J Anat* 142: 499–513, 1975.
75. Heasman J, Hynes RO, Swan AP, Thomas V, Wylie CC: Primordial germ cells of Xenopous embryos: The role of fibronectin in their adhesion during migration. *Cell* 27: 377–380, 1981.
76. Kuwana T, Fujimoto T: Locomotion and scanning electron microscopic observations of primordial germ cells from the embryonic chick blood in vitro. *Anat Rec* 209: 337–343, 1984.
77. Fujimoto T, Yoshinaga K, Kono I: Distribution of fibronectin on the migratory pathway of primordial germ cells in mice. *Anat Rec* 211: 271–278, 1985.
78. Kuwana T, Miyayama Y, Kajiwara Y, Fujimoto T: Behavior of chick primordial germ cells moving toward gonadal primordium in vitro: Scanning electron microscopic study. *Anat Rec* 219: 164–170, 1987.
79. Jeon KW, Kennedy JR: The primordial germ cells in early mouse embryos: Light and electron microscopic studies. *Dev Biol* 31: 275–284, 1973.
80. Blackler AW: Contribution to the study of germ cells in Anura. *J Embryol Exp Morphol* 6: 491–503, 1958.
81. Gruenwald P: The development of the sex cords in the gonads of man and mammals. *Am J Anat* 70: 359–397, 1942.
82. Merchant-Larios H: Germ and somatic cell interactions during gonadal morphogenesis. In: Ultrastructure of Reproduction. J Van Blerkom, PM Motta (eds), Boston, The Hague: Martinus Nijhoff, p 19–30, 1984.
83. Byskov AG: The anatomy and ultrastructure of the rete system in the fetal mouse ovary. *Biol Reprod* 19: 720–735, 1978.
84. Byskov AG: Sexual differentiation of the mammalian ovary. In: Biology of the Ovary. PM Motta, ESE Hafez (eds), The Hague, Boston: Martinus Njihoff, p 3–15, 1980.
85. Upadhyay S, Luciani JM, Zamboni L: The role of the mesonephros in the development of indifferent gonads and ovaries of the mouse. *Ann Biol Anim Biochem Biophys* 19: 1179–1196, 1979.
86. Zamboni L, Bézard J, Mauléon P: The role of the mesonephros in the development of the sheep fetal ovary. *Ann Biol Anim Biochem Biophys* 19: 1153–1178, 1979.
87. Wartenberg H: Structural aspects of gonadal differentiation in mammals and birds. *Differentiation* 23 (suppl.): S64–S71, 1983.
88. Wartenberg H: The influence of the mesonephric blastema on gonadal development and sexual differentiation. In: Development and Function of Reproductive Organs. Int. Congr. Ser. No. 559. AG Byskov, H Peters (eds), Amsterdam, Oxford: Excerpta Medica, p 3–12, 1981.
89. Wartenberg H: Germ cell migration induced and guided by somatic cell interaction. *Bibl Anat* 24: 93–110, 1983.
90. Motta PM, Makabe S: Development of the ovarian surface and associated germ cells in the human fetus. A correlated study by scanning and transmission electron microscopy. *Cell Tissue Res* 226: 493–510, 1982.
91. Haffen K: Sexual differentiation of the ovary. In: The Ovary, 2nd edition. Vol. 1. S Zuckerman, BJ Weir (eds), New York, London: Academic Press, p 69–112, 1977.
92. Wartenberg H: Development of early human ovary and role of the mesonephros in the differentiation of the cortex. *Anat Embryol (Berl)* 165: 253–280, 1982.
93. Gondos B: Germ cell differentiation and intercellular bridges. In: Ultrastructure of Reproduction. J Van Blerkom, PM Motta (eds), Boston, The Hague: Martinus Nijhoff, p 31–45, 1984.
94. Motta PM, Makabe S: Elimination of germ cells during differentiation of the human ovary: an electron microscopic study. *Eur J Obstet Gynecol Reprod Biol* 22: 271–286, 1986.
95. Makabe S, Motta PM: Fetal ovary. In: Atlas of Human Reproduction by Scanning Electron Microscopy. ESE Hafez, P Kenemans (eds), London: MTP Press, p 129–134, 1982.
96. Gondos B: Oogonia and oocytes in mammals. In: The Vertebrate Ovary, RE Jones (ed), New York: Plenum Press, p 83–120, 1978.
97. Baker TG, Franchi LL: The fine structure of oogonia and oocytes in human ovaries. *J Cell Sci* 2: 213–224, 1967.
98. Franchi LL, Mandl AM: The ultrastructure of oogonia and oocytes in the foetal and neonatal rat. *Proc R Soc B* 157: 99–114, 1962.
99. Hertig AT, Adams EC: Studies on the human oocyte and its follicle. I. Ultrastructural and histochemical observations on the primordial follicle stage. *J Cell Biol* 34: 647–675, 1967.
100. Ruby JR, Dyer RF, Gasser RF, Skalko RG: Intercellular connections between germ cells in the developing human ovary. *Z Zellforsch* 105: 252–258, 1970.
101. Gondos B, Bhiraleus P, Hobel CH: Ultrastructural observations on germ cells in human fetal ovaries. *Am J Obstet Gynecol* 110: 644–652, 1971.
102. Szollosi D: Changes of some cell organelles during oogenesis in mammals. In: Oogenesis. JD Biggers, AW Schuetz (eds), Baltimore: University Park Press, p 47–64, 1972.
103. Wartenberg H: Spermatogenese–Oogenese: ein cytomorphologischer Vergleich. *Verh Anat Ges* 68: 63–92, 1974.
104. Gondos B, Westergaard L, Byskov AG: Initiation of oogenesis in the human fetal ovary: ultrastructural and squash preparation study. *Am J Obstet Gynecol* 155: 189–195, 1986.
105. Russe I: Oogenesis in cattle and sheep. *Bibl Anat* 24: 77–92, 1983.
106. Gondos B: Germ cell degeneration and intercellular bridges in the human fetal ovary. *Z Zellforsch* 138: 23–30, 1973.

107. Stegner HE, Wartenberg H: Elektronenmikroskopische untersuchungen an Eizellen des Menschen in verschiedenen stadien der Oogenese. *Arch Gynak* 199: 151–172, 1963.
108. Weakley BS: Electron microscopy of the oocyte and granulosa cells in developing ovarian follicles of the golden hamster (Mesocricetus auratus). *J Anat* (London) 100: 503–534, 1966.
109. Weakley BS: Light and electron microscopy of developing germ cells and follicle cells in the ovary of the golden hamster: twenty-four hours before birth to eight days post partum. *J Anat* (London) 101: 435–459, 1967.
110. Zamboni L, Gondos B: Intercellular bridges and synchronization of germ cell differentiation during oogenesis in the rabbit. *J Cell Biol* 36: 276–282, 1968.
111. Gondos B, Zamboni L: Ovarian development: the functional importance of germ cell interconnections. *Fertil Steril* 20: 176–189, 1969.
112. Ruby JR, Dyer RF, Skalko RG: The occurrence of intercellular bridges during oogenesis in the mouse. *J Morph* 127: 307–340, 1969.
113. McReynolds HD, Blanchet LJ, Bowman DC, Hakim AA, Kiely ML: Intercellular bridges in ovaries of the newborn gerbil. *J Morphol* 180; 29–35, 1984.
114. Gondos B, Connell CJ: Transmission and scanning electron microscopy study of the developing testis. In: Proc VIII Intl Cong Animal Reprod. and Artificial Insemination. Vol. 3. Krakow: p 50–53, 1974.
115. Eddy EM, Kahri AI: Cell associations and surface features in cultures of juvenile rat seminiferous tubules. *Anat Rec* 185: 33–358, 1976.
116. Fawcett DW, Ito S, Slautterback DL: The occurrence of intercellular bridges in groups of cells exhibiting synchronous differentiation. *J Biophys Biochem Cytol* 5: 453–460, 1959.
117. Gondos B: Intercellular bridges and mammalian germ cell differentiation. *Differentiation* 1: 177–182, 1973.
118. van Wagenen G, Simpson ME: Embryology of the Ovary and Testis. Homo Sapiens and Macaca Mulatta. New Haven, London: Yale University Press, 1965.
119. Quattropani SL: Microscopy of ovarian cords and mesothelium in the fetal monkey (Macaca arctoides) and fetal human. *Anat Rec* 181: 454a, 1975.
120. Witschi E: Embryogenesis of the adrenal and reproductive glands. *Recent Progr Hormone Res* 6: 1–27, 1951.
121. Gondos B: Surface epithelium of the developing ovary. Possible correlation with ovarian neoplasia. *Am J Pathol* 81: 303–320, 1975.
122. Jeppesen TH: Surface epithelium of the fetal guinea-pig ovary. A light and electron microscopic study. *Anat Rec* 183: 499–516, 1975.
123. Merchant-Larios H: Rat gonadal and ovarian organogenesis with and without germ cells. An ultrastructural study. *Dev Biol* 44: 1–21, 1975.
124. Merchant-Larios H: Ovarian differentiation. In: The Vertebrate Ovary. RE Jones (ed), New York: Plenum Press, p 47–81, 1978.
125. Merchant-Larios H: Origin of somatic cells in the rat gonad: an autoradiographic approach. *Ann Biol Anim Bioch Biophys* 19: 1219–1229, 1979.
126. Pelliniemi JL: Ultrastructure of gonadal ridge in male and female pig embryos. *Anat Embryol* 147: 19–34, 1975.
127. Pelliniemi JL: Ultrastructure of the indifferent gonad in male and female pig embryos. *Tissue Cell* 8: 163–174, 1976.
128. Zamboni L: Modulation of follicles cell-oocyte association in sequential stages of mammalian follicle development and maturation. In: Ovulation in the Human. PG Crosignani, DR Mishell (eds), New York: Academic Press, p 1–30, 1976.
129. Guraya SS: Recent advances in the morphology, histochemistry and biochemistry of the developing mammalian ovary. *Int Rev Cytol* 51: 49–131, 1977.
130. Peters H: Folliculogenesis in mammals. In: The Vertebrate Ovary. RE Jones (ed), New York: Plenum Press, p 121–144, 1978.
131. Beck W: Postnatal development of ovarian surface epithelium. *Anat Anz* 147: 424–436, 1980.
132. Motta PM, Hafez ESE: Biology of the Ovary. The Hague, Boston: Martinus Nijhoff, 1980.
133. Byskov AG, Lintern-Moore SM: Follicle formation in the immature mouse ovary. The role of rete ovarii. *J Anat* 116: 207–217, 1973.
134. Byskov AG, Skakkebaek NE, Stafanger G, Peters H: Influences of ovarian surface epithelium and rete ovarii on follicle formation. *J Anat* 123: 77–86, 1977.
135. Zamboni L, Upadhyay S, Bezard J, Luciani JM, Mauleon P: The role of the mesonephros in the development of the mammalian ovary. In: Endocrine Physiopathology of the Ovary. RI Tozzini, G Reeves, RL Pineda (eds), Amsterdam, New York: Elsevier North-Holland, p 3–42, 1979.
136. Mitchell PA, Burghardt RC: The ontogeny of nexuses (gap junctions) in the ovary of the fetal mouse. *Anat Rec* 214: 283–288, 1986.
137. Grinsted J, Aagesen L: Mesonephric excretory function related to its influence on differentiation of fetal gonads. *Anat Rec* 210: 551–556, 1984.
138. Wartenberg H. Human testicular development and the role of the mesonephros in the origin of a dual Sertoli cell system. *Andrologia* 10: 1–21, 1978.
139. Makabe S, Motta PM: Germ cell dynamics in the surface layer of the fetal human ovary. A correlated study by STEM. In: 3rd International Congress of Cell Biology. S Seno, Y Okada (eds.), Tokyo: Japanese Society of Cell Biology, Abstract book, p 402, 1984.
140. Motta PM, Takeva S, Nesci E: Etude ultrastructurale et histochimique des rapports entre les cellules folliculaires et l'ovocyte pendant le développement du follicule ovarien chez les mammifères. *Acta Anat* 80: 537–562, 1971.
141. Motta PM, Makabe S: Morphodynamic changes of the mammalian ovary in normal and some pathological conditions. A scanning electron microscopic study. Biomed Res 2, suppl: 325–339, 1981.
142. Heller DT, Cahill DM, Schultz RM: Biochemical studies of mammalian oogenesis: metabolic cooperativity be-

tween follicle cells and growing mouse oocytes. *Dev Biol* 84: 455–464, 1981.
143. Ohno S, Smith JB: Role of fetal follicular cells in meiosis of mammalian oocytes. *Cytogenetics* 3: 324–333, 1964.
144. Baker TG, Scrimgeour, JB: Development of the gonad in normal and anencephalic human fetuses. *J Reprod Fert* 60: 193–199, 1980.
145. Motta PM: Superficial epithelium and surface invaginations in the cortex of mature rabbit ovaries. A note on the histogenesis of the interstitial cells. *Fertil Steril* 25: 336–347, 1974.
146. Byskov AG, Hoyer PE, Westergaard L: Origin and differentiation of the endocrine cells of the ovary. *J Reprod Fertil* 75: 299–306, 1985.
147. Guraya SS: Recent advances in the morphology, histochemistry, biochemistry, and physiology of interstitial gland cells in the mammalian ovary. *Int Rev Cytol* 55: 171–245, 1978.
148. Konishi I, Fujii S, Okamura H, Parmley T, Mori T: Development of interstitial cells and ovigerous cords in the human fetal ovary: an ultrastructural study. *J Anat* 148: 121–135, 1986.
149. Baker TG: A quantitative and cytological study of germ cells in human ovaries. *Proc R Soc B* 158: 417–733, 1963.
150. Gondos B, Hobel CJ: Germ cell degeneration and phagocytosis in the human fetal ovary. In: The Development of the Reproductive Organs and Functions in Female. H Peters (ed), Amsterdam: Excerpta Medica, 1972.
151. Baker TG: Oogenesis and ovulation. In: Reproduction in Mammals. 2nd edition. CR Austin, RV Short (eds), London: Cambridge University Press, p 17–45, 1982.
152. Jones EC, Krohn PL: The relationship between age, number of oocytes and fertility in virgin and multiparous mice. *J Endocrinol* 21: 469–495, 1961.
153. Peters H: The development of the mouse ovary from birth to maturity. *Acta Endocrinol* 62: 98–116, 1969.
154. Byskov AG, Rasmussen G: Ultrastructural studies of the developing follicle. In: The Development and Maturation of the Ovary and its Function. H Peters (ed), Amsterdam: Excerpta Medica, p 55–62, 1973.
155. Hiura M, Fujita H: Electron microscopic observations of the elimination of the oocyte through the peritoneal epithelium in the neonatal mouse ovary. *Cell Tissue Res* 182: 73–79, 1977.
156. Bonilla-Musoles F, Reanau J, Herandez-Yago J, Torres J: How do oocytes disappear? *Arch Gynaekol* 218: 233–241, 1975.

Author's address:
Dr. S. Makabe
Department of Obstetrics and Gynecology
Toho University School of Medicine
11-1, Ohmorinishi 6-chome
Ota-ku, Tokyo (143)
JAPAN

CHAPTER 3

Ultrastructure of fetal ovary including oogenesis

HUBERT WARTENBERG

1. Introduction

Ultrastructural aspects of the developing human ovary have been the subject of several studies during the last decades. Although most articles deal with the differentiation of the female germ cell (1–18), studies on the role of the somatic blastema and its differentiation into granulosa cells are rare (2,4,6). Even those papers which deal with the ovarian structure of nonhuman mammals concentrate on the cellular characteristics of oogonia (19–22) and premeiotic oocytes (19–26), the structure of intercellular bridges (27–32) and germ cell degeneration (30,33,34), but not on the supporting cell structure (23, 35–39).

This chapter describes the ultrastructure of the developing human ovary and focuses on the morphogenetic events during the initiation of ovarian growth and the origin of its blastemal content, the differentiation of the ovarian cortex, and the cellular interaction between granulosa and germ cells during oogonial proliferation, premeiotic differentiation, and early folliculogenesis. In order to describe adequately the fine structure of the fetal human ovary, a short survey of its early differentiation during the seventh and eighth weeks of gestation (Carnegie stages 20 to 23) is necessary; during this period the indifferent gonad changes stepwise to a sexually differentiated ovary. The basic morphogenetic process that transforms the indifferent gonad into an ovary occurs before commencement of the fetal period. These events are different from those in the young testis and enable us to recognize an ovary during its early developmental stages. The human ovary passes through a period of morphogenetic alterations which synchronously parallel early testicular differentiation. These events constitute an integral part of ovarian organogenesis, namely the formation of the cortex (40).

2. The origin of the ovarian blastema

The origin of the somatic content that constitutes the genital ridge is still an unresolved problem. Two basic concepts that have been discussed during the last 100 years are presently still under debate. These two concepts are (1) the *epithelial* theory, according to which the gonadal blastema derives from the coelomic epithelium, and (2) the *mesenchymal* theory, which holds an in-situ condensation of the local mesenchyme to be responsible for the origin of somatic cells (2,41–49). A variation of the latter concept, which was formulated a long time ago (43,50,51), has gained increasing importance during recent years (2,40, 52). This refined concept acknowledges the mesonephros as main contributor of cells to the formation of the gonadal blastema (2,53–58). In a previous volume of this series, Merchant-Larios (48) has summarized many arguments that support a synthesis of the two classical concepts, and that reject the proposal that the mesonephros furnishes cells to the gonad by cellular segregation. The counterargument is essentially that, in many species, disparity between the timetables of mesonephric and gonadal development prevents the cells segregated by the mesonephros from being available at the right period. According to Merchant-Larios's view, "mesenchyme participating in the formation of the gonadal medulla does not require prior organization as mesonephros" (48, p. 22) and "the process makes highly unlikely a de-differentiation of mesonephros to form the somatic part of the gonad" (48, p. 29).

It must be emphasized, however that none of these concepts take into account one crucial requirement: Which cell might function as an inductor and which cell could be inducted? Why does the genital ridge differentiate by proliferation of a very restricted area of the coelomic epithelium without reacting to a nonresident tissue which migrates into the region? A solution to this problem is offered by the neighbouring mesonephros and its disintegrating cells. Several LM and EM studies demonstrate the participation of cells deriving from the mesonephros in the formation of the genital ridge blastema (2,56–66). Cells of the ovarian rete blastema, which derive from the mesonephric tubulus system, are thought to be part of the follicular envelope of the early oocyte (59,67–70). Rete cells seem to initiate meiotic prophase in female germ cells (70–73) as well as mitotic proliferation in male germ cells (74).

Besides the question of cellular origin the main problem relates to the functional significance of the supporting cell. Two possibilities are conceivable: (1) A unitary concept that confines the cellular origin to one of the two potential sources, the coelomic epithelium *or* the mesonephros, and that limits the developmental destiny and functional importance of the supporting cell to one cell line, irrespective of the role the cell plays in the male or female organ; and (2) a dualistic concept that generally implies a separate origin of the two individual cells that intermingle in the genital ridge. These two cell lines would constitute a dualistic system of two distinguishable types of supporting cells that would not only derive from separate sources, but above all, could exercise function of a different kind. They could even function in opposite manners (53,55,63). The assumption that two antagonistic parts of the gonadal blastema may exist is an advanced concept of gonadal development that permits a combined view and, in a sense, a synthesis of the two other concepts, neither of which, due to inconsistent and overlapping results, presents a uniform idea on the mechanism of gonadal differentiation. This concept could possibly better explain two morphogenetic events: the sexual differentiation into male or female gonad and the cellular interaction that regulates germ cell differentiation and initiation of meiosis. Therefore, the following description of ovarian differentiation is based on this dualistic concept of cellular origin and diversity of supporting cell lines.

2.1. Two types of somatic cells constitute the genital ridge

There has never been any disagreement about the nature of the first thickening of the genital ridge. By convention, the primary structural signs have been described as the proliferation of the superficial coelomic epithelium (75–88). No consensus exists on the origin of those cells that are concentrated below the superficial layer. The most favoured concept is that of a growth in depth of the coelomic epithelium. The superficial cells continue their proliferation and generate epithelial cords, the *first ingrowth* described by many authors (49,87). Other investigators interpret this development as an in situ differentiation of the local mesenchymal cells. For decades this idea has been the alternative concept, certainly with less consequence. Neither concept, however, explains how the proliferation of the superficial epithelium or of the underlying mesenchyma could be conducted.

A further concept that has achieved some recognition during the last few years might answer this question. It refers to the role of the mesonephros in gonadal differentiation (53,55). If mesonephric cells participate in the formation of the early blastema of the genital ridge, these cells could assume the inductor role. They invade the region as a specific cellular system that, due to its local concentration, interacts with the stationary tissue. The cellular interaction between the nonresident, immigrating mesonephric cells and the coelomic epithelium would account for the local limitation of this process. Both cellular systems depend upon each other in order to become the gonadal blastema and in turn to attract the primordial germ cells.

In the human genital ridge (40,54,63), as well as in the ridge of other mammalian (54,63,89) and nonmammalian species (55,89), the epithelial cells differentiate into irregular cuboidal or columnar cells and intermingle with dense, dark staining cells that immigrate into the epithelium (Fig. 3-1,3-2). The difficulty of distinguishing these two cells in the mixed blastema makes it harder to convince critics of the dualistic system (51,90). In ordinary paraffin sections the two cells are indistinguishable. In one-micron sections of plastic-embedded material that has been prefixed in phosphate-buffered glutaraldehyde, a difference becomes evident: light, epitheloid cells stand out from dark, mesenchymallike cells (Fig. 3-2).

2.2. Segregation of cells from the mesonephros

Most authors of recent studies agree that the immigrating cells that are found within the gonadal blastema derive from the mesonephros (91). The contributory process, however, is not so easy to understand. For good reasons Merchant-Larios (48) has contested some points. Due to the fact that, in many mammalian species, the period of functional activity of the mesonephros does not correspond to the period of gonadal differentiation, a transfer of cells seems unlikely. His basic argument against the mesonephric contribution concerns the differential status: a cellular transfer from the mesonephros to the gonadal blastema would necessitate a *dedifferentiation* of the segregated cells, a mechanism that has not been detected under physiological conditions.

A mass of LM and EM results reveal a varying but mainly clear conception of the mechanism through which cells can be derived from this extragonadal source.

1. In some species, cells are segregated during the entire life of the organ, i.e., during the developmental, active, and regressional period of the mesonephros. First, cells are detached from the mesonephric vesicle, an early structure that develops after the Wolffian duct has become separated from the intermediate mesoderm (80). These cells are mesenchymallike and do not show any signs of differentiation. During further development of the mesonephric nephrons, cells are segregated from the primitive Bowman's capsule. Later, when the gonadal primordium is demarcated from the genital ridge, cells of the upper part of the mesonephros become incorporated into the gonad in great number. The nephrons of this region cease differentiating before or just upon attaining their full structure. During their early regression they deliver many cells to the gonad via the rete blastema.

Segregation and delivery of cells from a distinct section of the mesonephros to the gonad has been found in the human and the rabbit embryo (40, 53,54,62,63).

2. A similar process can be seen in small mammalian species (mouse, rat, hamster) with regard to the parallelism of the early development of the mesonephros and the genital ridge. Again, the upper part of the organ preferentially contributes cells to the developing gonad. In the final state, the S-shaped tubules are connected to the Wolffian duct while the other end is open or forms an incomplete vesicle from which cells segregate. The mesonephros persists in this rudimentary stage.

3. In some large mammals like cattle, sheep, and pigs, the differentiation of the mesonephros and the gonad do not show such strong parallelism. Mesonephric development starts sometime prior to onset of gonadal differentiation. The mesonephros exhibits a giant glomerulum that occupies the upper part of the organ. When this peculiar structure has grown to full size, gonadal development starts at the same level and cells are segregated in large numbers from the Bowman's capsule of the giant Malpighian corpuscle (57,58, 65,66,89). During further regression of the giant glomerulum, the release of cells continues.

It has to be emphasized, in particular, that the transfer of mesonephric cells is a long-lasting and/or discontinuous process and is not necessarily restricted to the regressive period of the organ. The indifferent gonad receives cells from the very beginning until it enters sexual differentiation, irrespective of the developmental state of the mesonephros. Cells enter the organ via the ovarian or testicular rete blastema during further growth of the ovarian cortex, or during differentiation of additional testicular cords (40,63).

In human gonads, the invasion of mesonephric cells must start in four- to five-mm embryos when the genital ridge is first formed (92), and it continues — including interruptions — for at least

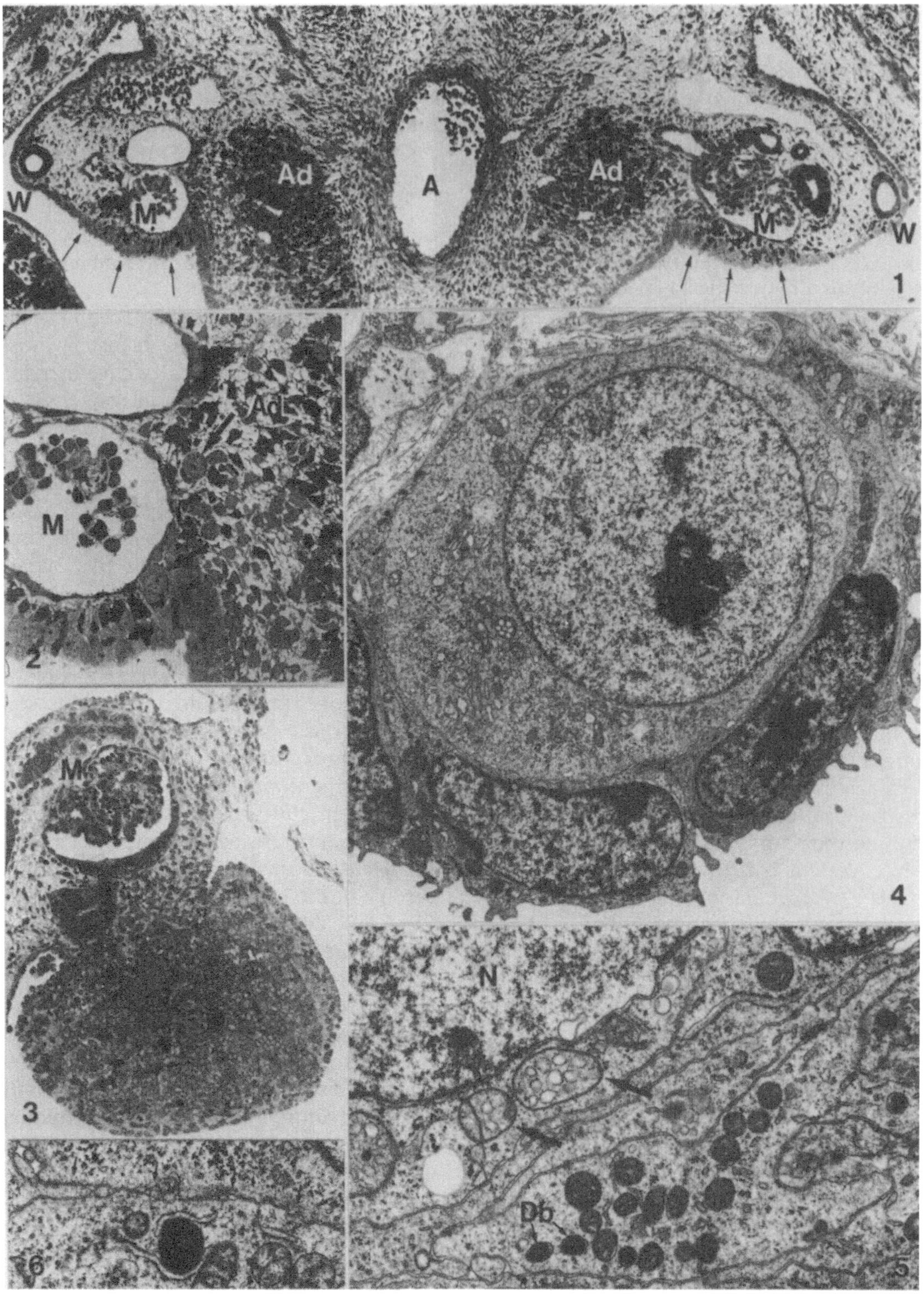
W
M
Ad
A
Ad
M
W
1
Ad
M
2
M
3
4
N
Db
5
6

five weeks until regression of the mesonephros terminates around the ninth week (93). Responsibility for cell release lies mainly with the upper part of the mesonephros. This genital section (123) comprises about 20% of the organ and starts regression early during the seventh week before it has reached its fully developed state. The mesonephric tubules of this region are destined to become the epididymic efferent tubules. The lower major part of the mesonephros continues differentiation and starts to regress some weeks later.

2.3. Altered function of segregated mesonepric cells and transdifferentiation

Comparative studies of the participation of the mesonephros in delivering cells to the gonad have shown that the mode of action is variable in different species and adapted to the developmental state of the mesonephros. Basically, these cells are derivatives of the intermediate mesoderm. Independent of the differential grade within the mesonephric system, these cells undergo a renewed utilization. The reintegration from the primary destination requires a fairly radical process of dedifferentiation or, better, *transdifferentiation*, because the cells change from one differentiated state to another. Phenotypically, these cells run through a state wherein they cannot be distinguished from plain mesenchymal cells (40,53, 62). If transdifferentiation, which in this case seems to be necessary as a physiological process, could be proven experimentally, an interesting mechanism in cell biology would be established.

At least two principles seem to underlie the differentiation of the gonadal blastema: (1) reutilization of cells that, during the development of the mesonephros, have only had a limited destination, and (2) the establishment of a cellular system in which an antagonistic behaviour seems to take place between derivatives of the *intermediate* mesoderm (mesonephros) and the *splanchnopleuric*, lateral plate mesoderm (coelomic epithelium). A kind of confrontation between these two tissue components is expressed first in the occurrence of rudimentary nephrostomata in some mammals. Roosen-Runge (94) has described these structures in rat embryos. In rabbit (I. Kinsky, personal communication) and rat gonadogenesis (39), these rudimentary nephrostomata herald the onset of genital ridge formation.

3. Sexual differentiation

3.1. The final structure of the blastema in the indifferent gonad

During the period of sexual differentiation, when the testicular cords occur in the male gonad, the ovary consists of a dense blastema that in the central part contains predominantly dark, fusiform cells, while in the outer zone, including the superficial layer, light, epithelioid cells prevail (Figs. 3-3,3-7). PGCs are found in a relatively small number. The majority of them are concentrated in the outer zone among the light stromal cells (Fig. 3-7).

The organ is connected to the mesonephric fold by means of a broad mesovarium. Through the wide hilar region, the ovarian blastema continues right to the Bowman's capsules of the mesonephric Malpighian corpuscles (Fig. 3-3).

On the TEM level, the germ cells are characterized by a high content of glycogen granules and a round nucleus with a very prominent nucleolus of reticular structure (Fig. 3-4). A few mitochondria are dispersed within the cytoplasm. They are round or oval-shaped, and are easily distinguishable from the mitochondria seen in the somatic cells (Fig. 3-5). Most PGCs are of irregular shape and show extended cytoplasmic

←

Figures 3-1–3-6. (1) Urogenital folds and adrenal primordia (Ad) of a 12-mm human embryo. Genital ridge epithelium (arrows) proliferates. M: mesonephric glomerula; W: Wolffian ducts; A: Aorta. X (2) Higher magnification of the right genital ridge with dark mesonephric and light epithelial cells and germ cells (arrows). M: mesonephric glomerulum; Ad: adrenal blastema. X 325. (3) Early ovary of a 21-mm embryo. M: mesonephros. X 135. (4) EM of an ectopic primordial germ cell (PGC) enclosed by cells of the coelomic epithelium. X 6300. (5) Mitochondria of a PGC (arrows) and adjacent supporting cells. Db: dense bodies; N: nucleus. X 17,600. (6) Subplasmalemmal filaments (arrows) in PGCs. X 15,500.
Note: All figures are taken from human ovaries. Data concerning age relate to ovulation age (OA).

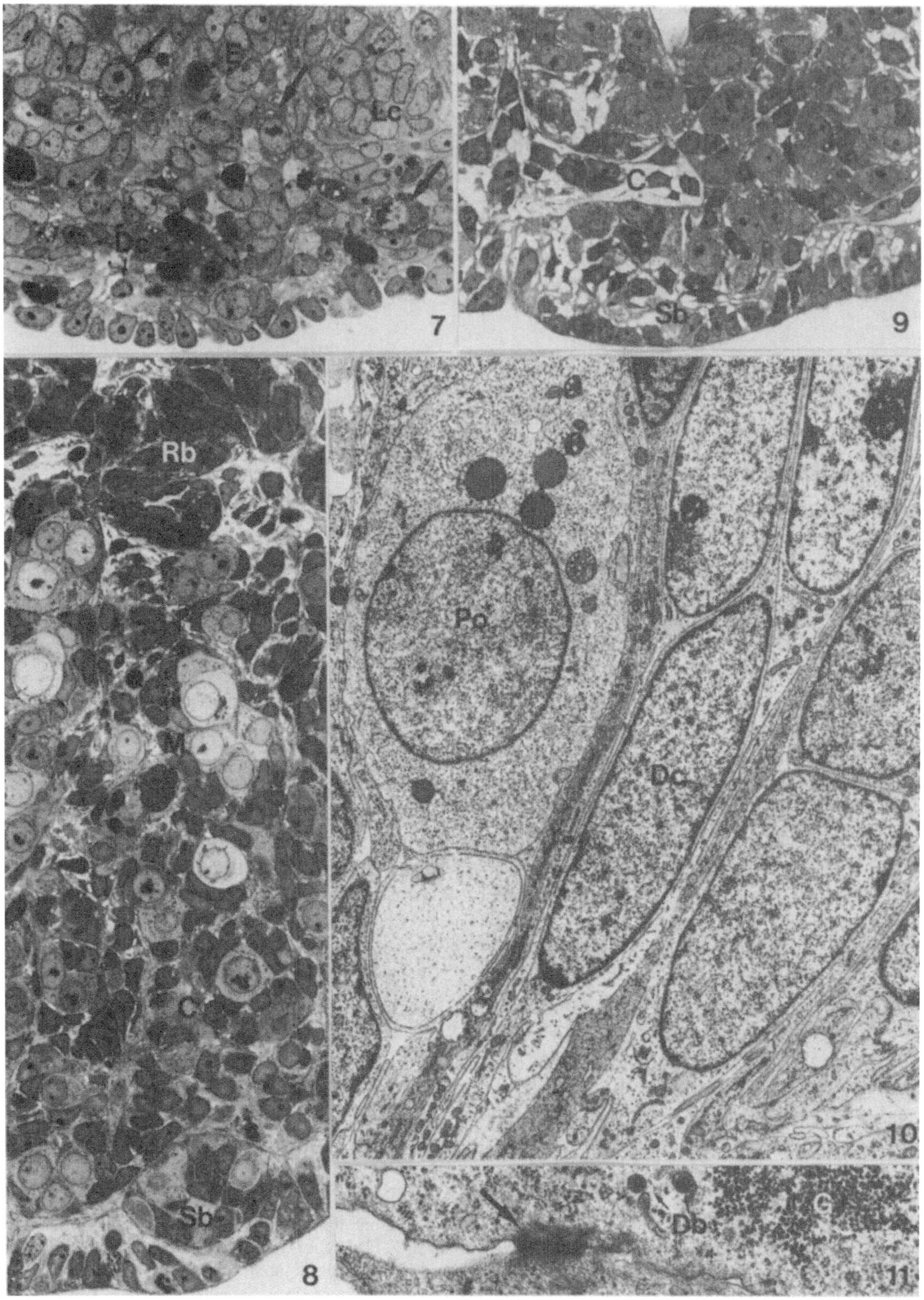
E
Lc
Dc
7
C
Sb
9
Rb
M
C
Sb
8
Po
Dc
10
Db
C
11

projections, suggesting an ability to migrate by amoeboid movement. In addition, these cells are characterized by a thin layer of fine filaments that are located beneath the plasmalemma (Figs. 3-5, 3-6).

3.2. Types of ovarian morphogenesis among mammalian species

Two morphogenetic events are presumed to be the basic requirements for the differentiation of the female gonad. One results in the rearrangement of the blastemal components into the female pattern during sexual differentiation: The initially intermixed blastemal parts tend to separate, with one accumulating in the outer zone and the other in the centre of the indifferent organ. This pattern has been found to be prevalent during the female sexual differential period in some mammals (40, 54,55,63,95) (man, rabbit, cattle). In male gonads, a different pattern dominates: Dark and light blastemal cells intermingle and dark cells of mesonephric origin penetrate the gonadal blastema. They regain dominance in the most superficial blastemal layer separating the larger part of the gonadal tissue, including the PGCs, from the surface epithelium (54,63). Male and female morphogenesis are dissimilar with respect to the degree of intermixture or separation of the two blastemal components during the onset of sexual differentiation. These distributional patterns are transient and, in the female gonad, a complete state of separation is never attained.

A second change of cellular distribution, which will be described in detail later, results from an invasion of the ovarian cortex by dark blastemal cells. The functional significance of this second cellular rearrangement must be attributed to the mechanism of germ cell activation. An increase in dark cells initiates germ cell proliferation and results in the rapid mitotic multiplication of the oogonia and onset of meiosis. Oogonial proliferation and initiation of meiotic prophase both depend upon this second morphogenetic event, the remodeling of cellular components in the developing ovarian cortex (40,55).

In the human ovary these events are chronologically separated: Sexual differentiation starts during the seventh week, cluster formation of proliferating oogonia can be found first during the 11th week, and the first meiotic prophase occurs at the end of the 12th and during the 13th week (9,21,33,40).

There is, however, a different type of ovarian morphogenesis among small mammals. In the mouse, for instance, sexual differentiation happens on day 12, while oogonial proliferation starts on the same day and is finished on day 15 when the first meiotic prophases appear (55,63). Consequently, the process of rearrangement of the blastemal content occurs in parallel with the process of female sexual differentiation, and the two patterns of cellular distribution overlap (55). Therefore, the cellular distribution that characterizes the period of sexual differentiation in the human female gonad cannot be seen in the mouse ovary, because this pattern is superimposed on the second pattern, which is necessary for the initiation of early meiosis.

4. Differentiation and growth of the ovarian cortex

The ovarian cortex is formed by a remodeling of the blastemal content during the eighth and ninth week (days 50 to 64) (40). This process is characterized by (1) an increasing looseness of the outer blastemal zone that finally comprises the entire anlage, including the central core (Figs. 3-7,3-9) and simultaneously, (2) an increasing invasion of the outer zone by dark, fusiform cells

←

Figures 3-7–3-11. (7) Blastema of an early ovary (day 42–46). A few PGCs (arrows) intermingle with light (Lc) and dark fusiform cells (Dc). E: erythrocytes in differentiating capillaries. X 550. (8) Differentiating ovarian cortex (day 47). Sb: superficial blastema; C: cortex; M: medulla; Rb: rete blastema. X 550. (9) Superficial blastema (Sb) and outer cortex (C) (day 51). Ovarian blastema is less compact and the number of germ cells is increased. X 550. (10) EM of dark fusiform cells of mesonephric origin (Dc) and a germ cell (Po: pre-oogonium) in the rete blastema. X 7,700. (11) Surface of a PGC exhibiting a desmosomal contact (arrow) with a supporting cell. G: glycogen granules; Db: dense bodies. X 25,000.

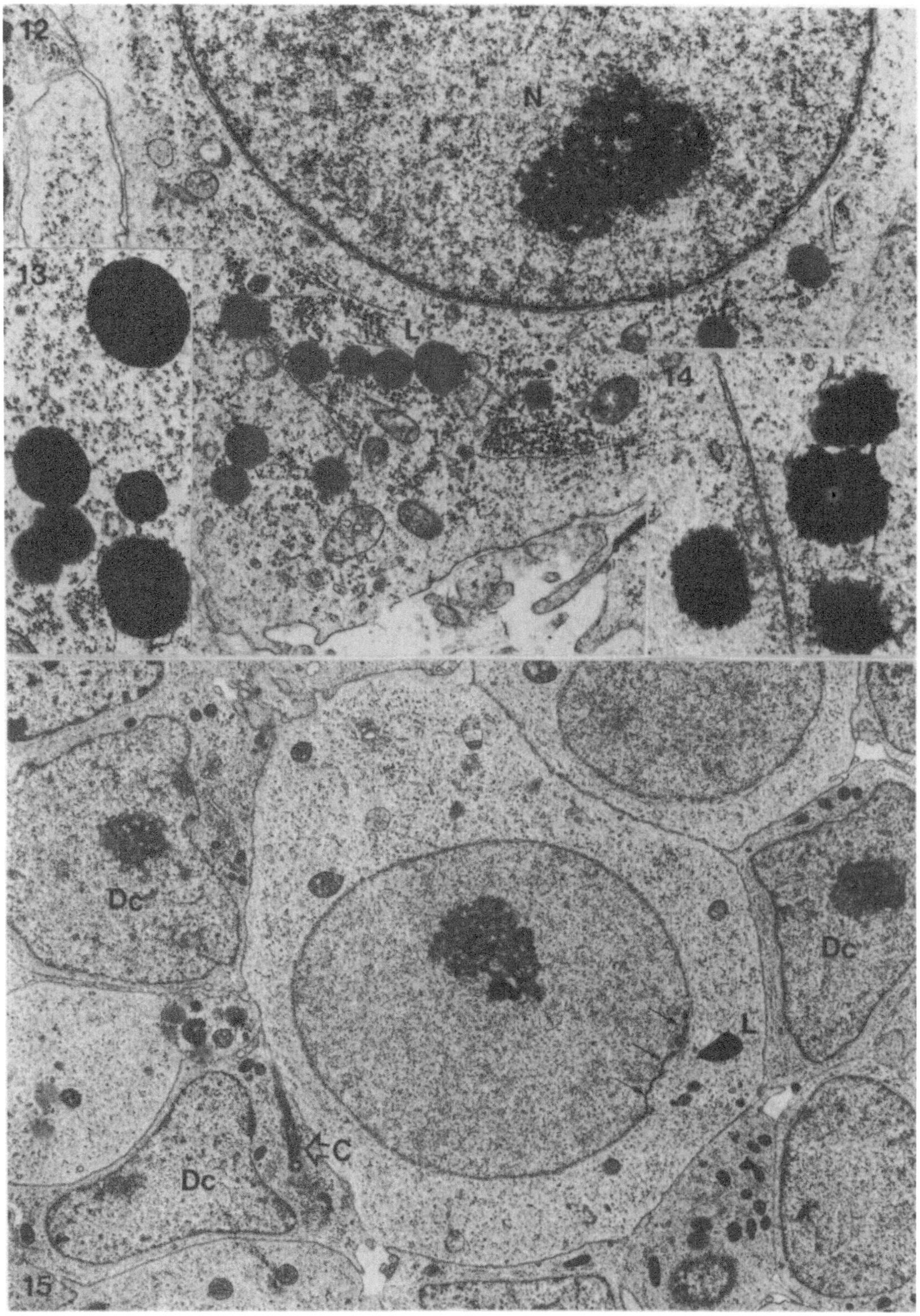
12
N
13
L
14
Dc
Dc
L
C
Dc
15

emanating from the central blastemal core (Figs. 3-8–3-10). Intercellular spaces, which first appear around day 50 below the superficial epithelial layer (Figs. 3-7–3-9), spread out in depth during week eight. These spaces permit the dark, fusiform cells to move peripherally, which results in a shift in the numerical proportion of the light and dark cells in favour of the latter. Coincidentally, with the increase of the dark blastemal content, the number of germ cells also suddenly starts to increase.

During the ninth week this process results in a condensation of the outer zone (see Fig. 3-16). The ovarian cortex is then formed. Increase in both thickness and density characterize the late period of cortical development. The inner core of the original gonadal blastema also changes. It decondenses in a slightly different way and finally separates into two compartments, the medullary area (Fig. 3-8) and the rete blastema (Figs. 3-8, 3-10). The ovarian medulla consists of remnants of the former dense blastemal core, and exhibits groups or strands of somatic cells that include single germ cells (Fig. 3-8). The number of germ cells seen in the Medulla is less than that noted in the cortex, and the structure of the medulla remains less dense. The basic cellular composition, however, is similar both in the medulla and in the cortex.

The border between cortex and medulla is accentuated by a concentration of dark blastemal cells (Fig. 3-8). Strands of these cells extend peripherally and subdivide the cortex into areas of different density. Even the superficial epithelium becomes interspersed with dark cells (Figs. 3-8,3-9). In addition, the light epithelial cells continue to proliferate, above all in the superficial layer of the cortex. In this area the invasion and increase of dark cells, together with the proliferation of light cells, result in a balanced proportion of the two types of cells. At the beginning of the 10th week (day 64), a mixture of dark and light somatic cells composes the superficial and deeper parts of the cortex and the medullary area.

During the onset of cortical differentiation, connective tissue fibres and basal lamina material are negligible. The scant occurrence of the extracellular matrix is conspicuous in the indifferent gonad (81,96). In the gonad of a 21-mm embryo, it is difficult to find collagen fibres and/or basal lamina material (see Figs. 3-21,3-22). The extracellular spaces below the superficial epithelium are empty (see Fig. 3-21). This changes during the eighth week: Basal lamina material occurs in several areas. (1) Fine filamentous material fills the extracellular spaces. (2) Subsequently a basal lamina is formed that lines the superficial layer of dark and light blastemal cells, including some germ cells (see Figs. 3-8,3-23). This basal lamina remains incomplete, so that the superficial layer *continues* through many openings into the deeper parts of the cortex. (3) Cords which are interspersed throughout the entire cortex show the same cellular composition as in the superficial layer and are *incompletely* surrounded by basal lamina material (see Fig. 3-24). This indistinct separation of the cortical cords from the interstitial tissue permits a continual *transfer* of cells from the interstitium into the cords or vice versa. During the ninth week and the subsequent period the basal lamina below the superficial layer shows an increased thickness. Finally an irregular stratum of extracellular homogeneous and dense material partially supports the superficial layer (see Figs. 3-23,3-25).

4.1. Cortex differentiation and second ingrowth; significance of the ovarian medulla and rete blastema

Descriptions of the essential process of differentiation of the ovarian cortex differ considerably, de-

←

Figures 3-12–3-15. (12) EM of an intermediate type of germ cell (pre-oogonium) with plaques of glycogen and lipidlike bodies (L) (day 57). N: nucleus with reticular nucleolus. X 15,500. (13 and 14) Lipidlike bodies exhibiting different grades of composition or disintegration and increasing density. X 25,500. (15) Advanced stage of pre-oogonium (day 59) without glycogen and reduced lipidlike bodies (L). Note the reduced heterochromatin beneath the nuclear membrane except at one area (arrows). Three dark supporting cells (Dc) with irregularly shaped nuclei, one of them exhibiting a cilium (C). X 7,700.

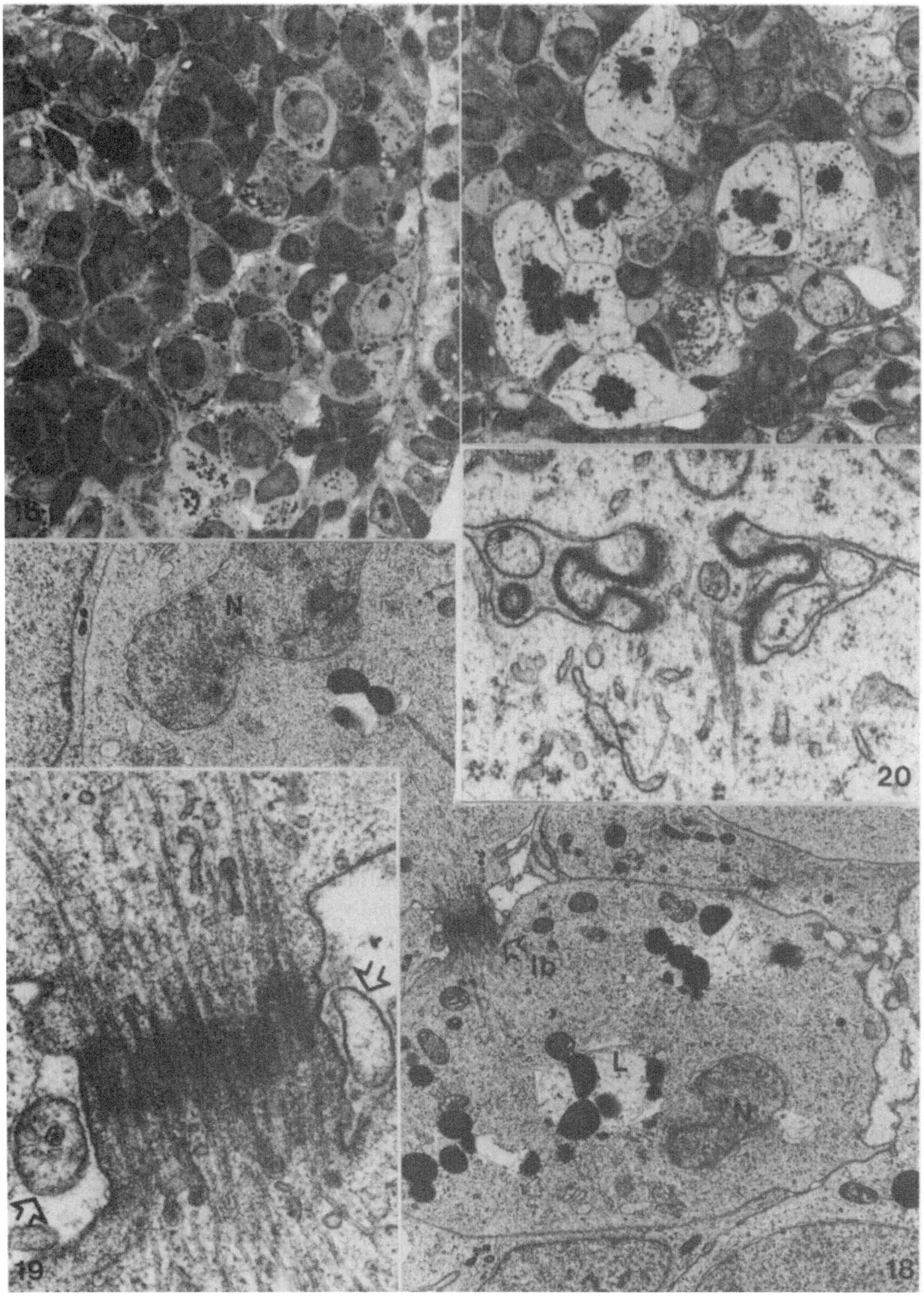

Figures 3-16–3-20. (16) Ovarian cortex (day 59). Most germ cells show accumulations of lipidlike bodies (=pre-oogonia). X 890. (17) Cluster of oogonia in synchronized mitosis (day 79). X 890. (18) Pre-oogonia in late telophase (day 59) connected by an intercellular bridge (Ib). L: lipidlike bodies; N: reconstructing nuclei. X 7,200. (19) Higher magnification of the intercellular bridge in Figure 3-18: Remnants of spindle microtubules form Flemming's mid-body. Processes of supporting cells (arrows) flank the bridge. X 40,000. (20) Intercellular bridge between two oogonia (day 79). X 40,000.

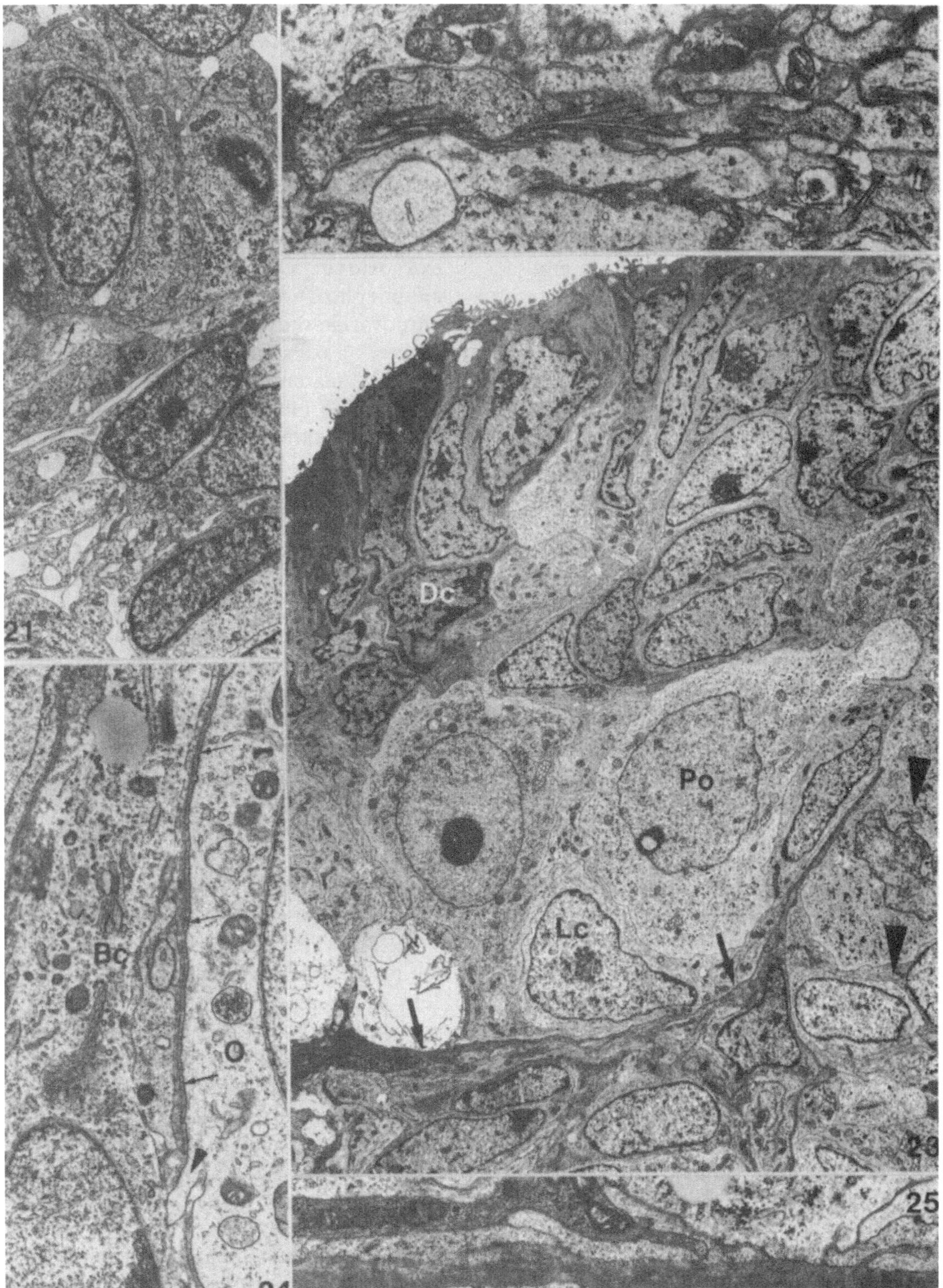

Figures 3-21–3-25. (21) Superficial blastema at onset of cortical differentiation (day 42). The epithelium is incompletely separated from the blastema below by a basal lamina (arrows) and some collagen fibrils. X 4,250. (22) Higher magnification demonstrating collagen fibrils between processes of blastemal cells. X 15,500. (23) Ovarian cortex (day 120). Superficial blastemal layer shows pre-oogonia (Po) and light (Lc) and dark cells (Dc). An irregularly thickened basal lamina (arrows) supports the superficial blastema which is connected to the deeper cortex (arrowheads). X 3,200. (24) Incomplete basal lamina (arrows) borders oogonium (O) which has contact (arrowhead) with a blastemal cell (Bc) of the interstitial space. X 15,500. (25) Higher magnification of the thickened basal lamina between superficial layer and deeper cortex. X 18,500.

pending on the mammalian species dealt with. Those who have studied large mammalian species, especially human ovaries (49,77–79,83,87), describe a *second ingrowth* of the superficial epithelium, forming *cortical cords* deriving from the coelomic epithelium. Cords which are located in the medullary area are believed to be remnants of a *first ingrowth* (97). Those investigators who have studied small mammalian species (rat, mouse, hamster) have not seen any significant structural changes (25,98–101).

According to our interpretation of the two morphogenetic events necessary for female sexual differentiation, and for the initiation of oogonial proliferation and onset of meiosis, these different views of ovarian differentiation can be explained. In small mammals a structure prevails during sexual differentiation that permits an early onset of meiosis. There is no second ingrowth and the structural situation is concealed by the meiosis-inducing cellular arrangement. This structure does not differ much from the composition during the indifferent period. Therefore, the maintenance and unchanged growth of the indifferent gonad have been postulated, a process which continues until development of the first follicles. The blastemal structure does not change much, and since it does not lose its connection to the surface, the cortex simply develops by growth of the preestablished blastemal arrangement.

The morphogenetic events of sexual differentiation and cortical growth are distinguishable in those mammals that exhibit an ontogenesis of long duration and an extended period of gonadal development. In these species a second ingrowth is necessary in order (1) to reestablish the blastemal structure, which in the female gonad has achieved a state of demarcation of the two blastemal parts, and (2) to guarantee a sufficient increase in ovarian volume, in order to establish an amount of cortical tissue sufficient to ensure the enormous multiplicative process of oogonial proliferation. Ovarian development and cortical differentiation differ basically among species by the different chronology of organogenesis and by different increases in ovarian volume and number of female germ cells.

In the ovary of small mammals, no medulla is formed that is equivalent to the structure in the human ovary; the developmental process in the rat or mouse ovary does not leave remnants of a first ingrowth (98). Therefore, the medullary structures are at most homologous to that structure in the human ovary that becomes separated from the medulla proper and that is positioned in the hilar region: the rete ovarii. In the rat, mouse or hamster (55,59,61,63,64,67,69–71), an intraovarian part of the rete ovarii connects the extraovarian rete tubules (the rudiments of the mesonephros) with the cortical blastema. The intraovarian rete consists of irregular strands and groups of cells of mesonephric origin (59). In the large mammalian ovary the amount of intraovarian and hilar rete cells is more pronounced, and forms extended reticular strands along the longitudinal axis of the ovary. These are connected to an area of undifferentiated blastema at the upper ovarian pole (central blastema), which in turn is connected to the regressing mesonephric system (40,50,53,63,93,102). The rete blastema in the human ovary (as in the human testis) serves as a pathway for mesonephric cells to become disseminated in the ovarian medulla and, finally, in the ovarian cortex.

4.2. Germ cell differentiation during cortical growth until onset of meiotic prophase

At the onset of cortical differentiation (seventh week) the number of germ cells is low (28.1 per 100 μm^2: see Figs. 3-3,3-7) and increases rapidly around day 50 (165.9 per 100 μm^2: see Figs. 3-8, 3-9). This increase is maintained at this level during the further growth of the cortex until the 12th week, when the first preleptotenal oocytes appear. The question that arises concerns which stage of germ cell differentiation coincides with this multiplying population of cells.

According to Gondos (19,21,33), primitive germ cells are found in the human ovary until the end of the 10th week, while a few days earlier (week 9.5), single oogonia are present, and some days later (week 10.5), groups of mitotically dividing oogonia appear. Oogonia are those germ cells that multiply by rapid proliferation and terminate in the meiotic prophase. The structural characteristics of these cells include a large and spherical nucleus with little chromatin and with one or two large reticular nucleoli (Fig. 3-15). The cytoplasm is reduced to a small rim containing few organelles (9,18,21). The most conspicuous structures are intercellular bridges that have a diameter of up to 1.0 µm and that interconnect a clonal association of oogonia (Fig. 3-20). The mitotic divisions are synchronized, and clusters of dividing oogonia exhibiting identical chromosomal configurations are a common feature (Fig. 3-17). By contrast, primordial germ cells (PGCs) are smaller and irregular in shape, and their cytoplasm may form long processes. In human PGCs the cytoplasm is permeated with glycogen particles, and a subplasmalemmal zone contains a dense rim of microfilaments (Fig. 3-6). Desmosomal junctions (Fig. 3-11) interconnect the PGCs with adjacent supporting cells. PGCs are single; no intercellular bridges connect them, and their mitotic rate is low.

In the human ovary during a period of three to four weeks (days 47 to 72), the mitotic rate of germ cells seems high, but the other characteristics of oogonia cannot be recognized. This intermediate stage of germ cell differentiation differs in some respects from oogonia as well as from PGCs. The following special features have been identified: (1) The intermediate type of germ cells are less dense than PGCs; they do not, however, show the very light appearance of proliferating oogonia (Figs. 3-4,3-12,3-15). Their shape is less irregular than that of the PGCs, but they still may have cytoplasmic processes, which oogonia never show. (2) Contact to the basal lamina, which oogonia rarely display, can be demonstrated at some sites by some — but not all — germ cells (Fig. 3-24). At the beginning of this four-week period, the germ cells still contain large amounts of glycogen that diminish during further differentiation. (3) In some areas glycogen granules show a reduced density, and an increasing number of lipidlike bodies appear (Figs. 3-12,3-13). The number of these lipid bodies reaches a maximum around day 60 (Fig. 3-16). Finally, they show signs of dissolution (Fig. 3-14) and within the 11th week, most germ cells are devoid of glycogen granules and lipid bodies (Fig. 3-15). (4) Although the mitotic rate must be quite high — the number of germ cells per area remains at a consistently high level during this period — the grouping of the germ cells does reach the same degree of compactness as observed during the 11th and 12th week (compare Fig. 3-16 and Fig. 3-17). Before the 70th day, cluster formation by germ cells remains an exception. A similar delay of the development of oogonial clusters has been found in bovine and ovine fetuses (102). (5) Intercellular bridges between germ cells have been first detected in a specimen from day 72 (Fig. 3-20). During the 8th to 10th weeks, no bridges could be found between germ cells in *interphase*. There are however intercellular bridges during the late *telophases* (Figs. 3-18,3-19). They show the pecularities of the *Flemming's mid-body* (103).

Summarizing the structural characteristics of the germ cells that populate the ovary shortly after sexual differentiation and during development of the cortex, one becomes aware that a special category needs to be inserted between the PGCs and the oogonium. This intermediate type of germ cell loses the structural attributes of PGCs (glycogen, cytoplasmic processes), yet it does not exhibit all the characteristics of oogonia. The intermediate germ cell seems to divide rapidly, but does not form oogonial clusters; no groups of synchronized mitoses can be observed and no intercellular bridges connect identical oogonia during their interphase. A distinct type of female germ cell must precede the oogonium.

The various cell stages might be defined in the following way:

- The intermediate type of female germ cell guarantees the increase of germ cell number by continued proliferation. Single mitoses occur separately. This type of germ cell should be termed *pre-oogonium*, because it accomplishes one of the oogonial characteristics: increased cell multiplication.
- The oogonium proper guarantees, by means of its synchronized mitosis and cluster formation, a final rapid but *limited* increase in number. In addition, the clonal arrangement of the oogonial cluster seems to initiate meiosis. After a definite number of synchronized mitoses, the oogonium becomes an oocyte and enters meiotic prophase.

4.3. Increase of dark supporting cells and oogonial proliferation

Coincident with preoogonial and oogonial proliferation, the penetration of the cortex by dark cells continues. At a final stage — during the 12th

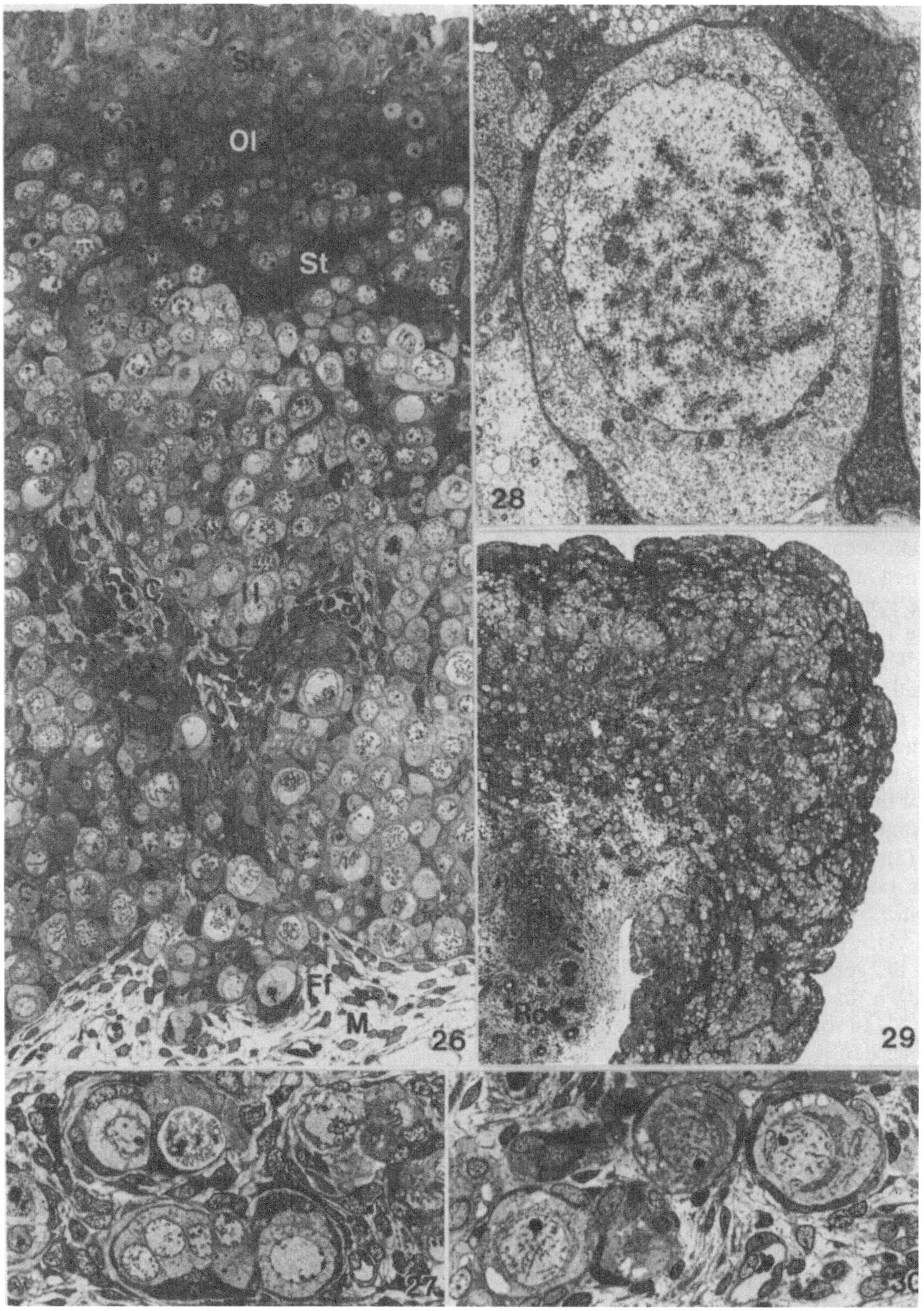
Ol
St
C
H
Ff
M
26
28
Ro
29
27

week — the thickness of the cortex has increased two- to three-fold. With the increasing number of dark immigrating cells, the mitotic activity of preoogonia and oogonia also increases. Light epithelial cells and dark mesonephric cells enclose the oogonia. But the rapid proliferation of the germ cells must depend on the cellular interaction between dark supporting cells and oogonia. If the intercellular contact between dark cells and oogonia exceeds that of the light cells, the germ cell starts to proliferate. The activation of the oogonial proliferation depends on the increasing number of dark cells in the immediate vicinity of the germ cell. These interactions between supporting cells and germ cells can be observed especially in the medullary area. Here, small groups lie in a loose mesenchymal tissue and the investment by dark (and some light) cells can be clearly seen (see Fig. 3-27). The oogonia multiply as in the cortex. Their proliferation, however, seems to be more advanced, and preleptotenal oocytes already appear during the 12th week.

5. Differentiation of the cortex during progress of meiotic prophase (13th to 16th week)

After the appearance of the first leptotenal stages during the 12th week and increasingly during the 13th week, the cortex changes. During the fourth month, the number of proliferating oogonia decreases and the number of oocytes exhibiting advancing stages of meiotic prophase increases. The cortex can be divided into three parts or layers that change in proportion during this period. The border between cortex and medulla becomes indistinct; cortical and medullary cords are difficult to separate. The three parts are as follows (Figs. 3-23–3-26):

(1) The superficial blastemal layer, which already has been accentuated during the preceding cortical differentiation. This layer consists of a variety of light and dark blastemal cells that enclose some germ cells (preoogonia), and is bordered by an incomplete basal lamina.

(2) The second layer corresponds to the cortical structure that has also developed during the preceding period and in which the proliferating preoogonia and oogonia assemble. This layer has attained its maximum size at the beginning of the fourth month and diminishes in the same measure as the third layer increases. At the end of this period, the zone with proliferating oogonia has been reduced to a thin layer.

(3) The third part becomes a voluminous layer filled with the advancing stages of oocytes in meiotic prophase. This layer stands out against the rest of the cortex due to its lighter appearance: Oocytes in prophase are large, empty-looking cells.

5.1. Distribution of the somatic cell content in the cortex after meiosis is initiated

Differences in the distribution of the light and dark supporting cells characterize the three (1–3) cortical layers. This cortical differentiation is very pronounced at the 15th week, and should be described in an appropriate stage (Fig. 3-26). (1) In the superficial blastemal layer, dark and light cells are intermingled. In some areas, however, the light cells dominate (Fig. 3-26). Dark cells are concentrated at those sites where the superficial layer is connected to the second part of the cortex through the interrupted basal lamina. They establish a connection with the somatic content of the (2) second layer. The proportion of dark cells increases centrally so that the deepest part of the oogonial layer and the adjacent part of the third layer (3) are prominent, due to their dense

←

Figures 3-26–3-30. (26) LM of the ovarian cortex (15 weeks). Sb: superficial blastema; Ol: 2nd outer dense layer of cortex with proliferating oogonia; Il: 3rd inner layer of cortex with oocytes in meiotic prophase; St: interlobular blastemal strands which contain capillaries (C) in the inner cortex; M: medulla; Ff: area of follicle formation at the cortical–medullary border. X 320. (27) Groups of oocytes in early folliculogenesis (week 20) (corresponds to Ff in Figure 3-26). X 530. (28) EM of an oocyte in late prophase of meiosis. Note mitochondria at the nuclear membrane and the thin cytoplasmic processes of dark supporting cells enveloping the germ cell. X 4,250. (29) Part of an ovary (week 18). Note the lobulation of the cortex and the increase of dark cells in blastemal strands. Medullary cords are reduced and the mesovarium contains the dense rete blastema (Rb) and tubules of the rete ovarii (Ro: epoophoron). X 50. (30) Primordial follicles (week 24). Note the light and dark granulosa cells. X 530.

structure (Fig. 3-26). Dense strands separate groupings of oocytes and thus give rise to a lobulation of the inner cortex. Within the lobules a few light and dark cells are interspersed between the leptotenal, zygotenal, and pachytenal oocytes (Fig. 3-26).

For a long time, the peculiar concentration of dark cells in the cortical areas has been considered by many authors to be an outgrowth of connective tissue and vessels (7,44,75,78,81). It is, however, only partially composed of connective tissue. Mesenchymal cells and capillaries do indeed invade the cortex, but they are preceded and joined by dark cells in large numbers. Because the dark cells can be traced as a continuing structural unit from the rete blastema through the medulla and through all parts of the cortex up to the ovarian surface, and because the formation can be traced chronologically from the first immigration of cells of mesonephric origin through all stages of increasing invasion of the developing cortex, this event can be conceived of as an integral process of cortical differentiation. This *outgrowth* of a distinct population of blastemal cells into the cortex actually underlies the often described process of *secondary ingrowth*. At least, more cells grow out than superficial cells grow in. The outgrowth makes mesonephric cells available in an appropriate number in the cortex and, secondly, triggers the meiotic prophase.

How is meiosis initiated and regulated? It might depend on a physiological clock and may be triggered by an internal factor of the germ cell itself (2). One factor triggering meiosis might pertain to the sequence of mitotic divisions that precedes the onset of meiosis: Oogonia finally undergo a number of rapid mitotic divisions that are largely synchronized. This oogonial proliferation results in the formation of a germ-cell cluster and intercellular bridges. The number of oogonial divisions differs among different species (21–104, 105). Even if these premeiotic mitoses prove to be an integral part of the meiotic process, it still does not explain how the entire process becomes initiated. Another factor that seems to play a decisive role in the onset of meiosis is the local distribution of the blastemal content. As has been shown before, the peripheral, nonlobulated areas of the cortex are completely crowded with cells of mesonephric origin, while the germ cells are in the final stage of clonal proliferation and enter meiotic prophase.

Byskov (71) has linked the initiation of meiosis with the local presence of rete cells, and has demonstrated experimentally the dependency of the trigger mechanism on the mesonephric cells (69,74). In some species, however, the oogonia are thought to be enclosed in the cortical cords for an extended period, during which time the germ cells are out of reach of the rete cells. These species exhibit a *delayed meiosis* (73) that starts long after sexual differentiation and that, moreover, might be inhibited by a high amount of estradiol produced by the ovaries during this period (21). Because estradiol production is low when meiosis starts, and because there is no experimental evidence of a functional relationship between estradiol and onset of meiosis (106), the role of the mesonephric cells must be of primary importance. In contrast to those species exhibiting delayed meiosis, the human ovary belongs to the group with an *immediate* type of meiosis (12,73,106). This type begins shortly after sexual differentiation. The human ovary during cortical differentiation does, however, structurally show a clearly retarded onset of meiosis, obviously without this being attended by an inhibitory mechanism deriving from a concomitant estradiol production (106). As discussed previously, the structural situation in the human ovary during sexual differentiation is not in a sufficiently advanced stage to initiate meiosis. Only after the differentiation of the cortex has reached a final stage, and when the invasion of dark, mesonephric cells has led to an appropriate accumulation of these cells, does an optimum arrangement at the 13th week permit the onset of meiosis. The preceding period (seventh to 12th week) is required for the multiplication of oogonia. In the human ovary, initiation of meiosis definitely depends on the mesonephric cell which, during the growth of the cortex, attains an appropriate position relative to the germ cell and is established in appropriate numbers.

5.2. Intercellular relationship between supporting cells and oocytes during meiotic prophase

During the 13th and until the 16th week, and in a more pronounced manner even until the 18th week, the cortex entirely changes its appearance (compare Figs. 3-26 and 3-29). This reorganization is due mainly to the increasing volume of the cortical layer which accommodates the increasing number of oocytes entering meiotic prophase. This change is accompanied by a strong concentration of dark mesonephric cells. On the EM level, a change of the intercellular relationship between germ cells and the covering supporting cells becomes obvious. Oocytes in the different stages of meiotic prophase either have contact with neighbouring germ cells — in most cases they are at the identical stage, some being still connected by intercellular bridges — or they are enveloped by dark supporting cells that cover the surface of the germ cell or fill the space between two germ cells with thin cytoplasmic lamellae (Fig. 3-28). The very dense sheets contrast with the very light cytoplasm and the large spherical nucleus of the leptotene, zygotene, or pachytene oocytes. In LM micrographs, only the perikarya of the dark supporting cells are visible between the empty-looking, enlarged oocytes (Fig. 3-26). Light supporting cells are found too, but in smaller numbers. The fine-structural characteristics of the prophase stages are as follows: (1) There is an increasing cytoplasmic proportion (nucleocytoplasmic ratio shifts) with few organelles. (2) The mitochondria have changed from the tubular type (in the oogonia) to the crista type; the round mitochondria are aligned along the outer face of the nuclear membrane in a single row (Fig. 3-28), primarily in the earlier stages (leptotene, zygotene). (3) If encountered in the plane of section, the Golgi area occupies a polar position close to the nuclear membrane (Fig. 3-28) and encircles the centriole. (4) Dense bodies that are membrane-bounded (37) are found at some sites. (5) The nuclei show the typical, well-known chromosomal configurations of the different stages of meiotic prophase (Figs. 3-26,3-28).

At the corticomedullary border the cortex is increasingly lobed. The structural situation has not changed much compared with the more peripheral cortex except for the increase in volume of the dark (and light) supporting cells between the germ cells in later pachytene. In these lobules, the dark cells push the germ cells apart (Fig. 3-27). In ovaries of the 17th and 18th week, in many of the lobules of the inner cortex, germ cells disintegrate. Often the entire lobulus is affected and no cell membranes can be detected between the germ cells. Germ cell *atresia* occurs primarily in this inner cortex and among the later pachytenal stages (1,21,107).

5.3. The ovarian medulla and its change to the definite state

The medulla, during differentiation and growth of the cortex, is the central area of the ovary which contains some remnants of the inner core blastema of the indifferent gonad. During the rapid increase of the cortical layers in the fourth month, the medulla shrinks somewhat. The germ cells within the medullary cords differentiate, just as they do in the cortex; some become atretic, but most finally develop into follicles. Because these medullary follicles are formed early, and soon become intermingled with follicles derived from the inner border of the cortex, it is difficult to distinguish the medullary and cortical follicles from one another. The definitive medulla appears, consisting of a loose mesenchyma, blood vessels and lymphatics (Zona vasculosa). This simple medullary *stroma* continues through the mesovarium into the tubular fold (mesosalpinx) and into the plica lata (Fig. 3-29).

5.4. The rete ovarii and the rete blastema

The only specific structure that persists in the medulla is the *rete blastema*. It can be found in the hilar region of ovaries during the 18th week. The dense concentration of epitheloid rete cells is restricted to the upper pole of the ovary. At this level the mesovarium is still a broad ligament that contains the rete blastema and, in a more extragonadal position, some tubular structures, the *rete ovarii* (= epoophoron) (Fig. 3-29). At lower levels of the ovary, the mesovarium is a long and empty duplicature and no rete blastema can be found. Later, in the 24-week-old ovary, some epitheloid structures remain from the former rete blastema. Germ cells may be enclosed in the rete blastema at any time. The cells that constitute the rete blastema are mainly of the dark, fusiform type. Earlier during ovarian development, they differ regarding the degree of density. Beside these mesonephric cells, a few epitheloid cells appear. During further development, remnants of the rete blastema form tubularlike structures that still consist of dark and light cells. These remnants persist, forming the hilus cells which can give rise to neoplasms in this region.

6. Folliculogenesis

The first primordial follicles appear during the 18th week of gestation (Fig. 3-30) (87) and are located at the inner border of the cortex or between the medullary cords. The onset of folliculogenesis, however, must be scheduled some days earlier in the 16th week, when an increasing number of oocytes enter diplotene (107). This is succeeded by an isolating process that separates the oocytes (some still in late pachytene) by means of the action of the dark supporting cells. This isolation must require a period of four weeks and is linked to a heavy wave of degeneration (107) that reduces the number of oocytes that can become enclosed in a follicle. Besides the increasing number of resting primordial follicles, some growing follicles can be detected at this stage.

After the increase of the follicles at the end of the fifth month, a homogeneous layer of follicles borders the cortex and the restricted medulla. During the fifth and the sixth month, the cortex changes continuously. The inner layer is incorporated in the lobular part and becomes reduced as the oocytes enter folliculogenesis. Finally, a small rim of the cortex is left that is composed of still-proliferating oogonia and meiotic prophases. Most of the former compact cortex is replaced by primordial follicles during the fifth month, and the residual layer partitions into a cortical remnant and cryptic invaginations. Oogenesis continues as long as the remnant persists, but generally it is terminated at the end of the seventh month. The superficial blastema changes into a simple cuboidal or low columnar epithelium that consists of a mixed population of light and dark cells. Beneath the ovarian surface epithelium (13,87), the *tunica albuginea ovarii* is formed by cells of the former cortical blastema. Deeper between the primordial follicles the *cortical stroma* differentiates. In both the tunica and the stroma, a fusiform type of cell predominates, which makes the cortical stroma cell-enriched and dissimilar from the common connective tissue stroma of other organs.

6.1. Origin of granulosa cells and fine structure of follicles in the fetal ovary

Granulosa cells are thought to be derivatives of the rete ovarii (67,69,70) and/or the coelomic epithelium (77,78,98). The former is a recently developed concept based on the observation that follicles are formed at the inner border of the cortex and, therefore, can easily be supplied with pregranulosa cells from the neighbouring rete blastema, particularly as these cells are engaged in the initiation of meiosis by means of the meiosis-inducing substances (72,108). In the human ovary, those cells that support the germ cell from the very beginning could continue their supportive function and differentiate into granulosa cells. The supporting cell population proliferates periodically, and is supplemented by additional cells from the regressing mesonephros via the rete blastema. The direct transfer of mesonephric cells from the rete to the granulosa layer occurs in the ovary of small mammalians (2, 67,70). In the human ovary, the continuing invasion of mesonephric cells during the entire period of cortical growth, and a subsequent in loco proliferation of the blastema, must be sufficient for granulosa formation at any site of the cortex. The same is true for the role of the *second* type of supporting cell: The light supporting cells participate in the formation of the granulosa envelope, too. This is in accordance with the former concept, which has been commonly accepted until recently (42,84,87). Could both types of supporting cells differentiate into granulosa cells? There are differing types of granulosa cells detectable in the human follicle (Fig. 3-30), and differing types have also been described in other mammalian species (24,25,109,110). They are dissimilar in the same manner and consist of dark granulosa cells that form a low covering and light ones that are less flattened.

The fine structural features of ovarian follicles have been demonstrated by several authors with convincing micrographs (4,10,12,15–17,26,102, 110–120). These studies are based mainly upon adult ovaries and to a lesser extent upon fetal

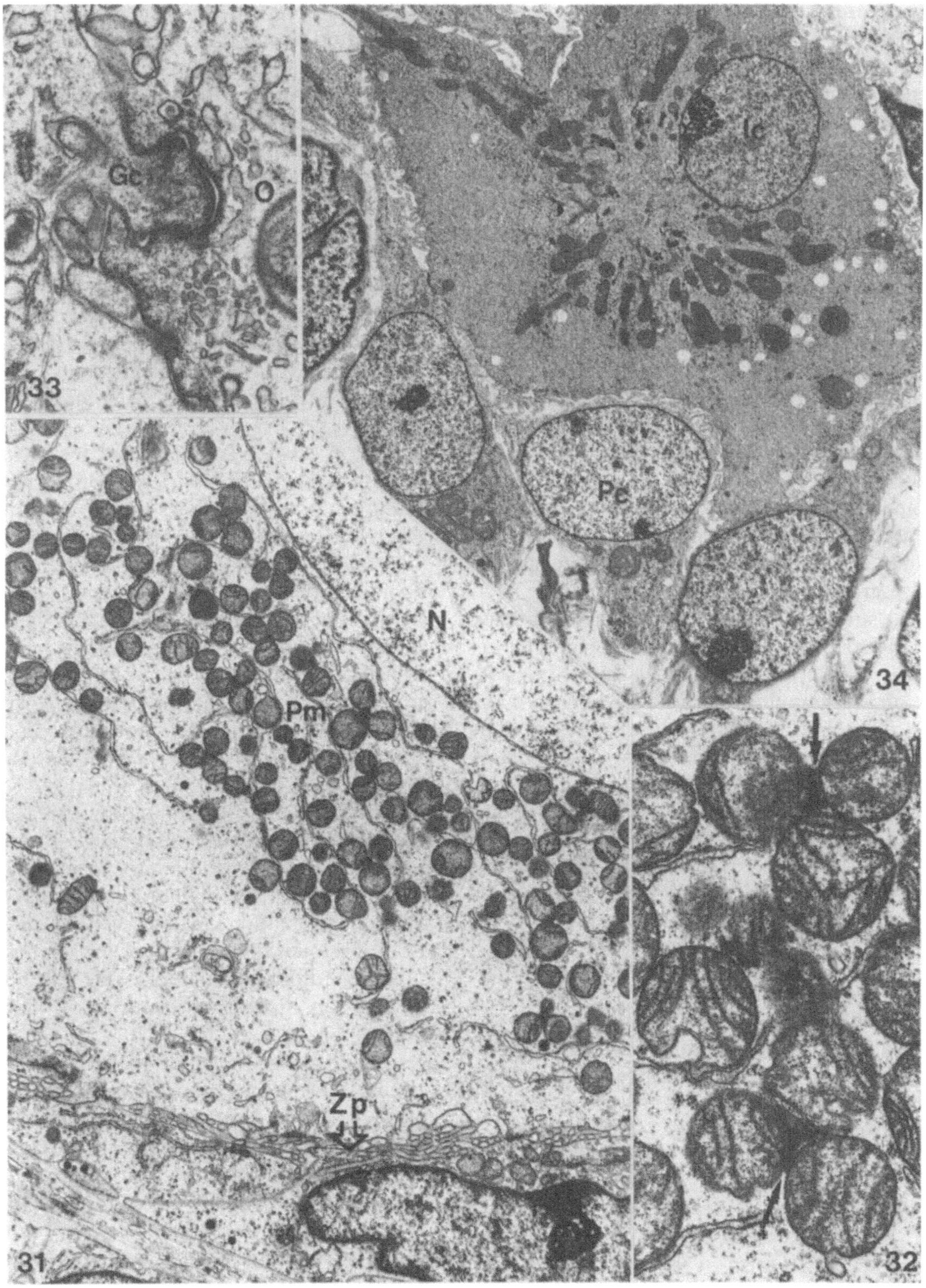

Figures 3-31–3-34. (31) Part of an unilaminar follicle (week 18). N = nucleus of oocyte; Pm = perinuclear cloud of mitochondria with concentric lamellae of ER; Zp = surface of oocyte interdigitating with granulosa cell membranes (area of future zona pellucida). X 7,700. (32) Higher magnification of mitochondrial cloud. Mitochondria show close contact with smooth ER and intermitochondrial cement (arrow = nuage). X 25,500. (33) Nexus between oocyte (O) and process of granulosa cell (Gc). X 25,500. (34) Ovarian interstitial cells (Ic) (week 15) and three pre-interstitial cells (Pc). X 3,250.

organs. Therefore, a summary description is sufficient. The oocyte of a resting follicle shows a large, spherical nucleus with two or three nucleoli, a less dense, spotted chromatin pattern, and a great number of nuclear pores that are distributed in irregular groups along the nuclear envelope. In the cytoplasm, the most conspicuous feature is the arrangement of the mitochondria. They are round or irregular in shape and show a peculiar arched formation of their cristae (Figs. 3-31,3-32); they are arranged in concentric layers that surround the nucleus and that alternate with concentric lamellae of the smooth ER. This perinuclear *cloud* of mitochondria can concentrate eccentrically and can include the Golgi area. During early follicle formation, the distribution of mitochondria and ER is not as regular as seen in the more advanced state. *Dense bodies* (37) appear regularly and *nuage* (121) occurs in two structural versions: (1) dense, intermitochondrial cement (18) and (2) nucleolarlike bodies or aggregations of dense granular material.

At the surface of the oocyte, the plasma membrane interdigitates with that of the granulosa cells (Fig. 3-31). At several sites, fingerlike processes of the granulosa cells have contacts with or invaginate into the surface of the oocyte, and exhibit intercellular junctions of the desmosomal type (Fig. 3-33). Extracellular material indicating the formation of the zona pellucida does not show up during this early stage.

6.2. *Interstitial cells in the fetal human ovary*

Gondos and Hobel (122) have described interstitial cells in 12–20-week-old human fetuses. We detected these cells in specimens between the 15th and 20th week. The maximum number of interstitial cells has been observed during the 18th week (Fig. 3-34). No interstitial cells have been seen in a 23-week-old ovary. The fine structure of these cells relates to that of the fetal Leydig cells. The cytoplasm is filled with a vesicular smooth ER and the mitochondria are dense, showing the characteristic features of steroid cells (Fig. 3-34).

7. Conclusions

The differentiation of the human ovary shows some morphogenetic events that relate primarily to the role of two different components of the ovarian blastema: dark fusiform cells that originate from the regressing mesonephros and that enter the ovary via the rete blastema, and light epitheloid cells that are derivatives of the former coelomic epithelium. These different types of cells function as early supporting cells and regulate the oogenetic differentiation of the female germ cells. The blastemal arrangement changes and influences the mitotic proliferation of the germ cells, initiates the meiotic prophase, and contributes to early folliculogenesis. The process of cortex formation is governed by means of a time-table: During the seventh week, the former indifferent gonad changes to an early ovary and the cortex starts to differentiate. This morphogenetic process is accompanied by an increased germ cell proliferation. Cluster formation of multiplying oogonia starts during the 11th week, when the cortex is fully developed. During the 13th week, meiotic prophase commences and is combined with a heavy concentration of dark mesonephric cells in the outer layer of the cortex. Finally during the 18th week the first primordial follicles are formed. It has to be emphasized that in the human ovary, periods of germ cell differentiation overlap: The proliferation of preoogonia, which starts shortly after sexual differentiation around day 46, continues when the oogonial cluster formation appears during the 11th week. Accordingly, oogonial multiplication extends beyond the onset of meiotic prophase and folliculogenesis. Therefore, the cortex of an 18-week-old fetus exhibits all stages of germ cell differentiation; the quantitative proportion of the stages, however, differs and is expressed in a continued change of the structural composition of the cortex.

The morphogenesis of the human ovary between the seventh and 24th week can be understood by three aspects. (1) The single germ cell requires an alteration of its local environment in order to start an increased mitotic proliferation. This shift happens by a rearrangement of the blastemal components and causes cortical differentiation and growth. (2) Differentiation of the ovarian cortex is combined with an increase of *mesonephrogenic* cells. *Outgrowth* of dark blastemal cells and *ingrowth* of cortical cords are two aspects of the same morphogenetic event; this is essential to achieve a local situation that permits a continued multiplication *and* differentiation of the germ cell. (3) In addition to the classical view of zonal structure of the cortex, and the movement of germ cells and/or blastemal cells from the outer to the inner cortex, and the growth of an interstitial vascular tissue from the medulla to the outer cortex, another aspect is of equal consequence. The differentiation and growth of the cortex can be caused by local proliferation of blastemal *and* germ cells and a local advancement of oogenesis, a process that spreads peripherally by adding new generations of proliferating zones at the ovarian surface. The ovary grows at the outside and to a lesser extend *into* the medulla.

The studies have also revealed that the transition from the primordial germ cell to the oogonial stage requires an intermediate type of germ cell that multiplies rapidly but does not form oogonial clusters.

Acknowledgments

I express my appreciation to Mrs. A. Ihmer and Mrs. I. Sieburg for technical assistance and to Mrs. I. Buchhauser and Alexander Gardner for aid in preparing the manuscript. Part of the studies was supported by grants of the Deutsche Forschungsgemeinschaft.

References

1. Baker TG, Franchi LL: Fine structure of oogonia and oocytes in human ovaries. *J Cell Sci* 2: 213–224, 1967.
2. Byskov AG: Sexual differentiation of the mammalian ovary. In: Biology of the Ovary. *Developments in Obstetrics and Gynecology*. Vol 2. PM Motta, ESE Hafez (eds), The Hague-Boston-London: Martinus Nijhoff Publ, p 3–15, 1980.
3. Byskov AG: Gonadal sex and germ cell differentiation. In: Mechanism of Sex Differentiation in Animals and Man. CR Austin RG Edwards (eds), London: Academic Press, p 145–164, 1981.
4. Dvorak M, Tesarik J: Ultrastructure of human ovarian follicles. In: Biology of the Ovary. Developments in Obstetrics and Gynecology. Vol 2. PM Motta, ESE Hafez (eds), The Hague-Boston-London: Martinus Nijhoff Publ, p 121–137, 1980.
5. Gondos B, Bhiraleus P, Hobel CL: Ultrastructural observations on germ cells in the human fetal ovaries. *Am J Obstet Gynecol* 110: 644–652, 1971.
6. Gondos B: Differentiation and growth of cells in the gonads. In: Differentiation and Growth of Cells in Vertebrate Tissues. G Goldspink (ed), London: Chapman and Hall, p 169–208, 1974.
7. Gondos B: Surface epithelium of the developing ovary: Possible correlation with ovarian neoplasia. *Am J Pathol* 81: 303–313, 1975.
8. Gondos B: Cellular interrelationship in the human fetal ovary and testis. Eleventh International Congress of Anatomy. Advances in the Morphology of Cells and Tissues. New York: Alan R. Liss, Inc, p 373–381, 1981.
9. Gondos B: Germ cell differentiation and intercellular bridges. In: Ultrastructure of Reproduction. J Van Blerkom, PM Motta (eds), Boston, The Hague, Dordrecht, Lancaster: Martinus Nijhoff Publ, p 31–45, 1984.
10. Hertig AT, Barton BA: Fine structure of mammalian oocytes and ova. In: Handbook of Physiology, Endocrinology II. RO Greep, EB Astwood, (eds), Washington, DC: American Physiological Society, p 317–348, 1973.
11. Kellokumpu-Lehtinen P-L, Söderström K-O: Occurrence of nuage in fetal human germ cells. *Cell Tissue Res* 194: 171–177, 1978.
12. Lanzavecchia G, Mangioni C: Étude de la structure et des constituants de follicule humain dans l' ovaire foetal. I.- Le follicule primordial. *J Microsc* 3: 447–464, 1964.
13. Motta PM, Makabe S: Development of the ovarian surface and associated germ cells in the human fetus. *Cell Tissue Res* 226: 493–510, 1982.
14. Ruby JR, Dyer RF, Gasser RF, Skalko RG: Intercelullar connections between germ cells in the developing human ovary. *Z Zellforsch* 105: 252–258, 1970.
15. Stegner HE, Wartenberg H: Elektronenmikroskopische Untersuchungen an Eizellen des Menschen in verschiedenen Stadien der Oogenese. *Arch Gynäkol* 199: 151–172, 1963.
16. Stegner HE: Die elektronenmikroskopische Struktur der Eizelle. *Ergebnisse der Anatomie und Entwicklungsgeschichte* 39/6: 1–113, 1967.
17. Wartenberg H: Elektronenmikroskopische Untersuchungen am Ovar der Feten und Neugeborenen. *Verh Anat Ges Anat Anz: Ergänzungsheft zu Bd* 112: 195–205, 1963.
18. Wartenberg H: Spermatogenese — Oogenese: ein cytomorphologischer Vergleich. *Verh Anat Ges* 68: 63–92, 1974.
19. Baker TG, Franchi LL: The fine structure of oogonia and oocytes in the Rhesus monkey (Macaca mulatta). *Z Zellforsch* 126: 53–74, 1972.
20. Franchi LL, Mandl AM: The ultrastructure of oogonia and oocytes in the foetal and neonatal rat. *Proc Roy Soc B* 157: 99–114, 1963.
21. Gondos B: Oogonia and oocytes in mammals. In: The Vertebrate Ovary. Comparative Biology and Evolution. RE Jones (ed), New York and London: Plenum Press, p 83–120, 1978.
22. Odor, DL, Blandau RJ: Ultrastructural studies on fetal and early postnatal mouse ovaries. II. Cytodifferentia-

tion. *Am J Anat* 125: 177–216, 1969.
23. Pelliniemi LJ: Ultrastructure of the early ovary and testis in pig embryos. *Am J Anat* 144: 89–112, 1975.
24. Weakley BS: Electron microscopy of the oocytes and granulosa cells in the developing ovarian follicles of the golden hamster (Mesocricetus auratus). *J Anat* 100: 503–534, 1966.
25. Weakley BS: Light and electron microscopy of developing germ cells and follicle cells in the ovary of the golden hamster: twenty-four hours before birth to eight days post partum. *J Anat* 101: 435–459, 1967.
26. Zamboni L: Comparative studies on the ultrastructure of mammalian oocytes. In: Oogenesis. JD Biggers, AW Schuëtz (eds), Baltimore: University Park Press, p 5–45, 1972.
27. Gondos B, Zamboni L: Ovarian development: the functional importance of germ cell interconnections. *Fertil Steril* 20: 176–189, 1969.
28. Gondos B: Germ cell relationships in the developing rabbit ovary. In: Gonadotropins and Ovarian Development. WR Butt, AC Crooke, M Ryle (eds), Edinburgh: E & S Livingstone, p 239–246, 1970.
29. Gondos B: Intercellular bridges and mammalian germ cell differentiation. *Differentiation* 1: 177–182, 1973.
30. Gondos B: Germ cell degeneration and intercellular bridges in the human fetal ovary. *Z Zellforsch* 138: 23–30, 1973.
31. Ruby JR, Dyer RF, Skalko RG: The occurrence of intercellular bridges during oogenesis in the mouse. *J Morphol* 127: 307–340, 1969.
32. Zamboni L, Gondos B: Intercellular bridges and synchronization of germ cell differentiation during oogenesis in the rabbit. *J Cell Biol* 36: 276–282, 1968.
33. Gondos B, Hobel CJ: Germ cell degeneration and phagocytosis in the human foetal ovary. In: The Development and Maturation of the Ovary and Its Functions. H Peters (ed), Amsterdam: Excerpta Medica Int. Congress Series 267, p 77–83, 1973.
34. Odor DL, Blandau RJ: Ultrastructural observations on artresia in whole organ cultures of foetal mouse ovaries. In: The Development and Maturation of the Ovary and Its Functions. H Peters (ed), Amsterdam: Excerpta Medica Int. Congress Series 267, p 63–76, 1973.
35. Merchant-Larios H: Ultrastructural events in horse gonadal morphogenesis. *J Reprod Fert* Suppl. 27: 479–485, 1979.
36. Merchant-Larios H: The establishment and sexual differentiation of the horse gonad. *Bol Estud Méd Biol Méx* 31: 3–23, 1980.
37. Merchant-Larios H: Rat gonadal and ovarian organogenesis with and without germ cells. An ultrastructural study. *Dev Biol* 44: 1–21, 1975.
38. Pelliniemi LJ: Ultrastructure of the indifferent gonad in male and female pig embryos. *Tissue & Cell* 8: 163–174, 1976.
39. Satoh M: The histogenesis of the gonad in rat embryos. *J Anat* 143: 17–37, 1985.
40. Wartenberg H: Development of the early human ovary and the role of the mesonephros in the differentiation of the cortex. *Anat Embryol* 165: 253–280, 1982.
41. Baker TG: Oogenesis and ovarian development. In: Reproductive Biology. H Balin, S Glasser (eds), Amsterdam: Excerpta Medica, p 398–437, 1972.
42. Baker TG, Wai Sum O: Development of the ovary and oogenesis. *Clinics in Obstetrics and Gynaecology* 3: 3–26, 1976.
43. Carlon N, Stahl A: Les premiers stades du développement des gonades chez l'homme et les vertébrés supérieurs. *Pathol Biol (Paris)* 21: 903–914, 1973.
44. Haffen K: Sexual differentiation of the ovary. In: The Ovary. 2nd edition, Vol. I. S Zuckerman BJ Weir (eds), New York, San Francisco, London: Academic Press, p 69–112, 1977.
45. Jost A: Embryonic sexual differentiation (morphology, physiology, abnormalities). In: Hermaphroditism, Genital Anomalies and Related Endocrine Disorders. 2nd Edition. HW Jones, WW Scott (eds), Baltimore: Williams and Wilkins, p 15–45, 1971.
46. Jost A, Vigier B, Prépin J, Perchellet JP: Studies on sex differentiation in mammals. *Rec Progr Hormone Res* 29: 1–41, 1973.
47. Merchant-Larios H: Ovarian differentiation. In: The Vertebrate Ovary. RE Jones (ed), New York, London: Plenum Press, p 47–81, 1978.
48. Merchant-Larios H: Germ and somatic cell interactions during gonadal morphogenesis. In: Ultrastructure of Reproduction. J Van Blerkom, PM Motta (eds), The Hague, Dordrecht, Lancaster: Martinus Nijhoff Publ, 1984.
49. Zuckerman S, Baker TG: The development of the ovary and the process of oogenesis. In: The Ovary. 2nd Edition, Vol I. S Zuckerman, BJ Weir (eds), New York: Academic Press, p 41–67, 1977.
50. Gropp A, Ohno S: The presence of a common embryonic blastema for ovarian and testicular parenchymal (follicular, interstitial and tubular) cells in cattle, Bos taurus. *Z Zellforsch* 74: 505–528, 1966.
51. Jost A: Initial stages of gonadal development. Theories and methods. *Arch Anat Micr Morphol Expér* 74: 39–41, 1985.
52. Witschi E: Embryogenesis of the adrenal and the reproductive glands. *Rec Progr Hormone Res* 6: 1–23, 1951.
53. Wartenberg H: Human testicular development and the role of the mesonephros in the origin of a dual Sertoli cell system. *Andrologia* 10: 1–21, 1978.
54. Wartenberg H: Der Mesonephros und die Gonadenentwicklung. *Verh Anat Ges* 73: 385–401, 1979.
55. Wartenberg H: Structural aspects of gonadal differentiation in mammals and birds. *Differentiation* 23 (suppl): 64–71, 1983.
56. Upadhyay S, Luciani JM, Zamboni L: The role of the mesonephros in the development of the indifferent gonads and ovaries of the mouse. *Ann Biol Anim Biochim Biophys* 19: 1179–1196, 1979.
57. Zamboni L, Bezard J, Mauléon P: The role of the mesonephros in the development of the sheep fetal ovary. *Ann Biol Anim Biochim Biophys* 19: 1153–1178, 1979.
58. Zamboni L, Upadhyay S, Bezard J, Luciani JM, Mauléon

P: The role of the mesonephros in the development of the mammalian ovary. In: Endocrine Physiopathology of the Ovary. RI Tozzini, G Reeves, RL Pineda (eds), Amsterdam: Elsevier/North-Holland, p 3–42, 1980.

59. Byskov AG: The anatomy and ultrastructure of the rete system in the fetal mouse ovary. *Biol Reprod* 19: 720–735, 1978.
60. Carlon N, Pizant J, Stahl A: Mesonephric origin of the gonadal primitive medulla in chick embryos. *Anat Embryol* 166: 399–414, 1983.
61. Fraedrich J: Licht- und elektronenmikroskopische Untersuchungen über den Zusammenhang der Mesonephros- und frühen Gonadenentwicklung der weissen Maus. Dissertation, Med Fakul Bonn, p 1–58, 1979.
62. Kinsky I: Bildung des somatischen Gonadenblastems durch degenerierende Urnierenanteile des Kaninchens. *Verh Anat Ges* (Innsbruck) 73: 403–406, 1979.
63. Wartenberg H: Differentiation and development of the testes. In: The Testis. Comprehensive Endocrinology. H Burger, D deKretser (eds), New York: Raven Press, p 39–80, 1981.
64. Upadhyay S, Luciani JM, Zamboni L: The role of the mesonephros in the development of the mouse testis and its pathways. In: Proceedings of the Vth Workshop on the Development and Function of the Reproductive Organs, Copenhagen. Amsterdam: Excerpta Medica, p 18–27, 1981.
65. Zamboni L, Upadhyay S, Bézard J, Mauléon P: The role of the mesonephros in the development of the sheep testis and its excurrent pathways. In: Proceedings of the Vth Workshop on the Development and Function of the Reproductive Organs, Copenhagen. Amsterdam: Excerpta Medica, p 31–39, 1981.
66. Zamboni L, Upadhyay S: The contribution of the mesonephros to the development of the sheep testis. *Am J Anat* 165: 339–356, 1982.
67. Byskov AG, Lintern-Moore S: Follicle formation in the immature mouse ovary. The role of the rete ovarii. *J Anat* 116: 207–217, 1973.
68. Byskov AG, Rasmussen G: Ultrastructural studies of the developing follicle. In: The Development and Maturation of the Ovary and Its Functions. H Peters (ed), Amsterdam: Excerpta Medica Int. Congress Series 267 p 53–62, 1973.
69. Byskov AG, Skakkebaek NE, Stafanger G, Peters H: Influence of ovarian surface epithelium and rete ovarii on follicle formation. *J Anat* 123: 77–86, 1977.
70. Byskov AG: The role of the rete ovarii in meiosis and follicle formation in the cat, mink and ferret. *J Reprod Fertil* 45: 201–209, 1975.
71. Byskov AG: Does the rete ovarii act as a trigger for the onset of meiosis? *Nature* (London) 252: 396, 1974.
72. Byskov AG: The meiosis inducing interaction between germ cells and rete cells in the fetal mouse gonad. *Ann Biol Anim Biochim Biophys* 18 (2B): 327–334, 1978.
73. Byskov AG: Regulation of meiosis in mammals. *Ann Biol Anim Biochim Biophys* 19: 1251–1261, 1979.
74. Byskov AG, Saxén L: Induction of meiosis in fetal mouse testis in vitro. *Dev Biol* 52: 193–200, 1976.
75. Brambell FMR: The development of sex in vertebrates. London: Sidgwick & Jackson, 1930.
76. Fischel A: Über die Entwicklung der Keimdrüsen des Menschen. *Z Anat Entwicklungs-Gesch* 92: 34–72, 1930.
77. Franchi LL, Mandl AM and Zuckerman S: The development of the ovary and the process of oogenesis. In: The Ovary. Vol 1. S Zuckerman (ed), New York, London: Academic Press, p 1–71, 1962.
78. Gillman J: The development of the gonads in man, with a consideration of the role of fetal endocrines and the histogenesis of ovarian tumors. *Contrib Embryol Carnegie Inst Wash* 32: 81–143, 1948.
79. Gruenwald P: The development of the sex cords in the gonads of man and mammals. *Am J Anat* 70: 359–389, 1942.
80. Jirásek JE: Development of the genital system in human embryos and fetuses. In: Development of the Genital System and Male Pseudohermaphroditism. MM Cohen, Jr (ed) Baltimore: Johns Hopkins Press, p 3–23, 1971.
81. Jirásek JE, Principles of reproductive embryology. In: Disorders of Sexual Differentiation. JL Simpson (ed), New York, San Francisco, London: Academic Press, p 51–110, 1976.
82. Merchant-Larios H: Ovarian differentiation. In: The Vertebrate Ovary. Comparative Biology and Evolution. RE Jones (ed), New York and London: Plenum Press, Chapt 2, 1978.
83. Mossman HW, Duke KL: Comparative Morphology of the Mammalian Ovary. Madison, WI, London: The University of Wisconsin Press, 1973.
84. Peters H, McNatty KP: The Ovary. A Correlation of Structure and Function in Mammals. London: Granada Publishing, 1980.
85. Peters H: Intrauterine gonadal development. *Fert Steril* 27: 493–500, 1976.
86. Politzer G: Die Keimbahn des Menschen. *Z Anat Entwicklungs-Gesch* 100: 331–361, 1933.
87. van Wagenen G, Simpson ME: Embryology of the Ovary and Testis. Homo Sapiens and Macaca Mulatta. New Haven: Yale University Press, 1965.
88. Witschi E: Migration of the germ cells of human embryos from the yolk sac to the primitive gonadal folds. *Contrib Embroyl Carnegie Inst Wash* 32: 67–80, 1948.
89. Wartenberg H: Germ cell migration induced and guided by somatic cell interaction. *Bibliotheca Anat* 24: 93–110, 1983.
90. Ritzén, EM, Hansson V, French FS: The Sertoli cell. In: The Testis. H Burger, D deKretser (eds), New York: Raven Press, p 171–194, 1981.
91. Grinsted J, Aagesen L: Mesonephric excretory function related to its influence on differentiation of fetal gonads *Anat Rec* 210: 551–556, 1984.
92. Witschi E: Development of vertebrates. Philadelphia, London: Saunders Company, 1956.
93. Wilson KM: Origin and development of the rete ovarii and the rete testis in the human embryo. *Contrib Embryol Carnegie Inst Wash* 17: 69–88, 1926.
94. Roosen-Runge EC: Rudimental "genital canals" of the gonad in rat embryos. *Acta Anat* 44: 1–11, 1961.
95. Wartenberg H: The influence of the mesonephric blastema on gonadal development and sexual differentiation.

In: Development and Function of Reproductive Organs. Proceedings of the Vth Workshop on the Development and Function of the Reproductive Organs, Copenhagen. AG Byskov, H Peters (eds), Amsterdam-Oxford-Princeton: Excerpta Medica, p 3–12, 1981.

96. Jirásek JE: Morphogenesis of the genital system in the human. In: Morphogenesis and Malformation of the Genital System. RJ Blandau, D Bergsma, (eds) New York: Alan R. Liss, p 13–39, 1977.
97. Forbes TR: On the fate of the medullary cords of the human ovary. *Contrib Embryol Carneg Inst Wash* 30: 9–15, 1942.
98. Brambell FWR: The development and morphology of the gonads of the mouse. Part I. The morphogenesis of the indifferent gonad and of the ovary. *Proc Roy Soc Lond B* 101: 391–409, 1927.
99. Torrey TW: The development of the urogenital system of the albino rat. II. The gonads. *Am J Anat* 76: 375–397, 1945.
100. Jost A: A new look at the mechanisms controlling sex differentiation in mammals. *The Johns Hopkins Med J* 130: 38–53, 1971.
101. Jost A: Donnèes préliminaires sur les stades initiaux de la diffèrenciation du testicule chez le rat. *Arch d'Anat Micros Morph Exp* 61: 415–437, 1972.
102. Rüsse I: Oogenesis in cattle and sheep. *Bibliotheca Anat* 24: 77–92, 1983.
103. Allenspach AL, Roth LE: Structural variations during mitosis in the chick embryo. *J Cell Biol* 33: 179–196, 1967.
104. Mauleon P, Mariana JC: Oogenesis and folliculogenesis. In: Reproduction in Domestic Animals. 3rd Edition HH Cole, PT Cupps (eds), New York, San Francisco, London: Academic Press, p 175–202, 1977.
105. Tokarz RR: Oogonial proliferation, oogenesis, and folliculogenesis in nonmammalian vertebrates. In: The Vertebrate Ovary. Comparative Biology and Evolution. RE Jones (ed), New York, London: Plenum Press, p 145–179, 1978.
106. Byskov AG, Høyer PE, Westergaard L: Origin and differentiation of the endocrine cells of the ovary. *J Reprod Fertil* 75: 299–306, 1985.
107. Baker TG: A quantitative and cytological study of germ cells in human ovaries. *Proc Roy Soc Lond B* 158: 417–433, 1963.
108. Yding Andersen C, Byskov AG, Grinsted J: Partial purification of the meiosis inducing substance (MIS). In: Development and Function of Reproductive Organs. AG Byskov, H Peters (eds), Amsterdam, Oxford, Princeton: Excerpta Medica, p 73–80, 1981.
109. Igbal SJ, Weakley BS: The effects of different preparative procedures on the ultrastructure of the hamster ovary. I. Effects of various fixative solutions on ovarian oocytes and their granulosa cells. *Histochemistry* 38: 95–122, 1974.
110. Sprumont P: Ovarian follicles of normal NMRI mice and homozygous "nude" mice. 3. Ultrastructural comparison. *Cell Tiss Res* 188: 409–426, 1978.
111. Baca M, Zamboni L: The fine structure of human follicular oocytes. *J Ultrastr Res* 19: 354–381, 1967.
112. Hertig AT, Adams EC: Studies on the human oocytes and its follicle. I. Ultrastructural and histochemical observations on the primordial follicle stage. *J Cell Biol* 34: 647–675, 1967.
113. Hertig AT: The primary human oocyte: Some observations on the fine structure of Balbiani's vitelline body and the origin of the annulate lamellae. *Am J Anat* 122: 107–138, 1968.
114. Hope J: The fine structure of the developing follicle of the Rhesus ovary. *J Ultrastr Res* 12: 592–610, 1965.
115. Kress A: Vergleichend ultrastrukturelle Aspekte der Oogenese bei Mensch und Säuger. *Biol Rdsch* 23: 49–62, 1985.
116. Motta P, Didio LJA: Microfilaments in the granulosa cells during the follicular development and transformation in corpus luteum in the rabbit ovary. *J Submicr Cytol* 6: 15–27, 1974.
117. Odor L: Electron microscopic studies on ovarian oocytes and unfertilized tubal ova in the rat *J Biophys Biochem Cytol* 7: 567–574, 1960.
118. Odor LD: The ultrastructure of unilaminar follicles of the hamster ovary. *Am J Anat* 116: 493–522, 1965.
119. Wartenberg H, Stegner HE: Über die elektronenmikroskopische Feinstruktur des menschlichen Ovarialeies. *Z Zellforsch* 52: 450–474, 1960.
120. Zamboni L: Fine morphology of ovarian follicle maturation. In: Endocrine Physiopathology of the Ovary. RI Tozzini, G Reeves, RL Pineda (eds), Amsterdam: Elsevier/North-Holland Biomedical Press, p 63–99, 1980.
121. Eddy EM: Germ plasm and differentiation of the germ cell line. In: International Review of Cytology. Vol. 43. GH Bourne, JF Danielli (eds), New York: Academic Press, p 229–280, 1975.
122. Gondos B, Hobel CJ: Interstitial cells in the human fetal ovary. *Endocrinology* 93: 736–739, 1973.
123. Balinsky BI: An Introduction to Embryology. 3rd Edition. Philadelphia, London, Toronto: Saunders, 1970.

Author's address:
Anatomisches Institut
Universität Bonn
Nussallee 10
D 5300 Bonn 1
West-Germany

CHAPTER 4

The ovary and ovulation: A three-dimensional ultrastructural study

GIUSEPPE FAMILIARI, SAYOKO MAKABE & PIETRO M. MOTTA

1. Introduction

The ovary is a highly complex organ that performs a dual role: it functions as an organ of reproduction and also as a gland of internal secretion. Gametogenesis has also been termed *folliculogenesis* (1). The ovary's secretory activity, including the release into the circulation of non-steroidal hormones that act locally within the ovary, was given the name *hormonogenesis* by Hodgen et al. (1).

From a morphological point of view, the ovary consists of an outer zone, the cortex, that is grossly separated from an internal zone, the medulla. Both the cortex and the medulla are composed of connective tissue, and there is no clear line of demarcation separating these two areas (2).

Concentrated in the cortex of the ovary are numerous small, primordial follicles, as well as a smaller number of comparatively large follicles. As many as 50 follicles grow during each cycle, measuring up to 1 mm during the follicular phase in the human ovary, but only three or less reach a size of 8 mm in the mid-cycle (3). In the human ovary, fully grown follicles are approximately 2–3 cm in diameter and have steroidogenic enzymes in both the theca and the granulosa cells (4).

Basically, two periods of follicular development can be distinguished: a) the growth of small primary and secondary follicles, which occurs up to a certain size, independently of the female cycle; and b) the growth of larger follicles, which is influenced by gonadotropins (5,6,7).

Recently it has been hypothesized that the precise regulation of follicular growth and selection is accomplished primarily by specific ovarian factors that act directly on the ovaries, and that gonadotropins, at so-called *tonic* levels, play a secondary, permissive role in folliculogenesis. Specific ovarian factors would govern the progressive winnowing of the pool of developing follicles down to the size of the species-characteristic ovulatory quota. Such factors are operative, however, only when the amount of circulating gonadotropins is above a minimal tonic level, and near a *tonic set-point*. In addition, ovarian hormones inhibit the secretion of gonadotropins in a negative-feedback fashion, limiting circulating gonadotropin levels to an appropriate range, around the tonic set-point (1,8,9).

On the basis of this hypothesis, if gonadotropin levels are below the tonic set-point, folliculogenesis will be arrested as a result of inadequate stimulation; on the contrary, if gonatropin levels are above the tonic set-point, then the intraovarian mechanisms regulating the ovulatory quota are inactivated and superovulation occurs (1,9). Other authors showed that in a normal cycle, the intraovarian feedback is finely tuned so as to inhibit gonadotropins and allow continued growth of only one follicle. Inhibin, renin, activin, granulosa peptides and luteinizing inhibiting factor all are probably responsible to a greater or lesser extent for this tuning (10,11,12,13).

During each cycle, primordial follicles leave the resting pool and enter the trajectory of follicular growth (14); this process has been termed *recruitment* (8). Recruitment of primordial follicles is not dependent on gonadotropins, but may only be enhanced by these hormones (5,6,7,15). Growing follicles are subjected to atresia and may relinquish this trajectory at any point. Afterwards, a characteristic species-specific number of follicles is selected and progresses towards ovulation in a typical cycle. The character of the selective mechanism remains uncertain (1,9).

What is certain, however, is that the selective process operates in primates with great precision; in fact, a spontaneous multiple ovulation is extremely unusual. The follicle destined to ovulate plays a key role in regulating the ovulatory quota; it is functionally dominant and inhibits the development of other competing follicles in both ovaries. The survival of the dominant follicle probably depends on a process of acquisition or retention of certain metabolic properties that enable the dominant follicle to resist atresia (1). The selected follicle becomes dominant about seven days before ovulation and it must maintain its dominance during the week preceding ovulation. Hypothetically, one mechanism might be that the dominant follicle secretes a substance (selectron) that acts directly on the ovaries to inhibit the development of potentially competitive follicles (1).

Furthermore, it has been shown that the dominant follicle's ability to synthesize estradiol under the influence of LH and FSH (16) is crucial to its continued development. The rise in serum levels of FSH and LH that occurs a few days before menstruation stimulates the further development of a small antral follicle. Within seven days, the favored follicle predominates over the other, asynchronously developing follicles, probably by inhibiting the secretion of FSH (16) through a negative-feedback mechanism.

The production of inhibin by the dominant follicle (11), probably together with other endocrine factors, may also affect interactions between gonadal steroids and the gonadotropins, resulting in follicular maturation, which culminates in ovulation that may be considered a step in the course of the evolution of the luteofollicular complex (17).

The results presented in this paper in part arise from experimental studies using laboratory mammals, and were correlated to original observations performed in humans using 1) small ovarian biopsies obtained in cases of various types of pathology, and 2) oocytes and granulosa cells, during the course of in vitro fertilization programs.

In an in vitro fertilization program, various protocols of hormonal stimulation, using clomiphene citrate, hMG, an FSH/LH combination or pure FSH, produce bilateral ovarian hyperstimulation, enhancing oocyte recovery as well as the probability of fertilization, even in endocrinologically normal women. Such treatment inhibited the midcycle LH surge. HCG is then usually required for the final maturation and ovulation of the follicles. (1,9,18–26).

Hodgen (19) proposes that

> ...the emergence of multiple follicles on both ovaries after administration of exogenous gonadotropins to monkeys or women is not only the result of augmenting the availability of gonadotropins, per se, but is also an indirect result of overriding the first-tier ovarian mechanism of follicle selection.

This article will deal mainly with ultrastructural modifications as viewed by Scanning Electron Microscopy (SEM) and Transmission Electron Microscopy (TEM) involved in 1) the growth of ovarian follicles, 2) the concomitant growth and maturation of oocytes and 3) the process of ovulation.

2. Primary follicles

2.1. *The oocyte*

The increase in size of the oocyte is the first morphological sign of follicular growth. In the growing oocyte cytoplasmic structures are randomly scattered. The mitochondria are characterized by a dense matrix, and annulate lamellae are rarely observed. The Golgi complex, characterized by smooth vesicles and tubules (27), moves from the original juxtanuclear position towards the periphery of the oocyte (28). Multivesicular bodies are present in the cytoplasm and seem to be of the heterolysosomal type (29). In a recent study on human ovaries in cancer patients treated with cytostatic drugs, oocytes showing similar ultrastructural features of atresia were frequently observed. Thus, the pool of resting follicles capable of undergoing further development appeared to be diminished in these patients (30).

During the primary follicle stage of maturation, cortical granules begin to develop; these are spherical structures limited by a smooth membrane and filled with intensely electron-dense material. Numerous motile microvilli containing microfilaments extend from the cell membrane and reach as far as the zona pellucida (31).

Between the oolemma and the follicular cell processes, two prominent types of cell contacts are present: the desmosome and the gap junction. The regions of heterologous cell interaction appear to be randomly distributed over the entire surface of the oocyte (32,33) (Fig. 4-1a–d).

2.2. *Follicular cells*

The growth of the primary oocyte is accompanied by thickening of the follicular wall. Follicular cells become cuboidal and, after initiation of mitotic activity, the cells are gradually arranged into increasingly numerous layers. The nucleus enlarges, becoming irregular in shape. A large nucleolus is present (31) (Fig. 4-1b).

In some cell regions gap junctions, tight junctions and desmosomes occur (31). Even in hypophysectomized rats, homocellular gap junctions between adjacent granulosa cells as well as heterocellular gap junctions between granulosa cells and oocytes were encountered in follicles ranging from primordial to multilaminar stages. However, the size and frequency of nexus contacts are significantly decreased if compared to preantral follicles from intact animals (34).

As soon as the oocyte becomes separated from the follicular cells through deposition of the zona pellucida, follicular cells form long processes that traverse the zona (31).

Metabolic coupling (via gap junctions) with follicular cells appears to be an important factor for oocyte growth. In fact, oocytes collected at the middle of their growth phase degenerate in vitro if cultured in the absence of follicular cells (35). In such a culture system, oocytes and follicular cells are metabolically coupled (36).

2.3. *The zona pellucida*

During the period of follicular growth an amorphous material rich in mucopolysaccharides is deposited between the surfaces of the oocyte and the surrounding follicular cells. Patches of zona material first appear around oocytes that are surrounded by a single layer of cuboidal follicular cells (37). These patches enlarge and coalesce to form a continuous layer that separates the oocyte from the follicle cells encircling it (Fig. 4-1b).

The cellular origin of the zona pellucida is still a matter of considerable controversy (37). Some studies seem to favour the involvement of the cumulus cells in its biosynthesis (38), while others conclude that it is formed by the oocyte (39,40,41). Some authors believe that the zona is secreted by both the oocyte and the follicle cells (31,42).

The zona pellucida of young follicles contains a nonfibrous mucopolysaccharide material (43). Studies using ruthenium red (RR) as a stain for proteoglycans show a weak reactivity in the zona pellucida of growing follicles (44). More recent studies using RR combined with detergents have clearly shown the three-dimensional fibrillar structure of the zona pellucida and the presence within this complex network of coarse fibrils of proteoglycan granules disposed as nodal-points (45). Greve and Wassarman (46) suggest that the zona pellucida is a highly organized extracellular coat in which glycoproteins are assembled into filaments possessing a recognizable structural repeat.

2.4. *Theca folliculi*

Stromal elements tend to become concentrically arranged around the growing follicle. In this way, another component of the follicular wall, the theca folliculi, is formed. Early theca cells resemble fibroblasts. Desmosomes and small gap junctions are found between the theca cells (47).

Theca cells are separated from the follicular cells by a basement membrane in which proteoglycans are regularly arranged on both the internal and external sides (48) (Fig. 4-1c)

3. Developing antral follicles

3.1. *The Oocyte*

In the developing antral follicle the oocyte no longer increases in volume. The nucleus is eccentrically located and is ovoid (49). Diplotene chromosomes become more distinct (50). The

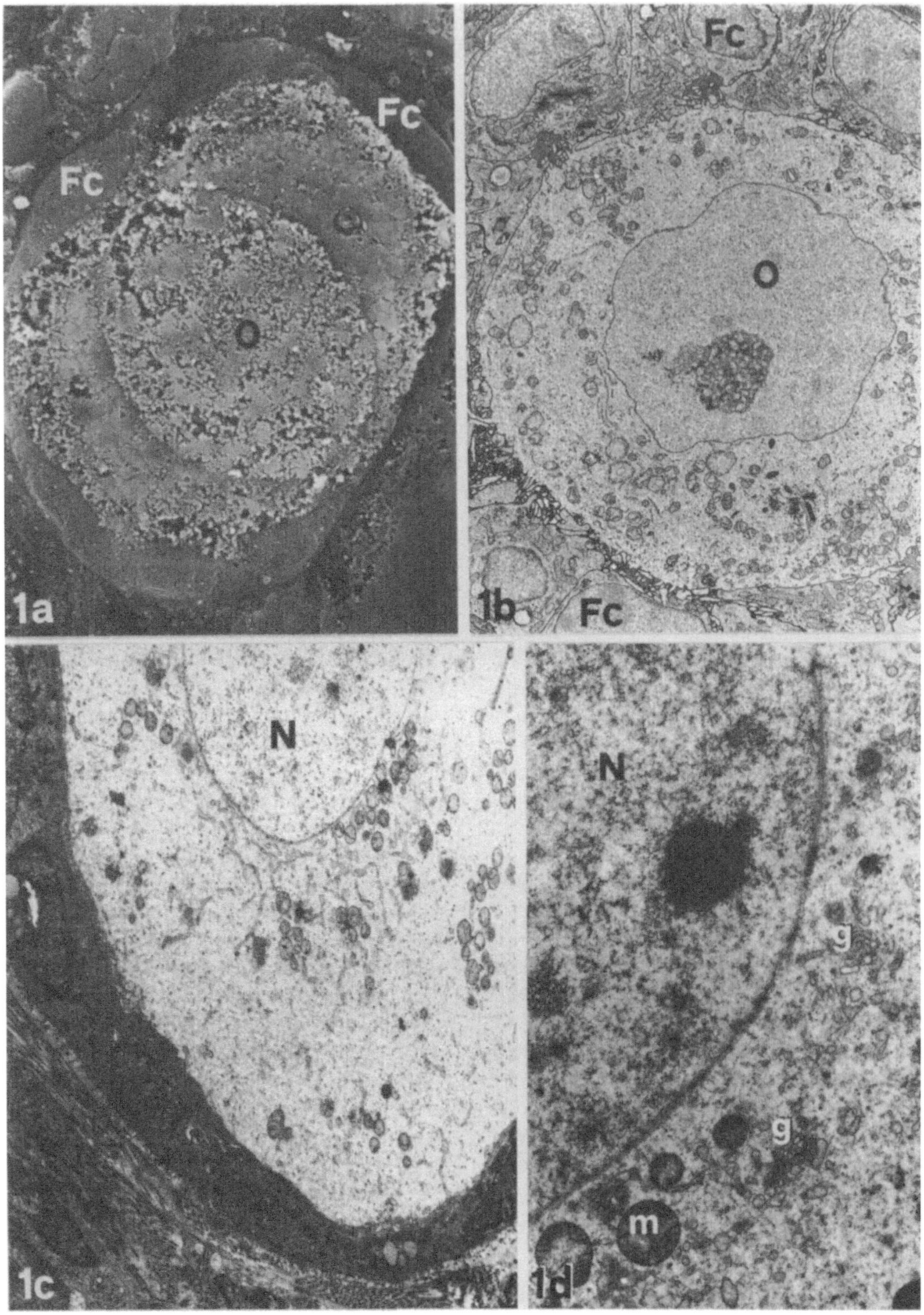
Fc
Fc
O
1a
Fc
O
1b
Fc
N
1c
N
g
g
m
1d

remainder of the nucleus appears somewhat empty at this stage, hence the term *germinal vesicle*.

Cytoplasmic structures are randomly scattered. The smooth endoplasmic reticulum is in the form of small vesicles. The mitochondria are evenly distributed and, usually, they are closely associated with the smooth endoplasmic reticulum. Multivesicular bodies are also present (51). The Golgi complex occurs predominantly in proximity to the cell membrane. Cortical granules become more numerous and are located just beneath the cell membrane (52) (Fig. 4-2a–d).

Cortical granules originate from Golgi complexes during maturation of the oocyte (53). Their synthesis takes place in two steps: the first occurs in oocytes of small antral follicles, in which there is a high level of Golgi activity, and the second is observed at the time of resumption of meiosis (54,53). Proliferation of cortical granules during the final maturational phase is accompanied by their centrifugal migration. Some authors (55,56) showed the presence in small antral follicle oocytes of cortical bands of microfilaments, likely composed of actin (57). Such microfilaments probably are related to the premature migration and activation of the cortical granules prior to the resumption of meiosis, at which time these bands of microfilaments become disorganized (53) (Fig. 4-2c,d).

As the oocyte completes its first maturational stage, Golgi complexes become progressively scarce and apparently disappear. In the preovulatory oocytes, Golgi complexes were never observed.

A narrow electron-lucent area, the perivitelline space, develops between the cell membrane and the zona pellucida. The oocyte's microvilli project into this space and appear to fill it. Numerous small gap junctions are present between the oocyte and the cumulus cells with which it is in close contact (32,33,58) (Fig. 4-2b).

Intercellular communication between granulosa cells and the oocytes via gap junctions may be related to the nutritional dependency of the oocyte on the granulosa cells. The nutritional dependency of oocyte growth on the granulosa cells may reflect, in part, an economization and compartmentalization of metabolic function, thereby liberating the oocyte from numerous housekeeping functions (36).

Furthermore, the intercellular communication between these two cell types could allow the passage of small molecules from follicle cells to the oocyte. It has been proposed that cAMP inhibits resumption of meiotic maturation (59,60) and that the origin of oocytic cAMP is in the granulosa and cumulus cells (61,62,63).

The transfer of follicle cell cAMP to the oocyte seems to be vital for maintenance of meiotic arrests (66). It is likely that cumulus cell cAMP enters the oocyte via the gap junctions connecting the two cell types (64). Therefore, cAMP might be regarded as a paracrine regulator of oocyte development (65).

In this stationary stage, the oocytes persist until they either degenerate into atretic follicles or resume meiosis a few hours before ovulation. Very little is known about the mechanisms that regulate the prophase meiotic arrest, which include contact with follicular cells (33), transfer of cAMP from follicular cells to oocytes (64,66), changes in the oocyte surface properties (67) and release of an oocyte-maturation inhibitor (68) into the follicular fluid.

An important functional change occurring in this period in oocytes released from the ovary and cultured in vitro is the acquisition by the oocytes of meiotic competence (69–71).

←

Figure 4-1. Primordial and primary follicle.

a) SEM. Human primordial follicle. The oocyte (o) is surrounded by a single layer of closely apposed, flattened follicle cells (Fc). X 2,000.

b) TEM. Mouse primary follicle. The oocyte (o) is surrounded by a single layer of cuboidal follicle cells (Fc). Cytoplasmic structures of the oocyte are randomly scattered. X 2,000.

c) TEM. Human primordial follicle. Note the simple organization of the peripheral cytoplasm of the oocyte and the condensation of the organelles in the region immediately around the nucleus (N). X 2,300.

d) TEM. Human oocyte in the quiescent stage. Note the presence of the nucleus (N), mitochondria (m), the Golgi complex (g), and cisternae of the endoplasmic reticulum. X 18,000.

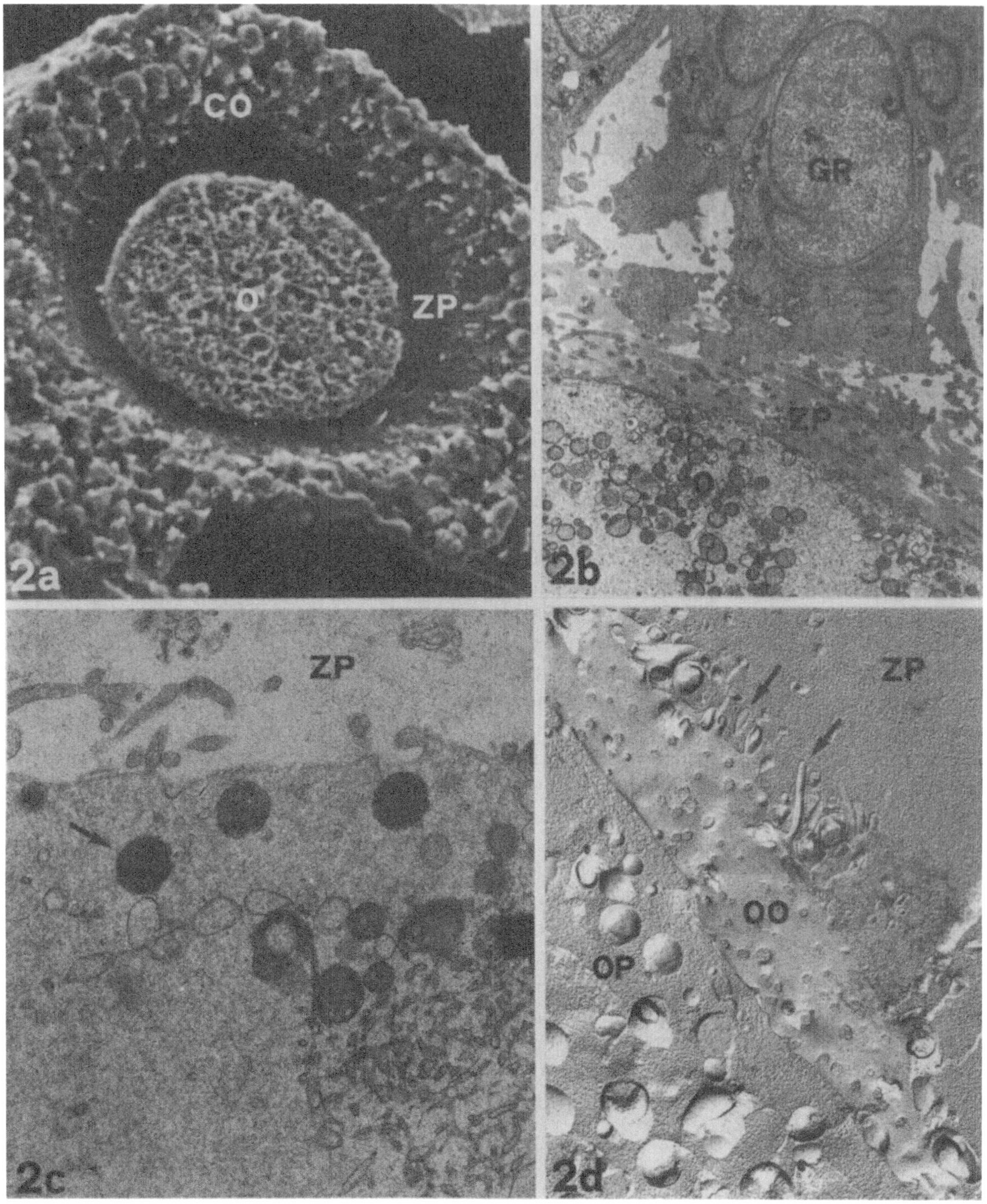
CO
O
ZP
2a
GR
ZP
2b
ZP
2c
ZP
OO
OP
2d

Recently, Channing et al. (72) reported that the activity of oocyte-maturation inhibitor (OMI) declined as a result of follicular maturation in women who ovulated after stimulation with menopausal gonadotropin (hMG) and human chorionic gonadotropin (hCG).

These results, which have been confirmed in other studies (11,73,74), suggest that, in addition to OMI, other factors may cooperate in inhibiting oocyte maturation.

Such studies further suggested that OMI is produced by the dominant follicle in response to FSH stimulation whereas the preovulatory surge in LH inhibits the production of this factor (11, 74). The level of OMI during induction of ovulation might also be considered as a protein marker of granulosa cell function having clinical and therapeutic importance (74).

3.2. Follicular cells

Fluid-filled spaces between granulosa cells first appear when these cells become dependent upon gonadotropins. As the follicle grows, these small areas gradually coalesce, forming a progressively enlarging cavity (antrum) filled with a liquid (liquor folliculi). The liquor folliculi originates from two sources: the plasma, and the granulosa cells of the follicle (75). When stained with RR, the liquor folliculi shows a fibrillar-granular appearance. In fact, it is formed by rounded, negatively charged particles united by a thin microfilamentous network (44) (Fig. 4-3a).

The liquor folliculi of mammals is rich in proteoglycans that are produced mainly by the granulosa cells (76,77). The pattern of their secretion depends on the stage of follicular maturity and is correlated with the hormonal profile in the follicular fluid (78–82).

During this period small fluid-filled cavities surrounded by rosettes of granulosa cells (Call–Exner bodies) appear among the granulosa cells. The content of these closed cysts, limited by a basal lamina and often in continuity with a delicate network of strands of similar material, may be released into the large follicular cavity at any time (83,84). Call–Exner bodies can even be seen floating free within the liquor folliculi filling the antrum of large preovulatory follicles (85). In contrast to studies carried out by others (86), it was indeed very easy for us to identify Call–Exner bodies by SEM (87). This technique, in fact, offers the possibility of obtaining a general view of the granulosa layer and permits, through the use of stereo views, a complete three-dimensional reconstruction of the follicular wall and, consequently, the exact distribution and location of such structures (88).

Follicular cells have a spherical nucleus; chromatin distribution is regular, and the reticular nucleolus is large and conspicuous. During the latter stages of follicular growth, the presence of RER cisternae containing an electron-dense material and clusters of Golgi vacuoles (31) is typical in the cytoplasm of granulosa cells (Fig. 4-4b–c). Most granulosa cells have an irregular, polyhedral appearance and a relatively rough surface. Cells of this type generally predominate near the basal membrane delimiting the follicle. Other granulosa cells, especially in areas that face the follicular cavity, are elongated or flattened. In these antral follicles, numerous granulosa cells possess large, irregular evaginations and microvilli. An increasing number of blebs and long, slender

←

Figure 4-2. Large Graafian follicle — Oocyte.

a) SEM. Dog Graafian follicle. The oocyte (o) is surrounded by the zona pellucida (ZP) and cumulus oophorus cells (co). The cytoplasm of the oocyte contains numerous vacuoles, some of which correspond to mitochondria, while others are likely Golgi membranes. X 900.

b) TEM. Rabbit Graafian follicle. The zona pellucida appears to be composed of an amorphous material traversed by long processes from the granulosa cells. ZP: zona pellucida; O: oocyte; GR: granulosa cells. X 7,000.

c) TEM. Human Graafian follicle. Cortical granules (arrows) are present beneath the plasma membrane of the oocyte. ZP: Zona pellucida: X 15,000.

d) Freeze-fracture. Mouse Graafian follicles. The medially fractured oocyte shows various fractured organelles in the ooplasm (op); they are probably cortical granules or membranes of mitochondria. It is possible to see the face of the oolemma (oo) in which numerous fractured or intact (arrows) microvilli are present. ZP: Zona pellucida. X 10,000.

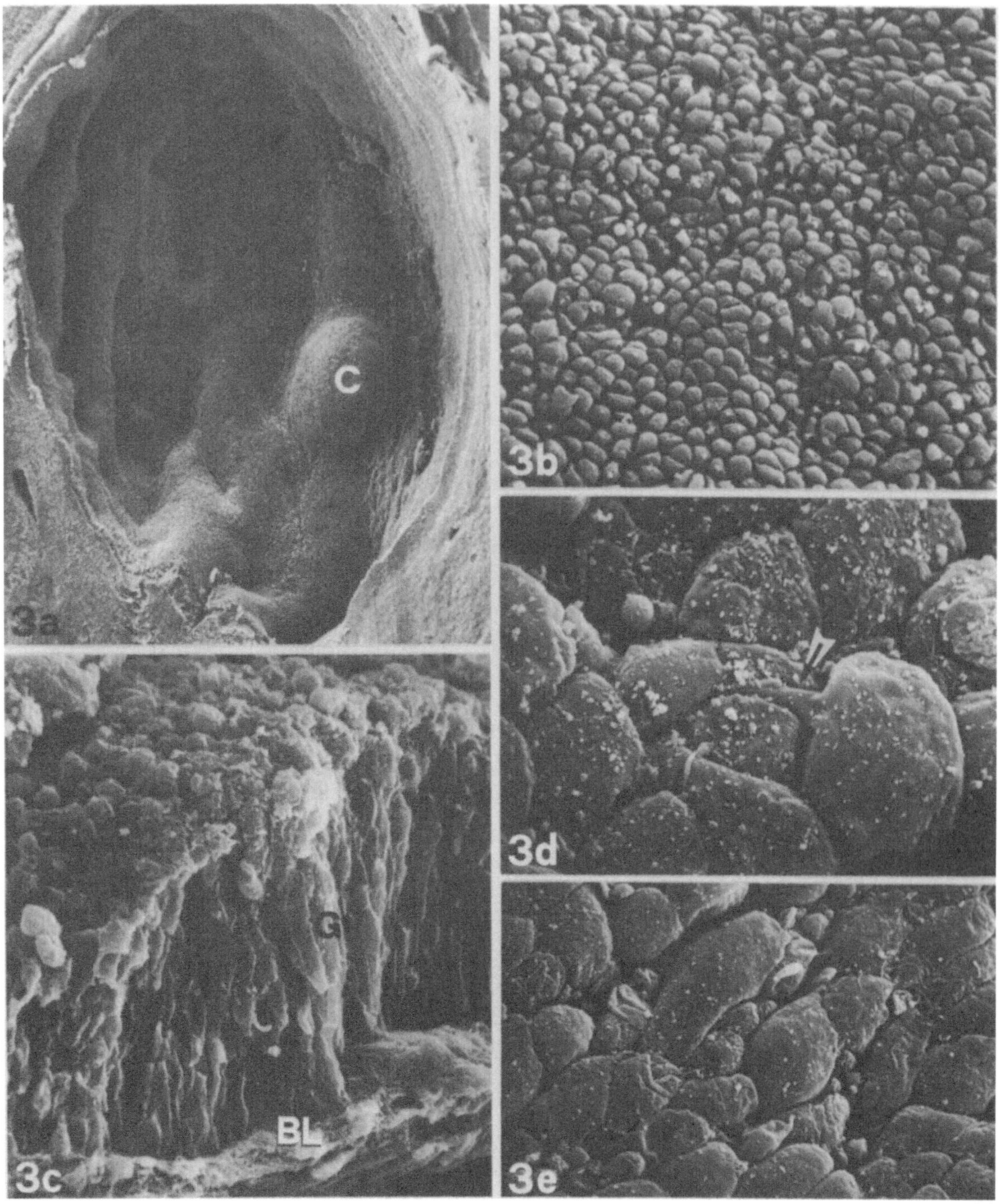

Figure 4-3. SEM. Human large Graafian follicle. Granulosa cells.
a) Note the presence of cumulus oophorus (c). X 70.
b) Note the cellular arrangement of the granulosa cells. X 300.
c) Granulosa cells (G) form a multilayered structure. BL: Basal Lamina. X 450.
d,e) Granulosa cells appear as polyhedral elements provided with cytoplasmic evaginations (arrows) and short microvilli. d: X 2,000; e: X 3,500. a, b, e from Makabe et al., 1982 (143).

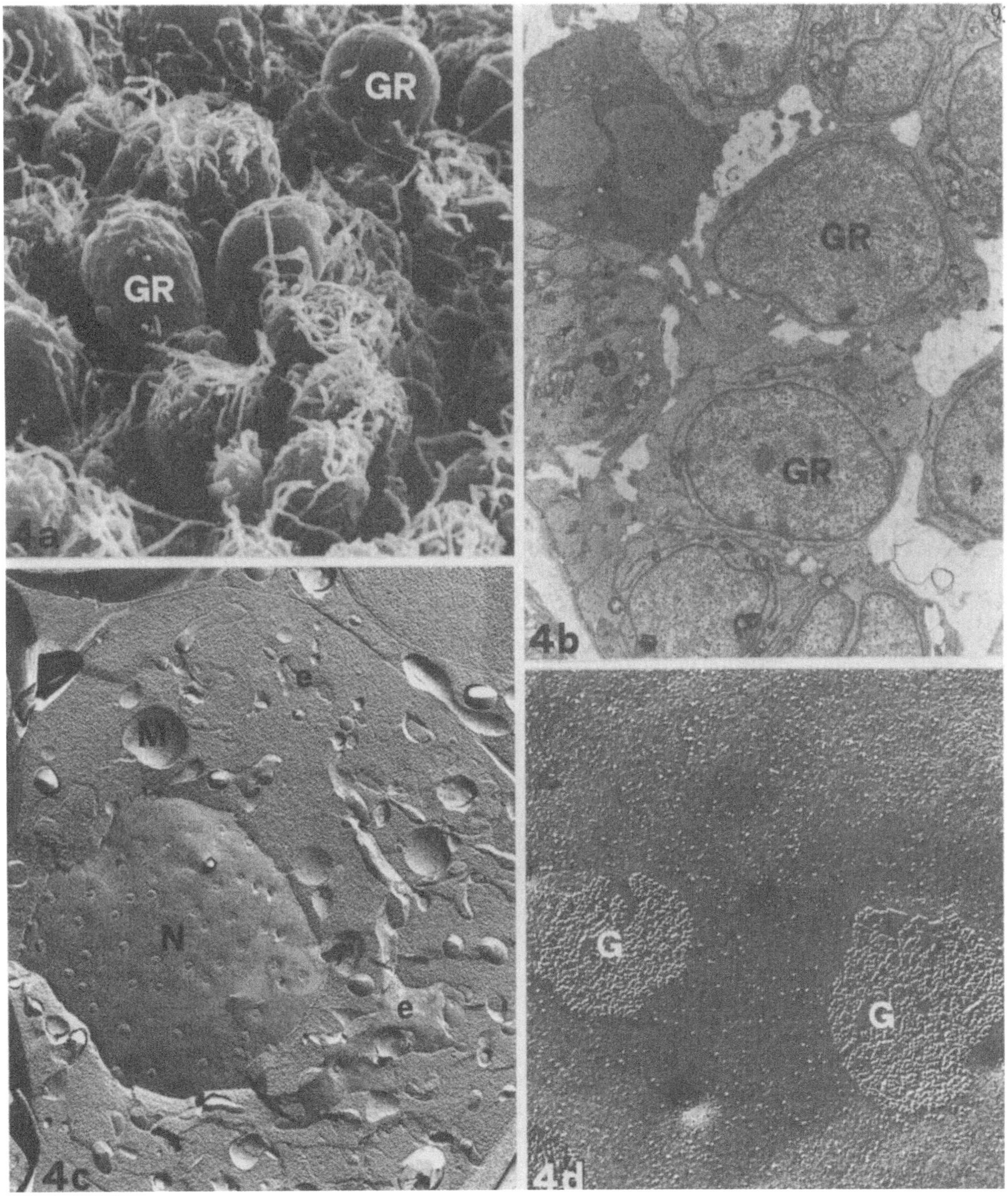

Figure 4-4. Large Graafian follicle. Granulosa cells.

a) SEM. Human Large Graafian follicle. Granulosa cumulus cells (GR) appear as spherical elements provided with many tortuous slender microvilli and blebs.

b) TEM. Rabbit Large Graafian follicle. Granulosa cells (GR) are polyhedral elements that contain ribosomes, mitochondria, Golgi vesicles and elongated and extended elements of the rough endoplasmic reticulum. Ameboid processes are packed with microfilaments. X 7,800.

c) Freeze-fracture. Mouse Large Graafian follicle. Medially fractured granulosa cell shows various fractured organelles: mitochondria (M) and endoplasmic reticulum (e). N: Nucleus. X 23,000.

d) Freeze-fracture. Mouse Large Graafian follicle. Cytoplasmic membrane of granulosa cell. Note the organization of particles and the presence of a gap junctions (G) on the P-face. X 75,000.

microvilli are also observed in many granulosa cells belonging to Graafian follicles and possibly undergoing precocious luteinization (89–91) (Fig. 4-3b,c,d,e;4a).

This characteristic increase of cell surface may depend on hormonal factors and, concomitantly, may be used as binding sites for an increasing quantity of gonadotropins (92).

In fractured follicles, the entire cell surface is covered by a thin layer of material having a fine reticular granular appearance. This material is most abundant on the surface of those cells that face the antrum, and is likely composed of condensed follicular fluid (87).

Contractile microfilaments are present in the cortical cytoplasm and in the large cytoplasmic projections of granulosa cells. These microfilaments are more abundant in cells exhibiting morphological evidence of possible contractile activity, such as cumulus cells, cells of the corona radiata, and granulosa cells largely detached from adjacent elements (93,94) (Fig. 4-4b).

Recent studies showed that the major constituent of the filamentous network in the cytoplasm of rat granulosa cells is actin (95). Actin is organized in large bundles in which individual filaments are aligned longitudinally; their polarity is generally away from the membrane with which the filaments are associated (95). Other studies showed that during early follicular growth actin is located at the base of those granulosa cells abutting the basal lamina. This transitional polarity of actin distribution is correlated with cell shape changes in the basal granulosa cells (96).

This contractile apparatus, together with the mechanical dissociation caused by the rapid accumulation of follicular fluid within the antrum, may be responsible for the progressive detachment of granulosa cells from adjacent elements in the stratum granulosum, of the cumulus oophorus from the granulosa layer and for the retraction of corona cell processes from the oocyte surface at the time of ovulation (17,87,93).

In addition, other studies have shown that in vitro differentiation of granulosa cells was associated with changes in cell shape and contacts, accompanied by simultaneous alterations in the organization and expression of cytoskeletal proteins (97). Further, these proteins may play an important role in the deposition and spatial arrangement of the extracellular matrix (basal lamina and follicular fluid in vivo) (97).

During this stage, different types of intercellular contacts exist among granulosa cells, including zonulae adherentes, tight junctions and gap junctions (31,34,98,99) (Fig. 4-4d).

The nature of intercellular contacts and the communication existent among granulosa cells has been examined by tracer and freeze-fracture techniques.

Gap junctions are randomly distributed between granulosa cells, and they increase in size and number with continued growth of the follicle, and especially during that stage of growth dependent upon pituitary gonadotropin stimulation (61). The presence and distribution of gap junctions in the ovary have been correlated with its hormonal environment. Homocellular gap junctions between granulosa cells persist in rat ovarian follicles following hypophysectomy. Gonadotropic and/or steroid hormones are therefore not required for the maintenance of gap junctions between these cells during early follicular development. However, replacement therapy with estrogen results in amplification of their surface area (34). On the contrary, high levels of androgens do not seem to affect the morphology or the surface area of the gap junctions between granulosa cells (100). On the basis of these and other studies, some authors believe that gap junctions may exert some role in the coordination of cellular activities during follicular growth and development (34, 98–100).

Only rarely have tight junctions been observed on granulosa cells (98,101). These junctions are *focal tight junctions* of limited length, and their significance is obscure. In the growing follicles of androgen-treated animals, their average length is significantly greater than that of healthy follicles in control animals (100).

At the periphery of large Graafian follicles in the mouse, focal cell contacts mediated by transmembrane intercalated glycoproteins were observed (102). These structures may be correlated to an increased adhesiveness of granulosa cells, probably controlled by gonadotropins (102).

Furthermore, solitary cilia were frequently observed in granulosa cells of growing follicles (83, 103). Althought they apparently represent rudimentary organelles, the possibility that they are

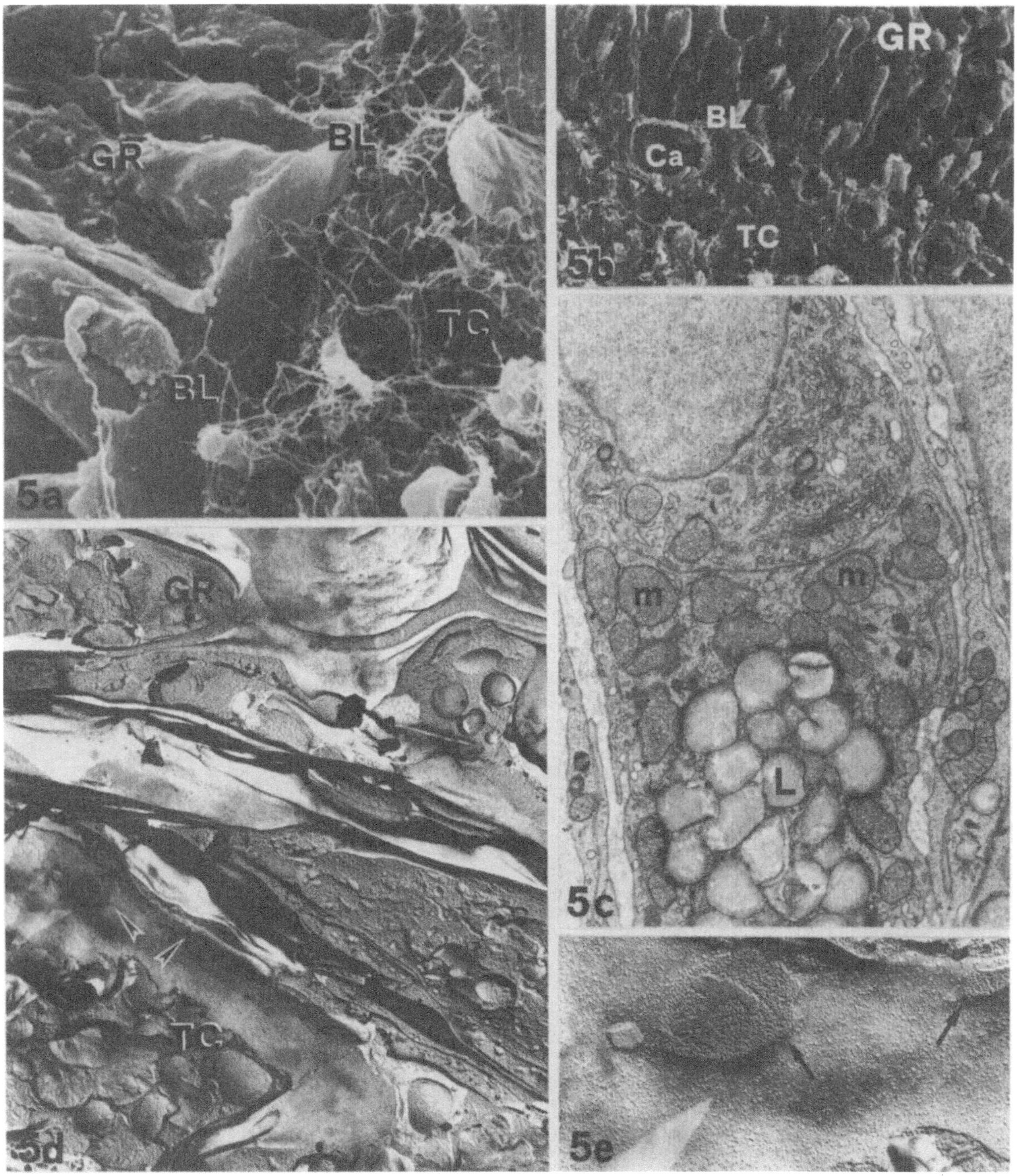

Figure 4-5. Graafian follicle. Theca interna cells.
a,b) SEM. Human Graafian follicle. Note the granulosa cells (GR), basal lamina (BL), capillary vessel (Ca) and theca cells (TC) surrounded by collagen fibrils. (a: X 7,000; b: X 500)
c) TEM. Mouse Graafian follicle. Typical, differentiated, theca interna gland cell. The cytoplasm of these cells contains numerous lipid droplets (L) and mitochondria (m) which possess tubular and/or villiform cristae. X 12,000.
d,e) Freeze-fracture. Mouse Graafian follicle.
GR: granulosa cells; TC: theca interna cells. Note the presence of typical gap junctions between theca interna cells (arrows). d: X 15,000; e: X 45,000. d, c, e, from Familiari et al., 1981 (28).

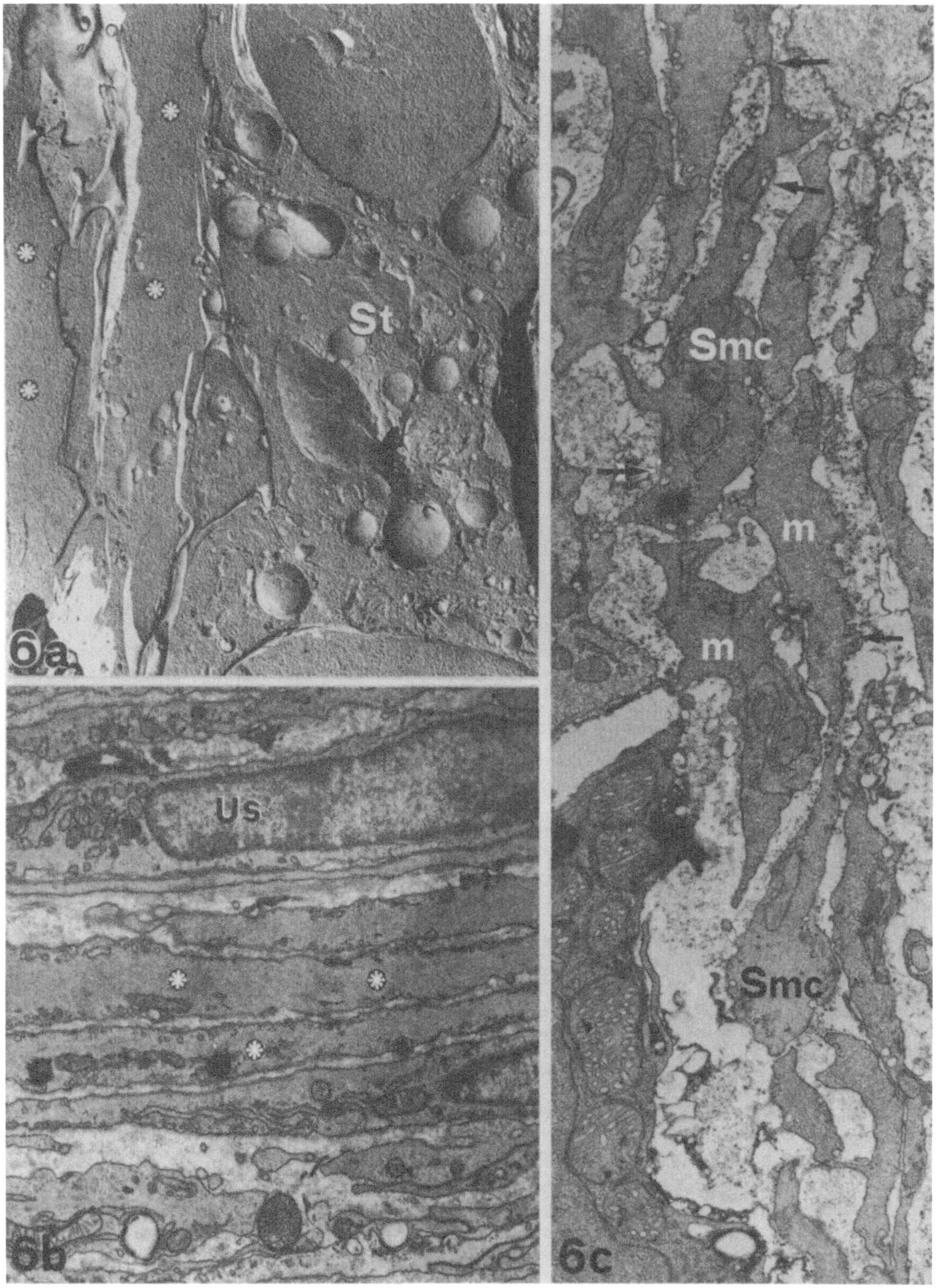
6a
St
6b
Us
6c
Smc
m
m
Smc

chemoreceptors serving a population of cells functionally coupled by communicating gap junctions (88) cannot be ruled out.

3.3. Theca cells

During development of the antral follicle, the theca becomes stratified and two parts can be differentiated.

The inner part is a richly vascular layer (theca interna) composed of epithelioid elements, and the outer part is a theca externa characterized by the presence of spindle-shaped cells similar in appearance to the cells of the surrounding stroma (Fig. 4-5a,b).

In the theca interna three different cell types are often recognized: fibroblastlike cells, steroid-secreting cells and transitional elements. In the antral follicle a great number of fibroblastlike cells enlarge and take on a polyhedral shape (104). These cells have all the morphological signs of steroid-secreting cells: a multilobulate nucleus and a cytoplasm containing a large number of lipid droplets, round mitochondria with tubular cristae and an endoplasmic reticulum with smooth membranes. A well-developed Golgi apparatus also occurs in this type of cell (Fig. 4-5c) (105–108).

Thecal cells are also attached by gap junctions. However, the contacts do not appear to be as great in number or proportions as those joining granulosa cells (47). Gap junctions are always present not only between two differentiated theca interna cells, but also between two immature elements, as well as between closely associated mature and immature cells (47). The theca interna cells are the major site of follicular androgen production (109,110). No significant differences in gap junction surface area is observed between androgen-treated and control animals, suggesting that the size of gap junctions is probably unrelated to the steroidogenic activities of theca interna cells (100) (Fig. 4-5d,e).

A few adrenergic nerve fibers are present in the theca interna of the guinea-pig ovary (111). Numerous capillaries serve the cells of the theca interna, completing the picture of an endocrine gland. These vessels form fine, basket-shaped vascular casts (112).

The theca externa is composed in part of cells that do not differ in any respect from the cells of the undifferentiated theca of primary follicles. E/M observations extended to several mammalian species, including man, show the existence of smooth muscle cells mainly concentrated in the theca externa of developing follicles (113).

Furthermore, these studies reveal that this contractile tissue consists of a variable population of cells, often described either as typical smooth muscle cells, or, in other instances, as intermediary and/or myoid elements (smooth muscle-like cells) that, however, always display varying amounts of myofilaments in their cytoplasm (114–118) (Fig. 4-6a–c).

The follicle's theca externa is richly supplied by adrenergic nerves. At this location, in fact, close contacts between the nerve terminals and the smooth musclelike cells were identified (111,119, 120).

4. Mature preovulatory follicles and ovulation

4.1. The oocyte

Morphological changes occurring in the preovulatory oocytes are probably initiated by the preovulatory surge in LH (121–125). Such alterations could be related to the increasing level of proges-

←

Figure 4-6. Large Graafian follicle. Theca externa cells.
a) Freeze-fracture. Mouse growing Graafian follicle. Elongated smooth muscle cells (*) are evident. St: steroidogenic cells of the theca interna. X 7,000.
b) TEM. Mouse growing Graafian follicle. Concentric layers of elongated smooth muscle cells (*) in the theca externa. The cytoplasm of these cells shows microfilaments, pinocytotic vesicles, mitochondria and dense bodies; undifferentiated smooth muscle cells (US) are visible as well. X 6,000.
c) TEM. Mouse atretic follicle. Irregularly shortened smooth muscle cells (smc) in the theca externa. Smooth muscle cells are provided with abundant bundles of myofilaments (m) and pinocytotic vesicles (arrows). X 18,000.
From Motta and Familiari, 1981. (118)

terone in the follicular fluid after the LH surge (126). Even though LH and HCG, which simulated, during in vitro fertilization, the natural hormonal stimulation that occurs in vivo, may be important in final oocyte maturation, studies involving perfused rabbit ovaries (126) would indicate that LH is not necessary.

In addition, high estradiol levels in the follicular fluid are essential to oocyte fertilization, while high androgen levels or high androgen/estradiol ratios were often associated with oocyte atresia (125). Premature luteinization of the follicle also appears to be associated with induced fertilizability of the associated oocyte (125).

Some nonsteroidal substances found in human follicular fluid, such as serum proteins, enzymes, placental proteins, granulosa cell proteins and glycosaminoglycans might play a part in ooctye maturation (75). Moreover, a variety of peptides may be involved in oocyte maturation. Although inhibin has been studied extensively (11,72–74), further studies are necessary to clarify the precise role in oocyte maturation exerted by some of these non-steroidal factors.

Oocyte maturation takes place in several stages, including the following: 1) the breakdown of the nuclear membrane germinal vesicle; 2) the events related to the first meiotic division, including extrusion of the first polar body; and 3) the formation of the second meiotic spindle and the arrest of meiosis at the second meiotic metaphase (125). The oocyte may be considered mature once it reaches the second meiotic metaphase (127).

Nuclear maturation is attended by cytoplasmic maturation, which is characterized by a general increase in the number of organelles that are scattered throughout the cytoplasm. The distribution and appearance of cytoplasmic structures do not seem to vary much compared with the oocyte of the growing antral follicle (128).

The recent introduction of IVF methods showed that events accompanying in vitro maturation of oocytes are basically similar to those normally occurring in vivo. Thus they offer the unique opportunity to study in great detail these crucial developmental events occurring in maturing oocytes. According to these studies, cortical granules are arranged in one or more layers just beneath the plasma membrane (51,54,56). These granules are always distributed uniformly throughout the cortical zone (129,130). Occasionally, the release of their contents may be observed; this phenomenon is thought to be a pathologic manifestation (130, 131) (Fig. 4-7a).

The mitochondria tend to aggregate with smooth endoplasmic reticulum and vesicles forming peculiar aggregates characteristic of preovulatory oocytes (51,130). These aggregates appear more evident in oocytes that have matured in vitro (51). It has been hypothesized that such complexes composed of mitochondria and cytoplasmic membranes represent the same functional apparatus, probably playing complementary roles in the production of substances related to fertilization (51). Indeed, these unusual aggregates might represent a rich pool of energy that could be harnessed for rebuilding the rapidly changing cytoplasmic membranes during cell cleavage (132), Golgi complexes are very small and rare, especially in oocytes that matured in vitro (51,53,54) (Fig. 4-7b).

By the use of TEM techniques in some mature human oocytes, early degenerative signs such as vacuolization of the cytoplasm, increased density of the cytoplasmic matrix and reduction in the number of microvilli could be observed (51,130, 132). Such aspects, always associated with an increased volume of the perivitelline space, seem to be related to failure of fertilization (51,130,132).

By SEM, immature hamster oocytes, at the germinal vesicle stage, have their surfaces uniformly covered by microvilli. Once meiosis has progressed to the first metaphase, the overlying surface shows the differentiation of a circular area with a low density of microvilli. Later, the first polar body emerges from this region, and the oocyte surface shows a cluster of cytoplasmic, conical projections at the point of extrusion of the polar body (133–135).

A number of studies have shown that the cortices of some mammalian oocytes are not structurally homogeneous but are highly polarized. The plasma membrane is a mosaic, and the cytoplasm overlying the meiotic spindle is devoid of cortical granules and consists of a filamentous layer containing actin (17,136,137). This cortical polarity is believed to facilitate gamete interaction and fusion, fertilization, cone development and polar body formation (17). An actin-based system may be responsible for the cortical localiza-

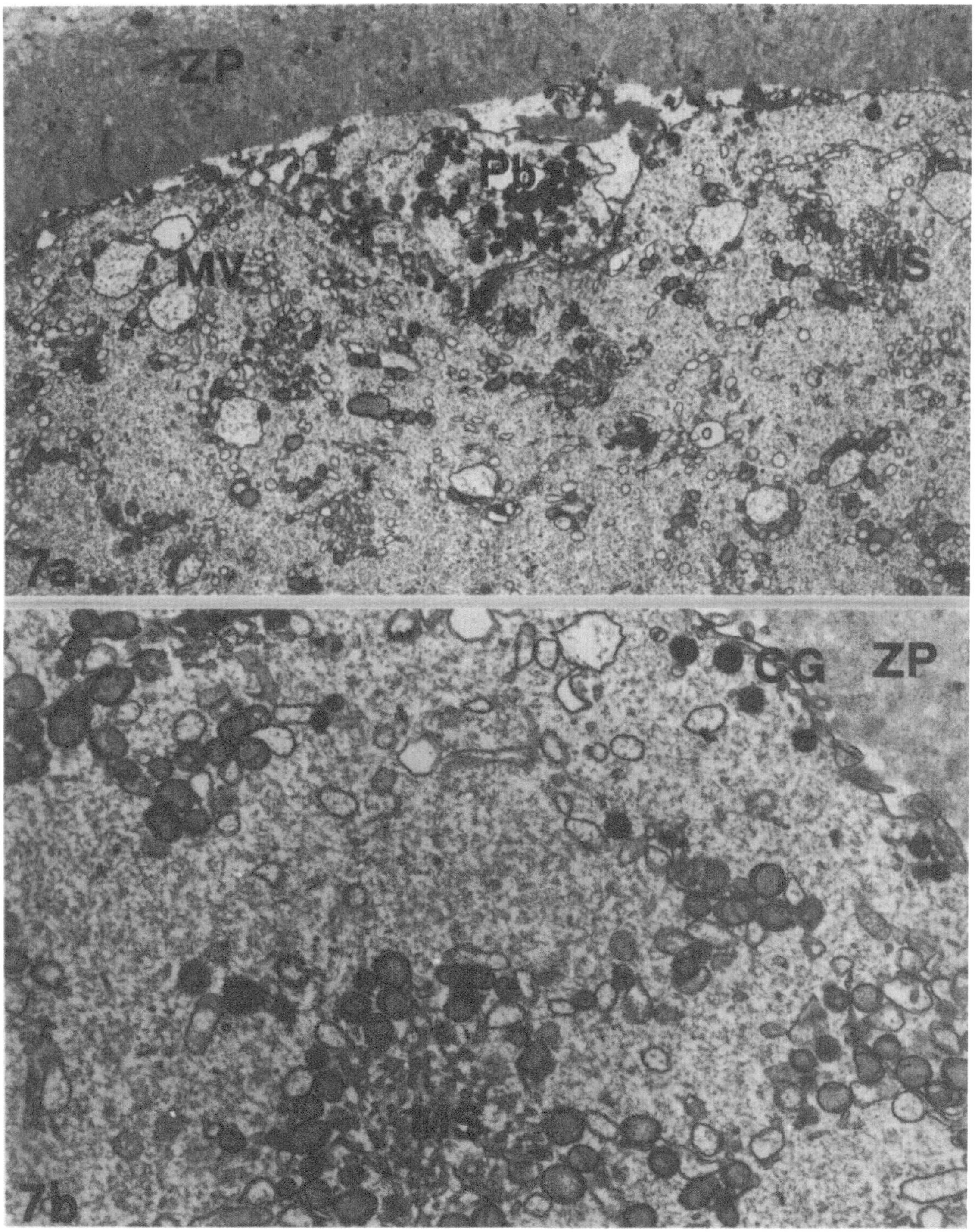

Figure 4-7. TEM. Human preovulatory oocyte.
Note the presence of mitochondria–smooth endoplasmic reticulum aggregates (MS) and mitochondria–vesicles complexes (MV). In a) the 1st polar body (PB) is observed. CG: cortical granules; ZP: zona pellucida. (a: X 3,500; b: X 9,100)
From: Motta et al., *Ann NY Acad Sci*, in press, 1988.

tion of the meiotic spindle in mammalian oocytes (137,138).

Recent studies on mature animal oocytes in the cumulus oophorus, using the freeze-fracture technique, showed that the P-face of the oolemma has randomly distributed intramembranous particles, and no marked difference is found between the intermicrovillous and microvillous regions of the oocyte with respect to the size and distribution of intramembranous particles. The E-face of the oolemma has fewer intramembranous particles than the P-face (139). Other observations show that microvillar membranes bear more intramembranous particles than those of the intermicrovillous regions (140,141).

Still other studies on preovulatory follicles showed that gap junctions present between oocytes and cumulus cells become small and decrease in number as the time of ovulation approaches (33, 58).

The density of intramembranous particles on areas of membrane between microvilli increases following fertilization. In addition, lipid-binding probes were applied to the oocyte membranes in order to study the distribution of specific lipids before and after fertilization; neither the distribution nor apparent quantity of these probes changed overall following fertilization, but this impression does not include specific sites of sperm–egg fusion (142).

In addition, decreasing small gap junctions are present between closely contacting cumulus cells (33,58).

Recently, new techniques for studying oocyte maturation have been developed. One such method is the use of specific fluorescent markers for cell surface, organellar and cytoskeletal components — a method that can be used for simultaneous multiple staining. Another new technique is coupling fluorescence staining, which is used with time-lapse video image intensification microscopy in order to study the relative disposition of these components in living oocytes (143).

Nuclear and cytoplasmic maturation in the preovulatory oocyte generally showed the same morphological pattern in both gonadotropin-stimulated and unstimulated animals (144,145).

In addition, the ultrastructural features of in-vitro and in vivo oocyte maturation have been compared; such comparisons have revealed only minor differences (56,144,145,146).

Hyttel et al. (145) showed a delay in the time of in vitro cortical granule release compared to that occurring in vivo, and little variation in the rearrangements of mitochondria and vesicles, in the clustering of the smooth endoplasmic reticulum and in the breakdown of the nucleus.

4.2. Zona pellucida

As observed by SEM, the zona pellucida of various mammals is characterized by numerous fenestrations that give it a somewhat spongy appearance (133,147). A zona pellucida with an outer surface having a "porous, net-like appearance" was also observed by Sundstrom (148) in human oocytes cultured in vitro. Few follicular cell processes traverse the zona pellucida.

Other studies (149,150) on in-vitro-cultured oocytes showed two different types of zona pellucida surface: a meshlike one, and the more compact, smooth one. The smooth type of zona pellucida was most common among immature and atretic cultured oocytes (149,150). Great numbers of spermatozoa were noted on oocytes showing the meshlike zona, in contrast to the presence of only few spermatozoa on oocytes having the more compact, smooth zona. It was therefore suggested that the condensation of the outer aspect of the zona pellucida may cause a disorientation of its sperm-binding sites that, through alterations in sperm attachment, ultimately lead to a reducation in oocyte fertilizability (149,150) (Fig. 4-8a–d).

4.3. Follicular cells

Due to the reduced occurrence of gap junctions, adjacent follicular cells lose their close interrelations. At this stage of maturation numerous annular gap junctions are present and may be the expression of an interiorization and segregation of junctional areas (61,151).

Granulosa cells still in close contact with the basal lamina of the follicle retain their polyhedral shape and appear larger than those in the interior layers. During this phase, many granulosa cells enlarge, and their surfaces often become highly infolded or expanded in voluminous pseudopodia. An increased number of blebs and slender microvilli are also noted in many granulosa cells of preovulatory follicles (89,152). Such alterations

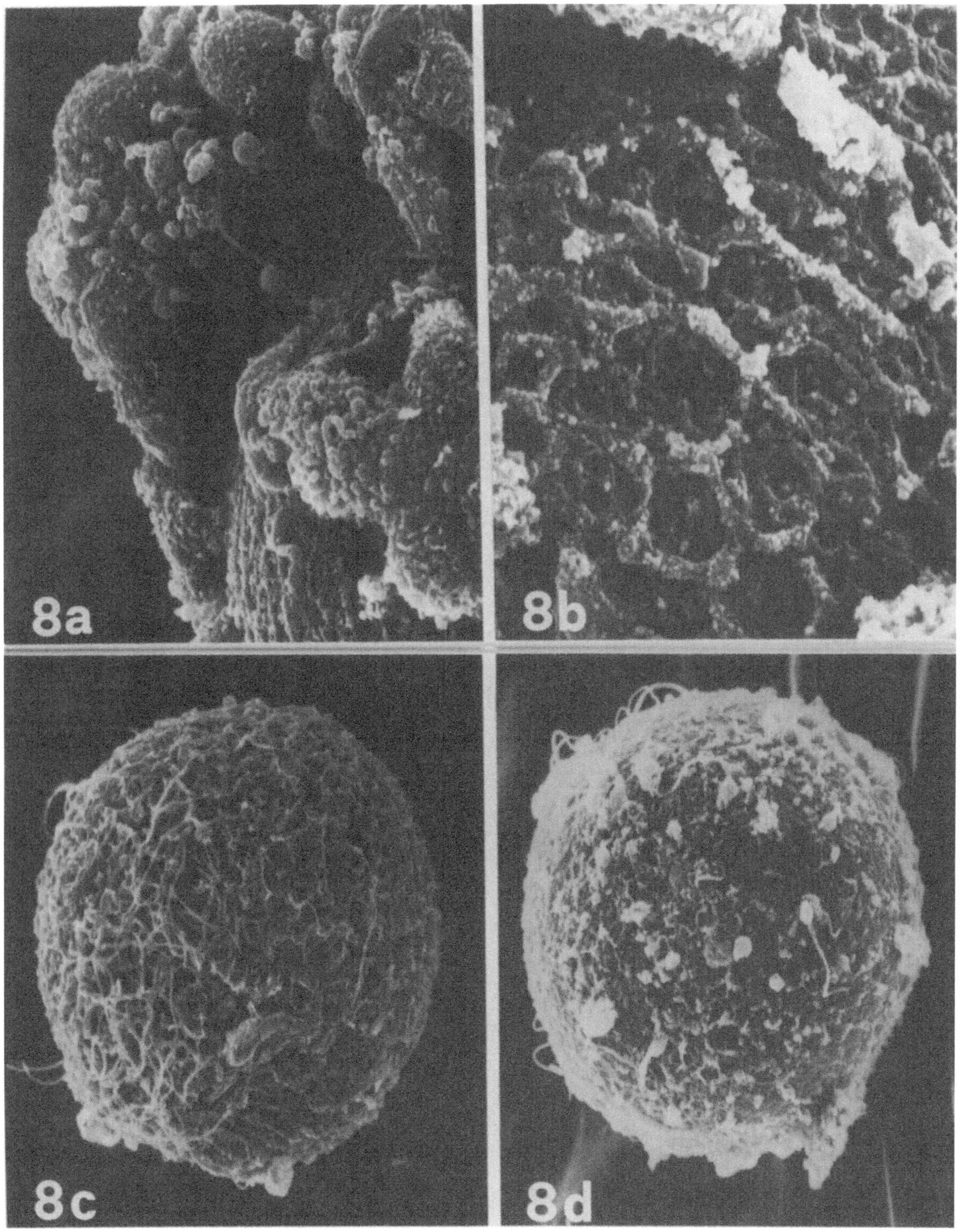

Figure 4-8. SEM. Human preovulatory oocyte (in culture).
a) The prevulatory oocyte is always surrounded by cumulus cells with morphological evidences of luteinization. X 4,000.
b,c) The preovulatory oocyte ZP shows a spongelike appearance. Note in Figure c the presence of numerous spermatozoa (b: X 1,300; c: X 7,000)
d) Atretic oocyte (in culture). The ZP has a more compact appearance. Note the presence of a few spermatozoa. X 1,300. From: Motta et al., *Ann N Y Acad Sci*, in press, 1988.

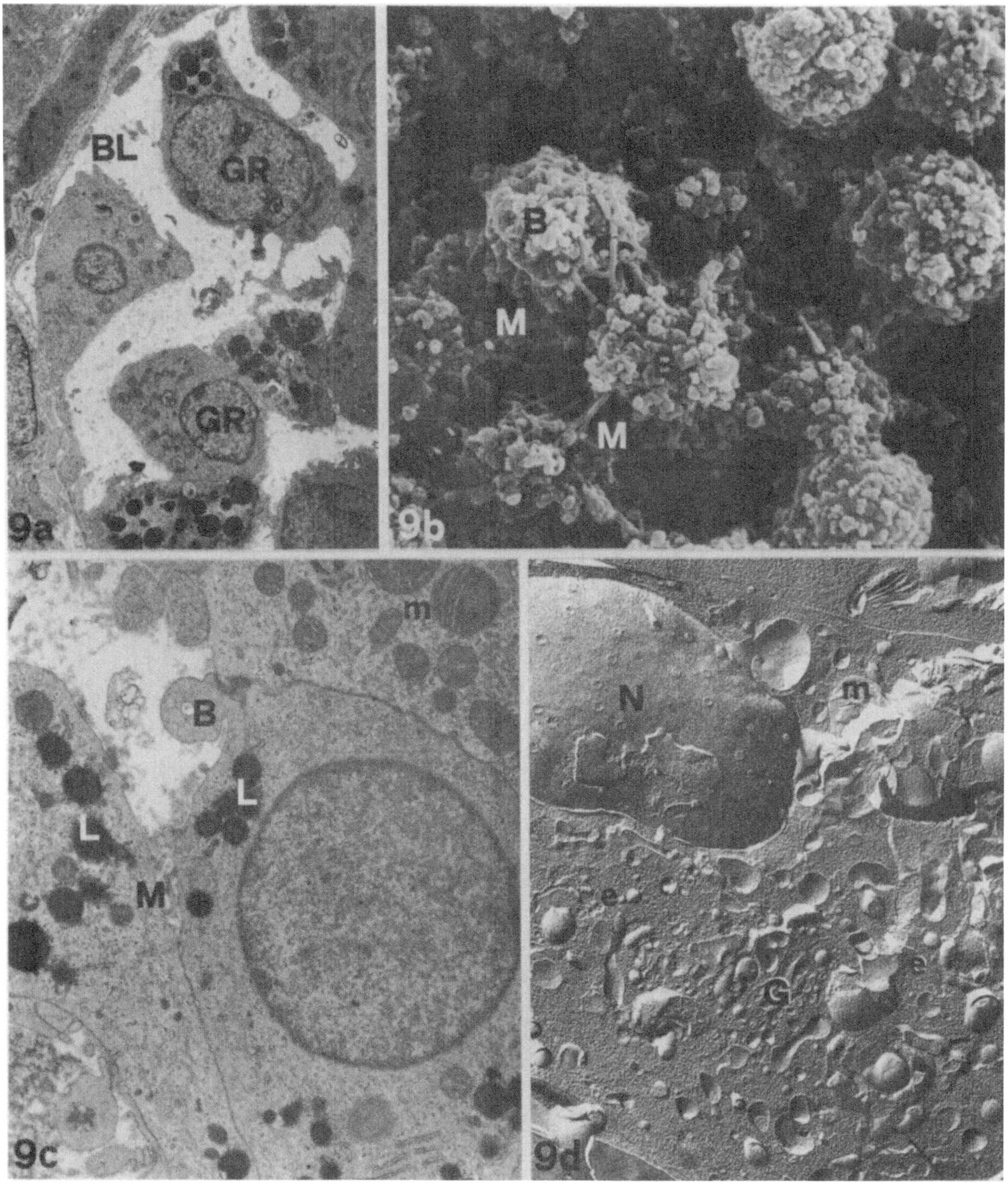
BL
GR
GR
9a
B
M
M
9b
m
B
L
L
M
9c
N
m
e
e
G
9d

are also seen in the preovulatory follicles of HCG-perfused ovaries (153,154) (Fig. 4-8a,4-9b)

Even though granulosa cells of preovulatory follicles stimulated in vitro by HCG showed the presence of relatively small amounts of smooth endoplasmic reticulum with respect to granulosa cells stimulated in vivo, the incomplete functional differentiation of the latter cells can probably be attributed to insufficient exposure to gonadotropins (155).

On the other hand, granulosa cells of the cumulus oophorus remain small and have irregular star shapes and smooth surfaces. These results indicate that, at this time, only the cells of the parietal layer of the granulosa in intact follicles within the ovary are candidates for luteinization, thus assuming all the biochemical and ultrastructural features related to these events. At this stage, it is typical to observe an increase of membranes of granular and smooth endoplasmic reticulum, mitochondria with tubular cristae, large Golgi complexes and lipid droplets. This morphological picture, together with the increased activity of the enzymes involved in steroid metabolism, is correlated to the steroidogenic activity of granulosa cells induced by the LH surge (156–158) (Fig. 4-9a,c,d).

Moreover, 3β-hydroxysteroid dehydrogenase activity is localized on the tubular or lamellar cristae and on the inner membrane of the mitochondria, as well as on the membranes of the smooth endoplasmic reticulum in the granulosa cells of human prevulatory follicles (158).

The possible association between granulosa cell ultrastructure and oocyte fertilizability was investigated during the course of an in vitro fertilization program in which clomiphene citrate and human menopausal gonadotropins were used to induce the growth of multiple follicles (159).

These studies showed that, even under these particular conditions, a relation exists between the morphology of the granulosa cells and the degree of oocyte maturation. In fact marked structural differences were noted among granulosa cells of similar-sized follicles containing both fertilizable and nonfertilizable oocytes. In particular, granulosa cells associated with non-fertilizable oocytes are tightly packed and provided with large functional gap junctions. Such granulosa cells possessed mitochondria with lamellar cristae, poor endoplasmic reticulum and Golgi complexes, and few lipid droplets and microvilli. On the other hand, granulosa cells associated with fertilizable oocytes are widely dispersed, contain small, annular gap junctions, mitochondria with tubular cristae, highly developed smooth endoplasmic reticulum and Golgi complexes, numerous electron-dense lipid droplets, and numerous long microvilli (159).

These ultrastructural patterns could be explained in terms of different LH/HCG responsiveness (159). Interestingly, such findings seem in agreement with morphological evidence of luteinization seen in granulosa cells of preovulatory and ovulatory follicles in nonstimulated cycles (160).

In addition, the viability of the oocyte in an in vitro fertilization program may be coupled with a predictable steroidogenic capacity of the surrounding granulosa cells (125). In particular, steroid levels in the follicular fluid can be used to predict the behaviour of the associated ooctyes (125). Further, it was noted that one or more proteins secreted by cultured human granulosa cells

←

Figure 4-9. Large preovulatory follicle. Granulosa cells.
a) TEM. Human large preovulatory follicle. Granulosa cells adjacent to the basal lamina (BL). Note dissociated granulosa cells (GR) with irregular ameboidal shape. X 5,500.
b) SEM. Human large preovulatory follicle. Granulosa cell–surface features typically include numerous long microvilli (M) blebs (B) and irregular extensions. X 6,000.
c) TEM. Human large preovulatory follicle. Granulosa cells adjacent to the basal lamina. Note the microvilli (M), blebs (B), the lipid droplets (L), and mitochondria (m) which have, in part, lamelliform, and, in part, tubular cristae and stacks of rough endoplasmic reticulum (*). These cellular organelles suggest that the initial luteinization occurs before ovulation. X 12,000.
d) Freeze-fracture. Mouse large preovulatory follicle. Granulosa cell. Medially fractured granulosa cell shows various fractured organelles; mitochondria (m), lipid droplets, endoplasmic reticulum (e) and Golgi complex (G). N: nucleus. X 23,000. c from Makabe et al., 1983 (98).

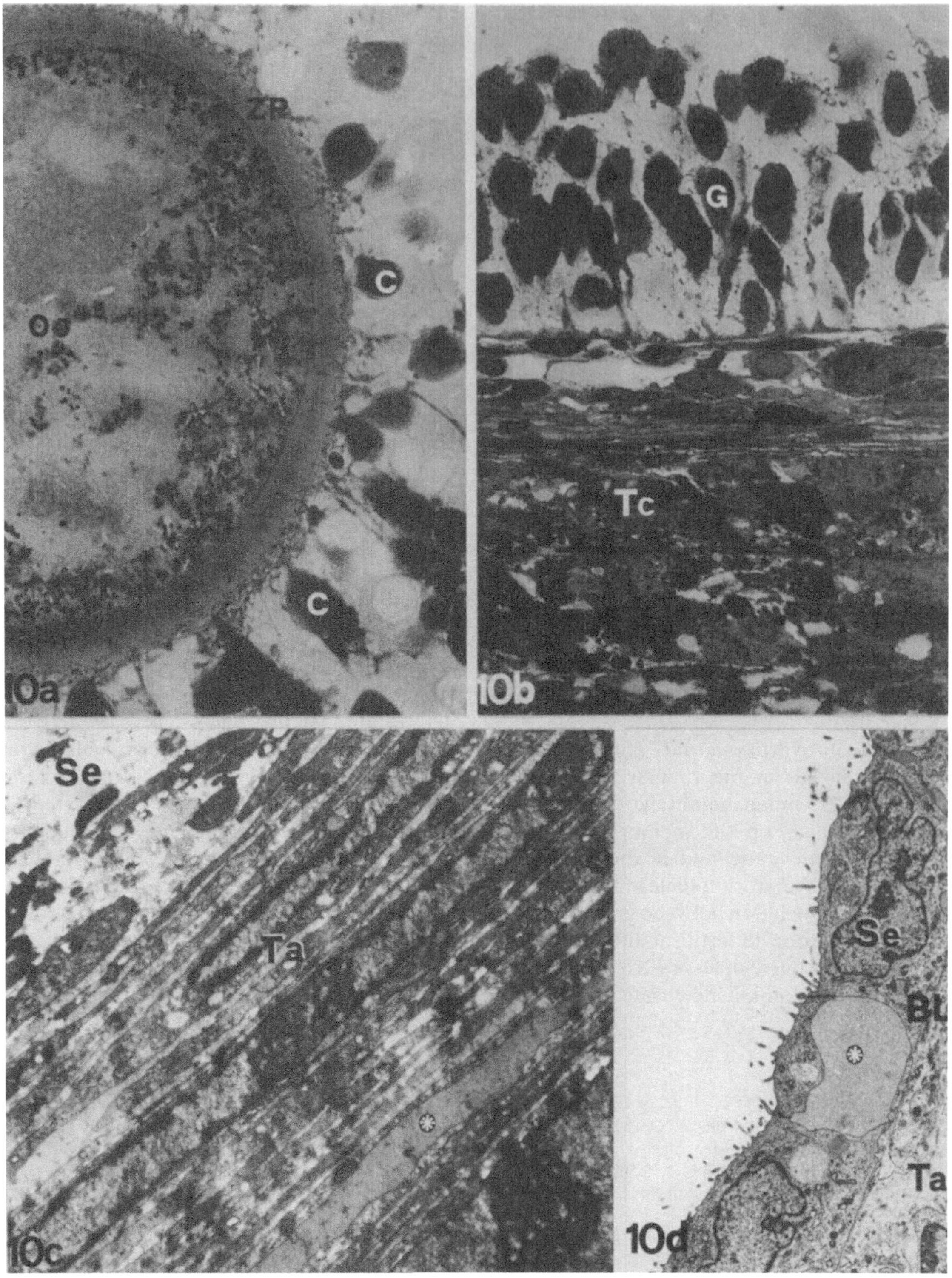
ZP
Oo
C
C
10a
G
Tc
10b
Se
Ta
10c
Se
BL
Ta
10d

might play a role in the gonadotropins' regulation of follicular function (162).

Rabbit granulosa cells taken from follicles near the time of ovulation show dense bodies that appear to be released into the follicular fluid by an exocytotic mechanism. Such unidentified material may be related to mechanisms involved in ovulation. A further possibility related to the cytolytic release of hydrolytic enzymes seems likely, since granulosa cells and many of the cellular elements of areas surrounding the follicle (theca interna) show a degenerative morphology at the time of ovulation. Alternatively, granulosa cells could produce activating agents that interact with locally synthesized or blood-borne substances to effect apical degradation of the preovulatory follicle. In this regard, Beers et al. (162) have shown that granulosa cells demonstrate plasminogen-dependent fibrinolytic activity, which indicates synthesis of plasminogen activator. The generation of the proteolytic enzyme, plasmin, may therefore mediate apical degradation (162,163) (Fig. 4-10c).

Other studies showed that *plasminogen activator* is not a primary proteolytic enzyme for follicular rupture, even if it has a "role in cumulus detachment and/or proliferation of granulosa cells during the ovulatory process" (164).

More recently, Espey et al. (165) showed that indomethacin, cycloheximide and colchicine were capable of inhibiting ovulation even while failing to reduce ovarian plasminogen activator activity. These authors concluded that plasminogen activator may not have a direct role in the final proteolytic degradation of ovulatory follicles, as previously thought, although it might contribute to the process by other mechanisms.

Because plasminogen activator is capable of mediating inflammatory reactions (166) and of inducing cell migration and tissue remodeling (167), it is possible that this substance instead plays a role in the intermediate stages of the ovulatory process (168).

By temporal and functional criteria the appearance of this activity correlates well with ovulation and can be demonstrated both in immature animals primed with hormones and in mature, cycling animals. In addition, granulosa cell cultures can be stimulated to release plasminogen activator by in vitro exposure to either luteinizing hormone or dibutyryl cyclic AMP (162).

In addition, the follicular basement membrane of preovulatory follicles is often penetrated by granulosa cell projections (151,153,169). These penetrating projections, occurring normally, may anchor the granulosa layer inside the follicle and may be considered a prerequisite for normal development of the corpus luteum (153).

The cumulus cells, which retain mucoprotein synthetic activity, are less sensitive to the gonadotropins and apparently are not transformed into luteal cells in vivo (170). Yet the cumulus cells respond to the gonadotropins (LH and FSH) by the deposition of an extracellular hyaluronidase-sensitive material that rapidly produces the mucification of the cumulus (171). These findings demonstrate that in the preovulatory follicles there are two populations of granulosa cells undergoing

←

Figure 4-10. The preovulatory follicle.
The oocyte, granulosa cells and the apex of the preovulatory follicle.
a) Light microscopy. Preovulatory rabbit follicle. The oocyte (oo) is surrounded by coronal cells (c) that have become detached from the granulosa wall, just prior to ovulation. Note the striations in the zona pellucida (ZP) which represent the ameboid processes of the coronal cells. X 900.
b) Light microscopy. Preovulatory rabbit follicle. Granulosa cells are dissociated from one another. Granulosa cells (G) appear to be undergoing precocious luteinization, as is indicated by the presence of dark bodies in the cytoplasm. Tc: theca interna cells. X 900.
c) TEM. Rabbit. Apical wall of a preovulatory follicle. The superficial epithelium (Se) has already been shed, leaving behind cellular debris. The connective matrix of the tunica albuginea (Ta) consists of dissociated fibers, degenerating cells and pools of a fluidlike material (*). X 13,500.
d) TEM. Superficial epithelium at the apex of a preovulatory follicle. A fluidlike material is accumulated in the connective tissue of the tunica albuginea (Ta) below the basal lamina (BL), and in the spaces between the cells of the superficial epithelium (Se). This material is so abundant in some regions as to cause extensive swelling of the intercellular spaces (*). A few microvilli are evident on the surface of the superficial cells, and spheroidal bodies that may correspond to lysosomes are present within the cytoplasm. X 4,500.

interesting specific differentiations (171). Hypothetically, this different activity of granulosa cells may depend upon the fact that the cells of the cumulus are closely related to the oocyte to which they are attached by gap junctions. These structures may inhibit the capability of the cumulus to luteinize, in contrast to the parietal granulosa cells that, while not related morphofunctionally with the meiotic oocyte, do undergo luteinization. An indirect support of this hypothesis can be found in the observation that progesterone synthesis in mammals can be induced by LH in cumulus cells cultured in vitro where such coupling between oocyte and cumulus cells might be altered (172).

These aspects have been corroborated by our own studies in humans, in which it has been shown that luteinization may occur in cumulus cells surrounding oocytes maturing in vitro where coupling between germ and follicle cells is lacking. Other observations indicate that luteinization might even continue just after fertilization (262,263). If these structural events are an expression of progesterone secretion by these cells, one might speculate that the activity of this hormone might have some effect on the fertilizability or healthy state of the ovum during its journey down the tube at the time of fertilization and cleavage.

Human cumulus cells represent the main source of the newly synthesized proteoglycans in the late preovulatory period, although their secretion by the oocyte was not negligible (82). These cumulus cell-secreted proteoglycans were associated with the cumulus extracellular matrix. The zona pellucida (82,173,174) may exert a role in the fertilization process since these molecules are constituents of the microenvironment in which fertilization occurs (82).

In clomiphene citrate-induced cycles, human cumulus cells may be seen that possess ultrastructural features commonly associated with luteinization (Nottola and Familiari, and Motta (263)), such as the presence of smooth endoplasmic reticulum, mitochondria with tubular cristae, lipid droplets and an increasing number of blebs and slender microvilli on their surface.

Just prior to ovulation, the granulosa cells of the cumulus and corona radiata tend to separate and create intercellular spaces into which the follicular fluid can penetrate (27). The number of follicular cell processes traversing the zona pellucida significantly decreases; these processes undergo degeneration and are believed to lose communication with their mother cells (17,87,93). Examination of the extent of ionic coupling in cumulus–oocyte specimens before and after ovulation indicates a progressive decrease as the time of ovulation approaches. This communication is terminated near the time of ovulation (33). The significance of this interrupted communication is not clear. Many classical and current studies suggest that physiological interaction between the mammalian oocyte and the surrounding granulosa cells is required to maintain the egg in meiotic arrest (63,66). On the other hand, some data show that the meiotic changes in the nucleus and cytoplasm of the oocyte precede by some hours the disruption of the intercellular communication between the oocyte and the cumulus (175,176). Other evidence suggests that the disruption of cumulus cell gap junctions in the rat provides a signal to the egg to resume meiotic maturation (Fig. 4-10a,b).

In fact, quantitative analysis of freeze-fractured rat cumulus–oocyte complexes reveals that the net area of cumulus cell gap junction membrane decreases about 15-fold, two to three hours following an ovulatory stimulus. This dramatic loss of gap junctions is temporally correlated with germinal vesicle breakdown and cumulus expansion, and suggests the possibility that cumulus junctional disruption plays a role in signaling meiotic resumption in the ovum (177). On the other hand, since several studies (178,179,180) suggest that the membrana granulosa plays a significant role in the control of meiotic maturation, it seems possible that the cumulus mass that anatomically interconnects the membrana granulosa with the oocyte can function as a bridge to convey membrane-generated inhibitory signals to the oocyte.

4.4. *Theca cells*

In preovulatory follicles, a great number of theca interna cells assume a clearly polyhedral shape and a steroidogenic structure; such cells are larger than those observed in previous stages. At this time of follicular development the gap junctions increase in number and size (47). No ultrastructural modifications can be seen in theca externa cells at this maturational stage (Fig. 4-10b).

During the hours preceding ovulation, in a number of mammalian species, a significant increase in vessel permeability was frequently reported (181). The capillary network of the theca interna becomes very developed and large gaps appear in the thinner cytoplasmic portions of the endothelial wall, whereas the capillary basal lamina undergoes fragmentation. Blood cells may appear free in the perivascular areas (17,182). When the blood follicular barrier is broken after the fragmentation of the basal lamina of the follicle, an *extrafluid* containing blood cells passes into the follicular cavity, enhancing the dissociation of the granulosa layer as well as intrafollicular pressure (17,181,183,184,185). Further, the passage of blood from the theca interna into the stratum granulosum may flush out the oocyte at the moment of ovulation (186). On the contrary, the apex of the follicle is avascular and an ischemic area becomes visible at this time (138).

In rabbit ovaries perfused in vitro after stimulation with human chorionic gonadotropin, perifollicular capillaries became dilated but interendothelial gaps were not observed (155), even though no blood components were present in the perfusate in this particular experimental situation. Also SEM observations also demonstrated an increased permeability of the perifollicular microvasculature just prior to ovulation both in vivo and in vitro (187).

In addition, the role of lymphatic capillaries in rabbit ovaries during ovulation was investigated (182). These studies showed that, like the blood capillaries, lymphatic capillaries may also dilate, producing gaps between adjacent cells. However, the edema around the lymphatic vessels disappeared whereas the blood capillaries continued to show large gaps in the endothelium. It was therefore concluded that the role of lymphatics may be consistent with the removal of edematous perifollicular fluid (182).

In the theca externa layer, evident alternations occur mainly at the follicular apex. In fact, at this location, the theca externa is reduced to a few randomly distributed fibrocytes and collagen fibers dissociated by the edema (Fig. 4-10c). On the other hand, all around the lateral and deeper parts of the follicle, myoid cells are present (113–117,188). Their presence in mammalian ovary, humans included, and their motor adrenergic innervation (111,119) suggest that neurogenic contraction of this muscle may play a role in follicular rupture and ovum ejection (17,90,189–193). In the case of the follicle at ovulation, it may be suggested that the activation of such perifollicular contractile tissue, even if incapable of producing alone the dehiscence of the follicle, may nevertheless be capable of squeezing the follicular wall. Then, if other factors do not simultaneously occur to cause a normal ovulatory process, the contraction of this tissue, instead of assisting in the rupture of the follicle, may favor its collapse and subsequent involution during atresia (118).

Even if smooth muscle cells, in atretic as well as in post-ovulatory follicles, are simply regarded as contractile cells favoring the transformation of the follicular complex into a corpus luteum or a corpus atreticum, it appears evident that their presence and contraction at some time during the follicle cycle is closely related to the life and involution of the luteofollicular complex (118,193). Other studies showed that smooth muscle cells persist in the basal hemisphere throughout the lifespan of the corpus luteum (194). During luteolysis most of the smooth muscle cells degenerate, while some persist as smooth muscle cells or transform into noncontractile cells (194). These observations support the suggestion that the smooth muscle cells of the theca externa play an active role in collapse of the ruptured follicle and compaction of the corpus luteum.

Although it is doubtful that the presence of a neuromuscular complex in the follicle is capable of producing alone the mechanical forces sufficient to rupture the follicular wall, nevertheless it may affect the incidence of ovulation. Contraction mediated by nerves in fact may have a role in the maintenance of a constant intrafollicular pressure in Graafian follicles. Contractile force may also produce alterations in the vascular wall with consequent ischemia or may simply have a trophic, modulatory function on the luteofollicular complex (195).

4.5. Apex of the preovulatory follicle

At the apical portion of the preovulatory follicle, dramatic morphological changes in the covering tissue layers can be observed by SEM and TEM.

Preovulatory follicles appear as large blisterlike structures swelling on the ovarian surface. The cells covering the basal areas of these follicles are polyhedral in shape and are provided with numerous microvilli and occasional solitary cilia such as those present in other regions of the ovarian surface (17,89,196). Moving up over the lateral zones of the apex, the superficial epithelium consists of elongated cells showing only a few microvilli. The cells covering the apex are extremely flattened and possess only rare microvilli, mainly assembled in the areas of intercellular contacts. Near the time of ovulation, a number of apical cells tend to degenerate and to desquamate (90). On the surface of the apex, a fluid material, extruding in the form of small droplets from the underlying intercellular spaces, is detectable by SEM (17,90) (Fig. 4-11a,c,d). When observed by TEM, the same fluidlike material infiltrates the stroma of the tunica albuginea, accumulates under the basal lamina and abundantly fills the dilated intercellular spaces of the superficial epithelium. The collagen fibers and fibrocytes of the apical perifollicular stroma are very rare and the connective matrix appears to be dissociated by the diffusion of the interstitial fluid (147). Martin and Miller-Walker (197) described the three-dimensional distribution of collagen fibrils over preovulatory follicles in the hamster. Approximately halfway up the follicle wall, the bundles fan out and form a meshwork of fibrils that covers the apex. As the time of ovulation approaches, the number of layers of the fibrils decreases over the apex until a tear forms in the weakened matrix (Fig. 4-10c,d).

An important role in determining the regressive phenomena that occur at the apical region of the follicle seems to be played by lytic enzymes acting in this zone (75). The mechanism by which collagen is removed from the follicle wall is not well known. Although numerous studies have implicated hydrolytic enzymes in this function (198–202), it is not clear if they degrade the collagen fibrils or the cross-links that bind them together into a matrix (188,197,203).

Although it has been recently suggested that proteolytic enzymes residing in the lysosomes or in other unidentified cytoplasmic sites of the superficial epithelium, as well as in other different cellular components of the follicular wall, are liberated at the appropriate time in these zones, these results still show some controversial aspects (17,198,204).

Espey (205) has hypothesized that connective tissue fibroblasts release zymogen into the extracellular space. Zymogen is subsequently activated either by itself or by other means, leading to degradation of the follicle wall. Further, he observed multivesicular structures in apical fibroblasts of preovulatory rabbit follicles and suggested that they might be a source of enzymes (206). Similarly, Okamura et al. (207) showed that the fibroblastic cytoplasm at the apex of the human Graafian follicle is well developed, is rich in lysosomelike granules and contains peripheral multivesicular structures, whereas intercellular collagen fibers are sparse. They suggested that the collagen fibers are digested by the content of the lysosomal granules and multivesicular structures, thus leading to follicular rupture. These morphological changes were related to the possible role of collagenase in follicular rupture. The role of collagenase, which might be activated by plasminogen activator (208), has been corroborated by many observations such as a decrease in follicular and ovarian collagen before ovulation (209, 210) and inhibition of ovulation after administration of collagenase inhibitors in vitro and in vivo (211,212). Finally, the preovulatory increase in collagenase activity after hCG administration was demonstrated (213), supporting the role of collagenolysis in ovulation.

The modified proteins that are released during this degradative process induce an acute inflammatory reaction that results in histamine release, leukocyte migration and concomitant activation of prostaglandins (185,214). This inflammatory response retards the repolymerization of the connective tissue matrix and thereby facilitates collagenolysis of the follicle wall. The connective tissue around the follicle wall, especially at the apex of the follicle, is degraded and, particularly, the thin region at the apex of the follicle is the area most susceptible to distension under the stress of a low intrafollicular pressure (185,214).

Observations made on perfused ovaries show that ovulation in this experimental condition is preceded by apical wall degeneration, which begins in the outer cell layers. This is compatible with the hypothesis that lysosomal enzymes are released extracellularly, close to the surface of

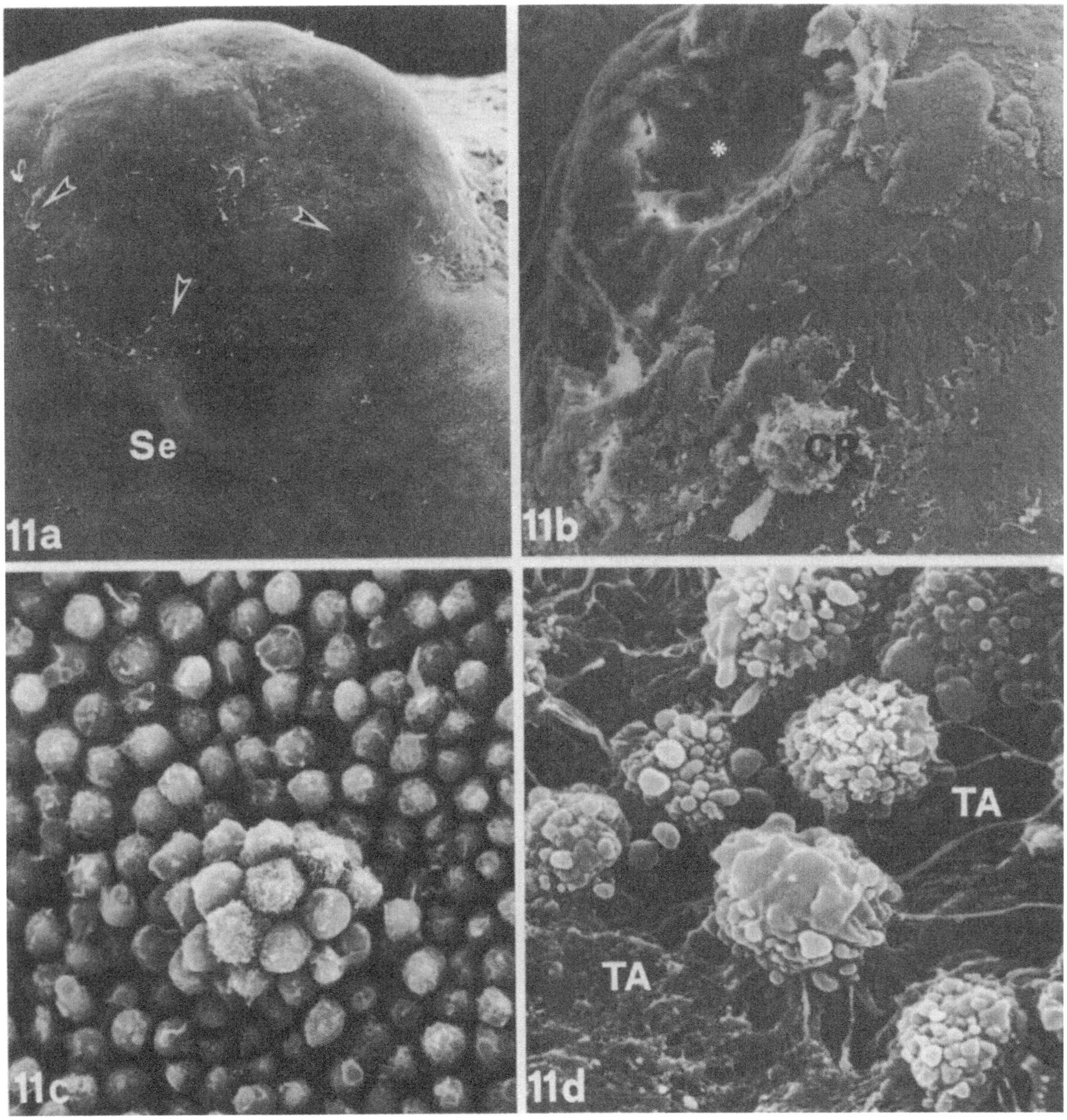

Figure 4-11. Preovulatory and ovulatory follicles.
The superficial epithelium and ovulation.
a) SEM. Apex of preovulatory rabbit follicle. The apex of the preovulatory follicle is not completely covered by the superficial epithelium (Se). In other regions, patches of the superficial epithelium are in the process of sloughing off (arrowheads). X 110. From Van Blerkom and Motta, 1979 (10).
b) SEM. Apex of a follicle just after ovulation in the rabbit. From the surface, it is possible to directly observe a ruptured stigma that is surrounded by a gellike mass of granulosa cells. The collapsed residual cavity of the follicle is clearly evident from the surface (*). An oocyte, covered by the cells of the corona radiata (Cr) is evident. X 130.
c) SEM. Human superficial epithelium in the lateral areas of preovulatory follicle. These cells are polyhedral elements which are covered with microvilli and isolated cilia. X 1,000.
d) SEM. Human superficial epithelium covering the apex of a human preovulatory follicle. The uncoupled, mulberry cells are connected by a thin extension, leaving exposed large areas of tunica albuginea (TA). X 4,000. a,b, from Motta and Van Blerkom, 1975 (147).

preovulatory follicles (198,199,215). A common inflammatory reaction with the accumulation of leukocytes previously found apically in vivo (188, 214) seems less probable in vitro (153,154).

Degradation of collagen fibrils seems to be less prominent in ovaries perfused in vitro with HCG than in ovaries similarly perfused in vivo, while the disappearance of surface epithelium at the apex of follicles was similar in ovaries so perfused both in vitro and in vivo (155).

Degeneration and sloughing of the surface epithelial cells are related to the formation of the stigma (147). A large, protruding stigma is observed with the SEM in the mouse, rabbit, rat, guinea pig and in humans (216). On the contrary, stigma formation in the hamster is unremarkable and difficult to detect (217). Before ovulation there is localized tissue disintegration in the stigma and a general dissolution of the entire follicular wall. When the membrana granulosa cells in the area of the stigma degenerate, the follicle basement membrane bulges from it; when this membrane ruptures, ovulation occurs (Fig. 4-11 a,b).

Ovulation has been described mostly as an oozing process following the rupture of a large stigma (147,216). Pendergrass and Reber (217), however, found an early portion of the process to resemble an eruption with a detectable column of cells standing away from the surface of the follicle in the hamster. This suggests an underlying propelling force, rather than an oozing process, that may result in a clump of cells adhering to the side of the follicle. A possible explanation of the difference is that, in animals in which a large stigma is formed, rupture of the stigma results in an opening of sufficient size to allow passage of the oocyte. In the hamster, however, an ejecting force is needed to expel the oocyte through the small stigma (217).

The release of the oocyte and companion cumulus cells at ovulation is often obscured by a bolus of dense precipitated material, composed of follicular and interstitial fluids, in which free granulosa cells and other cellular debris are suspended.

The repair of the ruptured follicle occurs immediately after ovulation and consists of a mass of fluids and reactive tissues that closes the opened stigma. This mass is then rapidly covered by numerous flattened cells originating by proliferation and consequent migration from adjacent undisturbed zones of the superficial epithelium (218). Probably due to the active proliferation of this epithelium, the ovarian surface overlying the nascent corpus luteum is irregular in some mammals due to the abundance of papillae as well as to the presence of crypts invaginated into the subjacent luteal mass (218).

These morphological observations were recently confirmed by autoradiographic studies performed by Osterholzer et al. (219). These authors showed that a significant portion of the ovarian surface epithelium cells undergoes DNA synthesis throughout most of the postovulatory period.

After ovulation, the granulosa cells of the ruptured follicle undergo luteinization, under the influence of gonadotropins. Factors in the follicular fluid prevent luteinization of granulosa cells prior to ovulation. Luteinized granulosa cells, plus the surrounding theca cells, capillaries and blood vessels, intermingle to form a corpus luteum.

The ultrastructure of human granulosa cells differentiating in lutein cells has been studied in an attempt to correlate morphological features with functional events (220).

Mitochondria with tubulovesicular cristae are characteristic of steroid-secreting cells and contain the enzymes necessary for cholesterol side-chain cleavage. During the early stages of lutein cell differentiation, endoplasmic reticulum increases in total amount, concomitant with a shift from rough morphology to a predominantly smooth reticulum during luteinization. Lipid droplets represent the storage site for cholesterol and for other steroid precursors or final secretory products (221). The immature granulosa–lutein cells exhibit numerous microvilli that gradually disappear, while blebbing activity becomes prominent. Inside these cells mitochondria, with both lamelliform and tubular cristae, rough and smooth endoplasmic reticulum and dense lipid droplets, become evident. However, mature granulosa–lutein cells show a few microvillous projections and some blebs and many ameboid cytoplasmic extensions. Mitochondria with tubular and/or villiform cristae, prominent smooth endoplasmic reticulum and lipid droplets are characteristic of these cells. The luteal cell surface in the regression stage have a mulberrylike appearance. The cytoplasm of the luteal cells is filled with numerous lipid droplets,

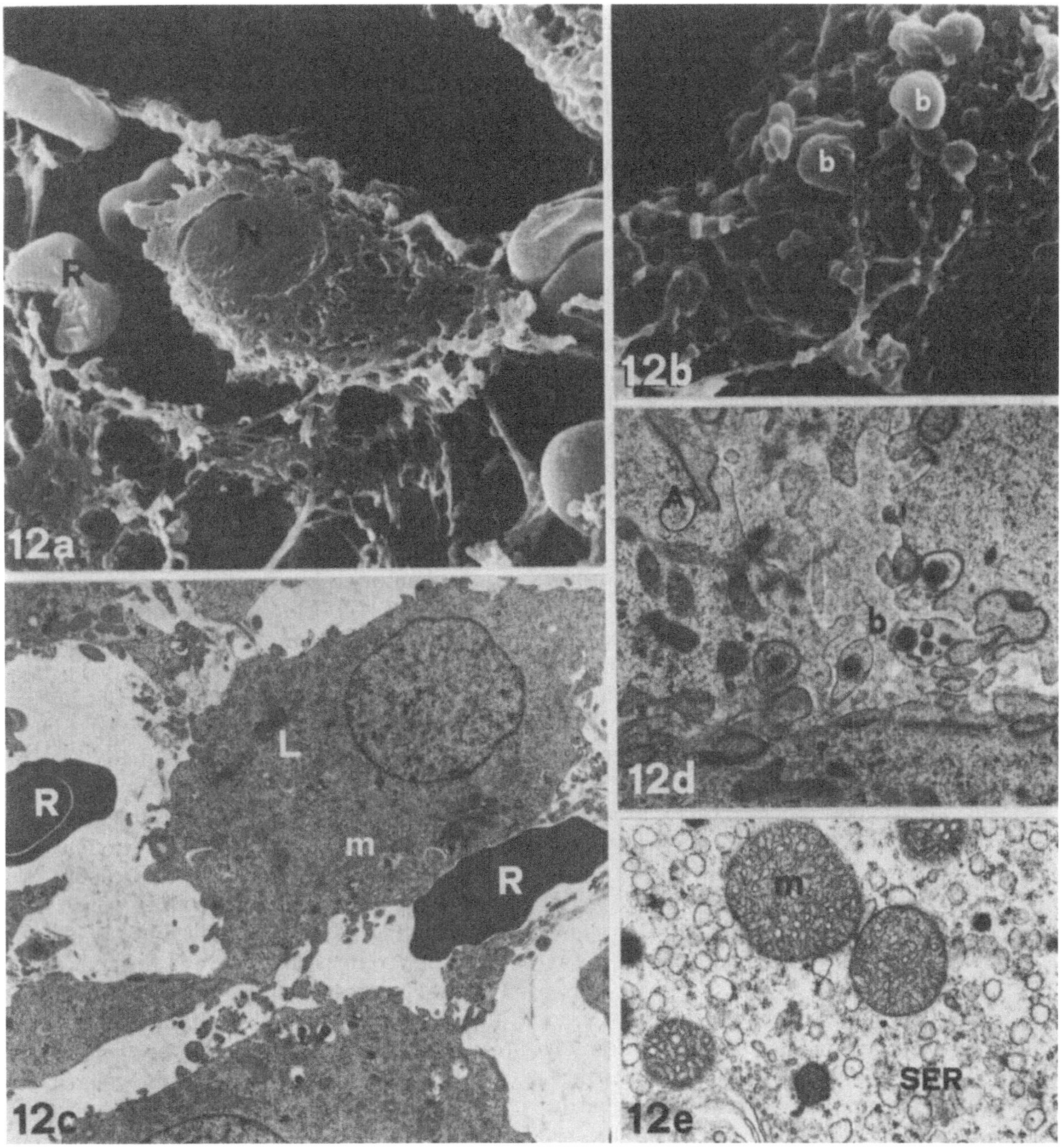

Figure 4-12. Postovulatory follicle. Granulosa–lutein cell.

a) SEM. Human granulosa–lutein cells. Sectional surface of a granulosa–lutein cell on the 3rd day of the luteal phase. N: nucleus; R: red cells. X 5,300.

b) SEM. Human granulosa–lutein cells. Part of the apical surface of a granulosa–lutein cell on the 3rd day of luteal phase. Note the prominent blebbing activity (b). X 9,500.

c) TEM. Mature granulosa–lutein cells taken from the fifth to seventh day of the luteal phase. The cells have become larger, with ameboidal cytoplasmic extensions and prominent organelles. Note the many large mitochondria with villiform cristae (m), lipid droplets (L) and smooth endoplasmic reticulum. R: red cells. X 7,000.

d) TEM. Mature granulosa–lutein cells. Part of three adjacent granulosa–lutein cells on the 3rd day of the luteal phase. Note several blebs (b), and annular nexus (A). X 15,000. a,b,d from Makabe, 1981 (91).

e) TEM. Mature granulosa–lutein cells. Note the mitochondria with larger villiform cristae (m) and smooth endoplasmic reticulum (SER). X 13,000. From Makabe et al., 1982 (143).

polymorphic lysosomes and autophagic vacuoles. A sectional surface of the luteal cells having the appearance of a honeycomb is also observed. Many connective tissues surround these luteal cells (220) (Fig. 4-12a–e).

The function and morphology of the differentiating corpus leteum cells were also studied in culture (222). In this case, too, the fine structure of these cells was characterized by the presence of abundant smooth endoplasmic reticulum, lipid droplets and junctions.

Stimulation with HCG and LH resulted in changes in the ultrastructure of cultured granulosa –lutein cells and of progestin secretion, leading to a redistribution of the organelles involved in steroid synthesis. (223). These antral changes seem to be mediated by cAMP and are probably related to a rearrangement of cytoskeletal microfilaments (223).

Oxytocin was recently localized in the corpus luteum cells (224–227).

The levels of circulating oxytocin increased to a maximum during the first part of the luteal phase and fell during luteolysis, reaching basal levels around the time of oestrus (228,229).

Luteal oxytocin might play a role in luteolysis (229).

Relaxin was also noted in the corpus luteum in various species (230,231).

5. Ovulation

The midcycle LH surge, or the simulation of it in in vitro fertilization through administration of HCG, is the focal point of a series of events that occurs within the preovulatory follicle and culminates in ovulation. LH inhibits the follicular production of androgens and estrogens and stimulates the luteinization process of granulosa cells and the production of progesterone (10,232).

The LH surge stimulates cAMP production (205) as well as the release of histamine, serotonin and bradykinin which mediate the initial phase of an inflammatory reaction (increased vascular permeability) probably involved in the ovulatory process (214).

Schmidt et al. (233) showed that histamine alone, like LH, was capable of inducing ovulation in the isolated perfused rat ovary, though the number of follicular ruptures was lower in histamine-perfused ovaries. Histamine might affect the ovulatory process by increasing permeability of perifollicular capillaries, producing interstitial edema (181), by influencing follicular contractility either directly or through some interaction with the transmitter released by nerve terminals or even by stimulating the production of progesterone (233).

Further, LH (234) and cAMP production (235–237) stimulates the production of prostaglandins, which have been shown to be essential mediators of ovulation (214,238,239). Prostaglandins stimulate collagenase activity in connective tissue cells by activation of thecal fibroblasts (214). Other prostaglandin-induced phenomena that have been implicated in follicular rupture include the mobilization of fibroblasts (240), enhancement of the inflammatory reaction (214) and follicular contractions (241). Moreover, studies carried out by Downs and Longo (242) suggest a role for prostaglandins in follicular rupture by two processes: a) the mobilization of granulosa cells, and b) the dissolution of thecal and peripheral connective tissue. It is possible that, at the time of ovulation, these events are coordinated with other prostaglandin-induced activities such as the production of plasminogen activator by the hyperemic, inflammed follicle, by systemic leucocytes and by proliferating fibroblasts (214), all of which activate local collagenase and produce a cascade of proteolytic enzymes that degrade the follicular connective tissue and cause ovulation. Prostaglandins are also capable of stimulating the secretion of ovarian vasopressin (243). Observations in animals have demonstrated an anovulatory effect for this hormone (244). Studies in humans have revealed that concentrations of vasopressin in the preovulatory follicular fluid are generally higher than those in the plasma (245). These observations lend strong support to a role for vasopressin in human ovulation.

More recently, a possible role of oxytocin in the ovulatory process was investigated (246). In fact, within the human ovary, the concentration of oxytocin is highest in the follicular fluid of the preovulatory follicle. The ovarian secretion of oxytocin, which is stimulated by prostaglandins (247), may promote ovulation by inhibiting the formation of follicular wall proteins such as collagen (246).

Immunochemical investigations (248) have indicated the presence of relaxinlike substances in granulosa cells of cavitary follicles in the human and pig ovary. Relaxin might activate collagenase and thereby play some role in the process of ovulation (249).

Human ovarian follicular fluid contains many other serum proteins, and, in particular, enzymes and protease inhibitors (alpha 1-antitrypsin, alpha 2-macroglobulin, antithrombin III, pregnancy-associated plasma protein A, and placental protein 5). Many of these proteins are thought to be derived from the circulation, whereas some evidence suggests that alpha 1-antitrypsin, pregnancy-associated plasma protein-A and placental protein 5 are produced by the ovarian follicle in humans (250,251).

Pregnancy-associated plasma protein-A is a large glycoprotein, produced by the granulosa and/or theca interna cells, that increases in concentration during the follicular phase, reaching high levels in the liquor folliculi of large preovulatory follicles immediately prior to ovulation (250, 251). Levels of placental proteins also increase during follicular development, peaking prior to ovulation (253). Alpha 1-antitrypsin is another antiprotease found in human serum. Bagdasarian et al. (254) demonstrated the presence of alpha 1-antitrypsin in ovarian stromal cells, which seems to suggest the synthesis of this enzyme by these cells.

More recently a statistically lower concentration of alpha 1-antitrypsin was found in follicular fluids from which mature oocytes were recovered than in those yielding immature oocytes (252). These antiproteases may contribute to the pool of intrafollicular protease inhibitors and to the maintenance of the follicular proteolytic homeostasis in the pre- and postovulatory phases (255).

When the balance between proteolytic enzyme and antiprotease activities shifts towards increased protease activity, spontaneous ovulation of the maturing follicle could occur. In fact, one of the mechanisms involved in ovulation is the activation of a proteolytic cascade that has been implicated in the enzymatic digestion of the collagen fibers of the follicular apex. Some follicular proteins, such as the protease inhibitors mentioned above, are produced during the periovulatory period and modulate this proteolytic activity "as part of a general immunosurveillance system" (251).

Other studies have shown that prorenin is produced by the mature follicle in response to LH/hCG (12). Prorenin, via the action of angiotensin II, might control oocyte maturation, ovulation, ovarian blood flow, or ovarian steroid biosynthesis (12).

In conclusion, the surge in LH either directly or indirectly activates a series of biochemical and biophysical events that contributes to follicular rupture and ovulation. These events can be summarized as follows:

a) Capillary permeability increases and is accompained by alteration of the endothelial wall in the perifollicular vascular bed (181,196); the consequent escape of fluid and blood cells into the interstitium thus leads to edema of the apical portion of the follicle (17,90).
b) Enzymatic digestion of the collagen fibers of the follicular apex that is avascular and ischemic takes place (17,75,188,197,198,199, 202, 205,214).
c) Degeneration and desquamation of apical cells (17,90,217) occur.
d) The perifollicular smooth muscle cells assisted by their motor adrenergic innervation (111, 119) contract and squeeze the follicle (90,118, 189,217).
e) According to the inflammatory process model for ovulation, prostaglandins stimulate the production of proteolytic enzymes capable of disrupting the follicular wall during ovulation (214).
f) The pool of follicular protease inhibitors may also be regarded as an important modulator system for the proteolytic apparatus, being active during the periovulatory period (250–255).
g) In addition, the role of other substances such as histamine (233), vasopressin (243,244,245), oxytocin (246,247) relaxin (248,249) and prorenin (12) has not been fully evaluated and should be considered as a part of a complex and highly ordered system whose coordinated activities culminate in ovulation.

All of these mechanisms most likely operate synergistically, with no single one being determinant of the ovulatory event.

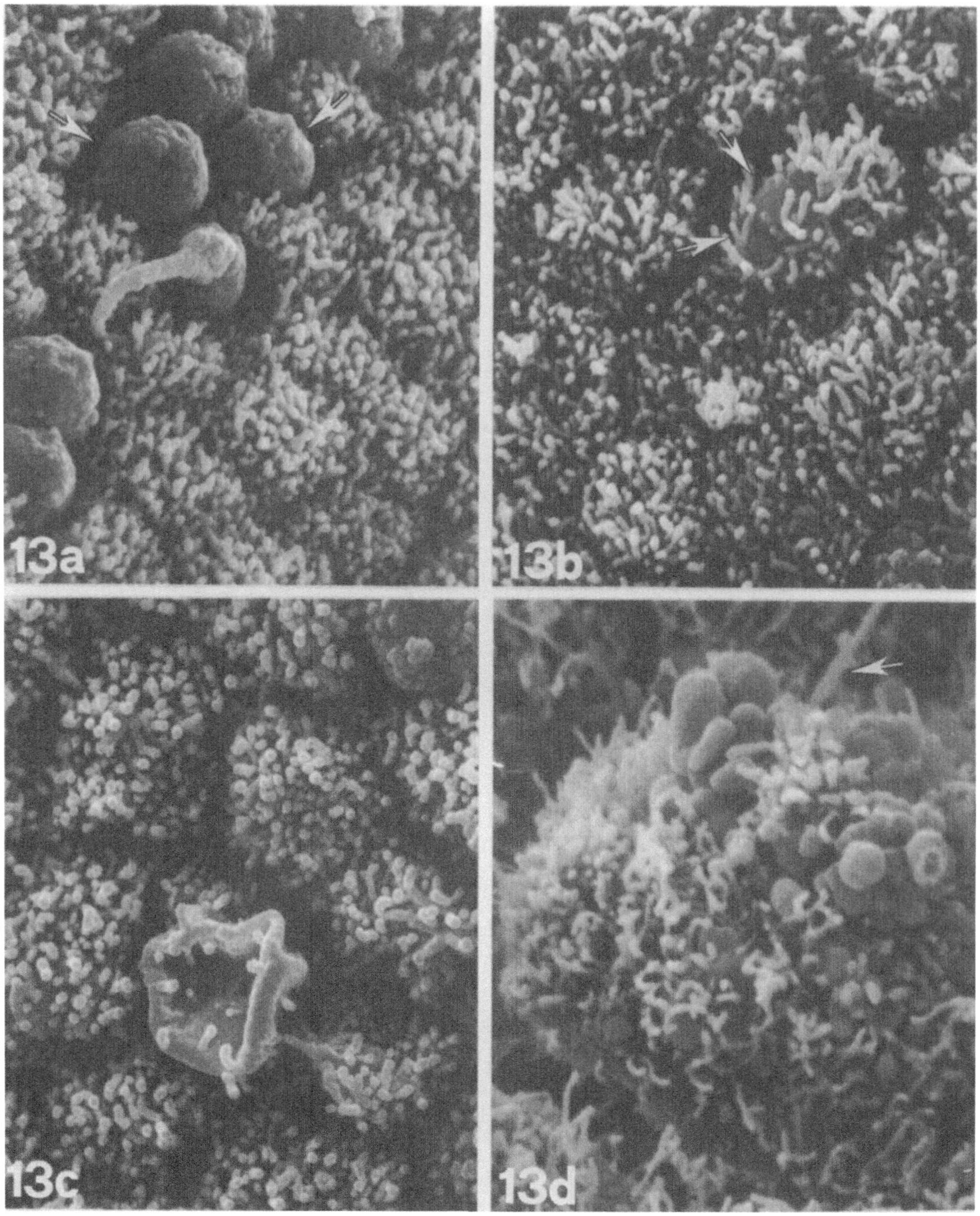

Figure 4-13. SEM. Anovulatory human ovaries. Superficial epithelium.

a,b) Luteinized unruptured follicle. On the follicle apex it is possible to see different types of superficial epithelial cells. Some have dense microvilli and some are completely without microvilli (arrows) suggesting that they may be at the S-stage of the cell cycle. a,b = X 3,500.

c,d) Polycystic ovary.

c) The superficial epithelium have large ruffles (lamellipodia). X 8,000.

d) On the apex of the cystic follicle, the superficial epithelium shows many blebs of various sizes, solitary cilia (arrow) and irregular microvilli. X 10,000.

From Makabe, S. In: Three Dimensional Microanatomy of Cells and Tissue Surfaces. Amsterdam: Elsevier, North Holland, Inc., 1981.

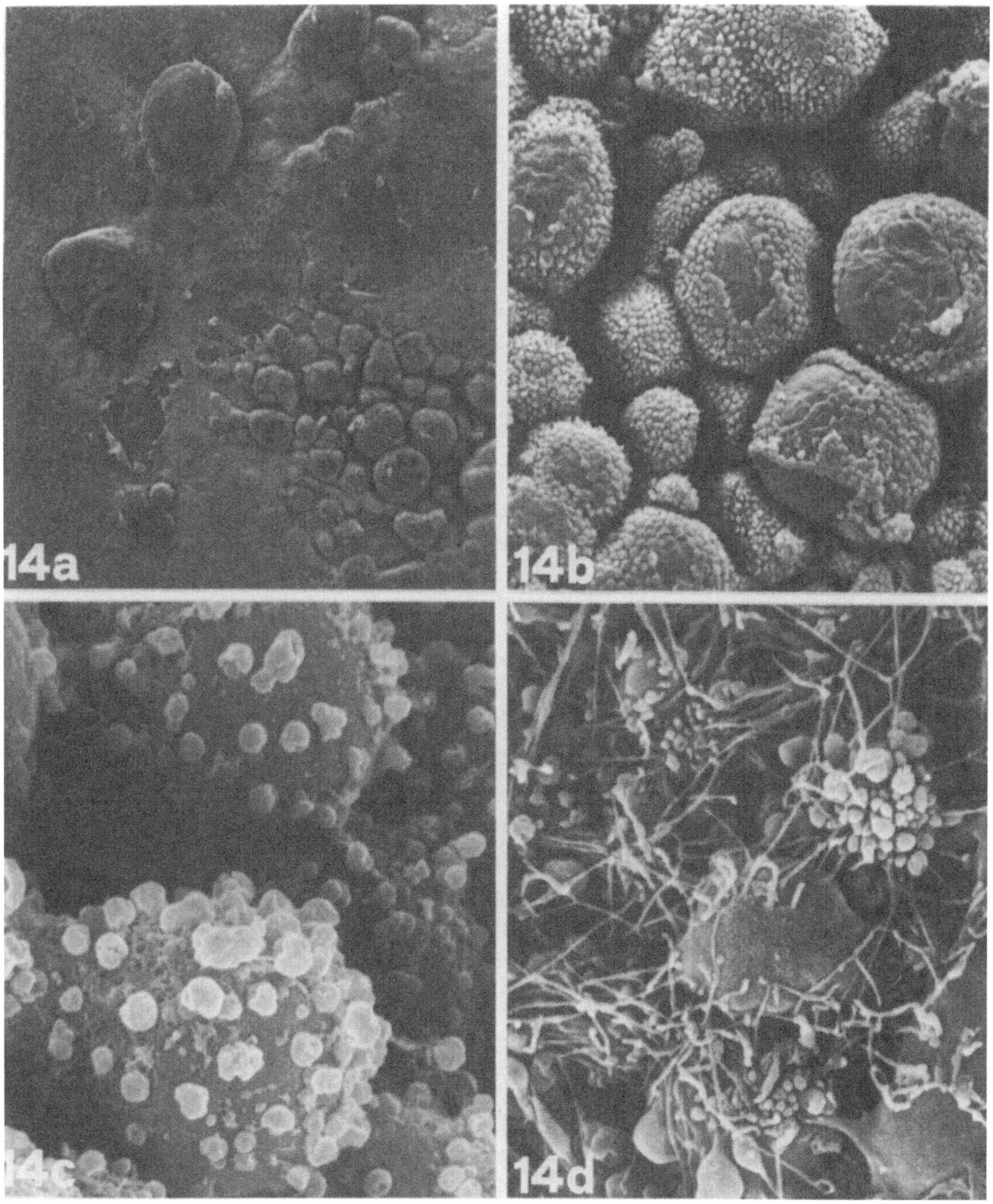

Figure 4-14. SEM. Anovulatory human ovaries (polycystic ovaries).

a,b) Superficial epithelium. Numerous round-shaped papillae covered with superficial epithelium are visible on the ovarian surface in the polycystic ovarian syndrome (a: X 300; b: X 1,100).

From Makabe, S. In: Three Dimensional Microanatomy of Cells and Tissue Surfaces. Amsterdam: Elsevier, North Holland, Inc., 1981.

c,d) Granulosa cells. In c, the granulosa cells appear as rounded elements, with scattered bleblike surface features. In d, granulosa cells are smooth-surfaced, flattened, and irregular in size and shape (c: X 9,000; d: X 4,000).

From: Makabe, S. In: Advances in the Morphology of Cells and Tissues. New York: Alan Liss, Inc., 1981.

At present the use of an appropriate combination of TEM, SEM and other ultrastructural methods facilitates a more consistent morphological reconstruction of the dynamic events leading to ovulation. However, although this study appears to provide a useful and easily interpretable ultrastructural image of the various mechanisms involved in ovulation, these morphological results will have to be integrated with biochemical and cytochemical information if a more correct evaluation of the relevant physiopathological phenomena is to emerge.

5.1. Pathology of ovulation

Many pathological conditions can lead to failure of ovulation and, consequently, to infertility. In this chapter a short account of the luteinized unruptured follicle (LUF) syndrome is provided, considering its highly frequent occurrence during ovarian stimulation protocols.

In 1975 Jewelewiez (256) described this condition, in which "luteinized unruptured follicles secrete progesterone, causing all the progestational changes without actual ovulation".

Comparative studies dealing with the hormonal characteristics of both the LUF syndrome and ovulatory cycles showed that both midcycle LH levels and midluteal progesterone levels were lower in LUF cycles than in control cycles, even though the duration of the luteal phase was not affected (257–259). In addition, laparoscopic and ultrasonic studies of this syndrome (258,260,261) reported evidence of failed ovulation.

Histologic studies (259) showed an entrapped ovum within a luteinized follicle consisting of spindle-shaped granulosa cells with small nuclei, and others showing increased cytoplasmic volume and vacuolization. Theca cells and perifollicular stromal luteinization were also noted in these follicles (259). Observed by SEM, the morphology of the superficial epithelium resembles that seen in polycystic ovaries (91). The superficial epithelium covers the entire surface of the organ; the cells are tightly packed, showing the presence of dense microvilli, solitary cilia, blebs, filopodia and ruffles (Fig. 4-13a–d) (91). Certain superficial epithelial cells display no microvilli, which would suggest that they are at the S-stage of the cell cycle (Fig. 4-13a) (91). Papillae occur frequently and are similar to those observed in polycystic ovaries; they often enlarge and become more numerous than in the normal ovary (Fig. 4-14a,b). Recently, SEM and TEM studies (Makabe, unpublished results) showed the presence of granulosa cells provided with numerous blebs, cytoplasmic projections, filopodia and ameboid evaginations analogous to structures appearing in polycystic ovaries (Fig. 4-14c,d). These morphological aspects suggest that some of the granulosa cells of unruptured follicles do undergo luteinization.

Both central and local factors might be responsible for failed ovulation. A disturbance of the LH surge is one possible cause of the LUF syndrome (258,259). Enzymatic disorders may also represent a local cause of the LUF syndrome; failure of plasminogen synthesis due to insufficient plasminogen activator activity might lead to ovulatory failure (258). It is also possible that in the LUF syndrome, the normal relationship between protease and antiprotease activity is defective, causing a lack of ovulation (252).

Some observations suggested that hCG might be useful in the treatment of the LUF syndrome (257).

Further ultrastructural and histochemical studies are needed to explain the etiology of this syndrome, which, in many respects, could provide a model for the study of ovulation and its alterations.

Acknowledgments

This work was supported by Grant N. 85.00612.04 from CNR and funds from MPI, Italy.

We are grateful to Mr. A. Familiari for his skillful technical assistance and Mrs. E. Vergnano for her patience in typing and retyping the manuscript.

Thanks are due to David Fynn for the excellent revision of the manuscript.

References

1. Hodgen GD, Kenigsburg D, Collins RL, Schenken RS: Selection of the dominant ovarian follicle and hormonal enhancement of the natural cycle. In: In Vitro Fertilization and Embryo Transfer. M Seppala, RG Edwards (eds), *Ann NY Acad Sci* 442: 23–37, 1985.

2. Mossman HW, Duke KL: Comparative morphology of the mammalian ovary. Madison, WI: University of Wisconsin Press, 1973.
3. Block E: Quantitative morphological investigations of the follicular system in women. *Acta Anat* 14: 108–123, 1952.
4. Breitenecker G, Friedrich F, Kemeter P: Further investigations on the maturation and degeneration of human ovarian follicles and their oocytes. *Fertil Steril* 29: 336–341, 1978.
5. Greenwald GS: Role of follicle-stimulating hormone and luteinizing hormone in follicular development and ovulation. In: Handbook of Physiology, Vol IV, American Physiological Society. RO Greep, EB Aswood, E Knobil, WH Sawyer, SR Geiger (eds), Washington: Baltimore Waverly Press, p 293–323, 1974.
6. Schwartz NB: The role of FSH and LH and of their antibodies on follicle growth and on ovulation. *Biol Reprod* 10: 236–272, 1974.
7. Richards JS: Maturation of ovarian follicles: actions and interactions of pituitary and ovarian hormones on follicular cell differentiation. *Physiol Rev* 60: 51–89, 1980.
8. Goodmann AL, Hodgen GD: The ovarian triad of the primate menstrual cycle. *Rec Prog Horm Res* 39: 1–73, 1983.
9. Hodgen GD: Physiology of follicular maturation. In: In Vitro Fertilization. HW Jones Jr, G Seegar Jones, GD Hodgen, Z Rosenwaks (eds), Norfolk: Pb Williams & Wilkins, p 8–29, 1986.
10. Seegar Jones GE, Acosta AA, Garcia JE, Rosenwaks Z: Specific effects of FSH and LH on follicular development and oocyte retrieval as determined by a program for in vitro fertilization. In: In Vitro Fertilization and Embryo Transfer. M Seppala, RG Edwards (eds), *Ann NY Acad Sci* 442: 119–122, 1985.
11. Baird DT, Messinis IE, Templeton AA, McNeilly AJ, Tsonis CG: Inhibin levels in gonadotropin-treated cycles. Fifth World Congress on In Vitro Fertilization and Embryo Transfer, Norfolk, Virginia ps-018, 1987.
12. Itskovitz J: Ovarian prorenin-renin-angiotensin system. Fifth World Congress on In Vitro Fertilization and Embryo Transfer, Norfolk, Virginia, ps-202, 1987.
13. Ying SY, Ling N, Guillemin R: Inhibins and activins. Fifth World Congress on In Vitro Fertilization and Embryo Transfer, Norfolk, Virginia ps-017, 1987.
14. Schwartz NB: The role of FSH and LH and of their antibodies on follicle growth and on ovulation. *Biol Reprod* 10: 236–241, 1974.
15. Lunenfeld B, Kraiem Z, Eshkol A: Structure and function of the growing follicle. *Clin Obstet Gynecol* 3: 27–234, 1976.
16. Baird DT: Factors regulating the growth of the preovulatory follicle in the sheep and human. *J Reprod & Fertil*, 69: 343–352, 1983.
17. Van Blerkom J, Motta PM: The Cellular Basis of Mammalian Reproduction. Baltimore: Urban & Schwarzenberg, 1979.
18. Acosta AA, Seegar Jones G, Garcia JE, Sandow B, Week L, Mantzavinos T: Correlation of human menopausal gonadotropin/human chorionic gonadotropin stimulation and oocyte quality in an in vitro fertilization program. *Fertil Steril* 41: 196–201, 1984.
19. Weeck LL, Wrotham JWE Jr, Witmyer J, Sandow BA, Acosta AA, Garcia JE, Jones JS, Jones HW Jr: Maturation and fertilization of morphologically immature human oocytes in a program of in vitro fertilization. *Fertil Steril* 39: 594–602, 1983.
20. Garcia JE, Jones GS, Acosta AA, Wright A: Human menopausal gonadotropin/human chorionic gonadotropin follicular maturation for oocyte aspiration: phase II, *Fertil Steril* 39: 157–163, 1983.
21. Marrs RP, Vargyas JM, Hoffman D, Yee B: Use of various ovarian stimulation methods to improve oocyte and embryo production for human in vitro fertilization. In: In Vitro Fertilization and Embryo Transfer. M. Seppala, RG Edwards (eds), *Ann NY Acad Sci* 442: 112–118, 1985.
22. Testart J, Lefevre B, Castanier M, Belaisch-Allart J, Guillet-Rosso F, Frydman R: Comparative effect of Clomipheme and Clomiphene/hMG on the preovulatory follicles and fertilizability of the oocyte. In: In Vitro Fertilization and Embryo Transfer. M Seppala, RG Edwards (eds), *Ann N Y Acad Sci* 442: 128–139, 1985.
23. Bayly CM, McBain JC, Clarke GA, Gronow MJ, Johnston WIH, Martin MJ, Speirs AL: Ovarian stimulation regimens in an in vitro fertilization program: a comparative analysis. In: In Vitro Fertilization and Embryo Transfer. M Seppala, RG Edwards (eds), *Ann NY Acad sci* 442: 123–127, 1985.
24. Russel JB, Polan ML, De Cherney AH: The use of pure follicle-stimulating hormone for ovulation induction in normal ovulatory women in an in vitro fertilization program. *Fertil Steril* 45: 829–833, 1986.
25. Yoshimura Y, Hosoi Y, Atlas SJ, Wallach EE: Effect of clomiphene citrate on in vitro ovulated ova. *Fertil Steril* 45: 800–804, 1986.
26. Templeton AA, VanLook P, Angell RE, Aitken RJ, Lumsden MA, Baird DT: Oocyte recovery and fertilization rates in women at various times after the administration of hCG. *J Reprod Fertil* 76: 771–778, 1986.
27. Balboni GC: Histology of the ovary. In: The Endocrine Function of the Human Ovary. VHT James, M Serio, G Giusti (eds), London: Academic Press, p 1–24, 1976.
28. Baca N, Zamboni L: The fine structure of human follicular oocytes. *J Ultrastruct Res* 19: 354–381, 1967.
29. Tesarik J, Dvorak M: Morphometric study of the occurrence of cytoplasmic structures in human and rat primordial oocytes. In: XXL Colloq Sci Fac Med Univ Carol et XIX Congr Morph Symp E Klika (ed), Prague: p 533, 1978.
30. Caggiati A, Familiari G, Marrapese M, Ermini M, Carenza L, Motta PM: Aspetti ultrastrutturali dell'ovaio in pazienti trattati con chemioterapia per linfomi Hodgkin e non-Hodgkin. *Fisiopat Rip* 3: 33–37, 1985.
31. Motta PM, Takeva S, Nesci E: Etude ulstrastructurale et histochimique des rapports entre les cellules folliculaires et l'ovocyte pendant le developpement du follicule ovarien chez les mammiferes. *Acta Anat* 80: 537–562, 1971.
32. Anderson E, Albertini DF: Gap junctions between the oocyte and companion follicle cells in the mammalian

ovary. *J Cell Biol* 71: 680–686, 1976.
33. Gilula NB, Epstein ML, Beers WH: Cell-to-cell communication and ovulation. A study of the cumulus–oocyte complex. *J Cell Biol* 78: 58–75, 1978.
34. Burghardt RC, Anderson E: Hormonal modulation of gap junctions in rat ovarian follicles. *Cell Tissue Res* 214: 181–193, 1981.
35. Bachvarova R, Baran M, Tejblum A: Development of naked growing mouse oocytes in vitro. *J Exp Zool* 211: 159–169, 1980.
36. Brower PT, Schultz RM: Intercellular communication between granulosa cells and mouse oocytes: existence and possible nutritional role during oocyte growth. *Dev Biol* 90: 144–153, 1982.
37. Kang Y: Development of the zona pellucida in the rat oocyte. *Am J Anat* 139: 535–566, 1974.
38. Hedrick JL, Fry GN: Immunocytochemical studies on the porcine zona pellucida. *J Cell Biol* 87: 143a, 1980.
39. Haddad A, Nagai MET: Radioautographic study of glycoprotein biosynthesis and renewal on the ovarian follicles of mice and the origin of the zona pellucida. *Cell Tissue Res* 177: 347–369, 1977.
40. Bleil JD, Wassarman PM: Structure and function of the zona pellucida: identification and characterization of the proteins of the mouse oocyte's zona pellucida. *Dev Biol* 76: 185–202, 1980.
41. Bousquet D, Léveillé MC, Roberts KD, Chapdelaine A, Bleau G: The cellular origin of the zona pellucida antigen in the human and hamster. *J Exp Zool* 215: 215–218, 1981.
42. Stegner HE, Wertemberg H: Elektronenmikroskopische und histopochemische Untersuchungen uber Struktur und bildung der zona pellucida menschlicher Eizellen. *Z Zellforsh* 53: 702–713, 1964.
43. Kang YH, Anderson WA, Chang SL, Ryan RJ: Studies on the structure of the extracellular matrices of the mammalian follicles as revealed by High Voltage Electron Microscopy and cytochemistry. In: Ovarian Follicular Development and Function. AR Midgley, WA Sadler (eds), New York: Raven Press, p 121–136, 1979.
44. Familiari G, Simongini E, Motta PM: The extracellular matrix of healthy and atretic mouse ovarian follicles studied by lanthanum nitrate and polycation ruthenium red. *Acta Histochem* 68: 193–207, 1981.
45. Familiari G, Magliocca FM, Macchiarelli G, Motta PM: I proteoglicani della zona pellucida in follicoli in sviluppo ed atresici. Studiò Ultrastrutturale. *Fisiopat Riprod* 3: 149–152, 1985.
46. Greve JM, Wassarman PM: Mouse egg extracellular coat is a matrix of interconnected filaments possessing a structural repeat. *J Mol Biol* 181: 253–264, 1985.
47. Familiari G, Correr S, Motta PM: Gap junctions in theca interna cells of developing and atretic follicles. In: Advances in Morphology of Cells and Tissues. E Vidrio, A Galina (eds), New York: Alan Liss, p 337–348, 1981.
48. Familiari G, Castellucci M, Motta PM: Structural features of follicular basement membrane in the mouse ovary. *Acta Anat* 120: 23, 1984.
49. Gonzales-Santander R, Clavero Nuez JA: The fine structure of the human oocyte. *Acta Anat* 84: 106–117, 1973.
50. Baker TG: Oogenesis and ovarian development. In: Reproductive Biology. H Balin, S Glasser (eds), Amsterdam: Excerpta Medica, p 398–437, 1972.
51. Sundstrom P, Nilsson BO, Liedholm P, Larsson E: Ultrastructural characteristics of human oocytes fixed at follicular puncture or after culture. *J In Vitro Fert Emb Transf* 2: 195–206, 1985.
52. Zamboni L: Fine morphology of the follicle wall and follicle cell–oocyte association. *Biol Reprod* 10: 125–149, 1974.
53. Sathananthan AH, U G, S g, Chia CM, Law HY, Edirisinghe WR, Ratman SS: The origin and distribution of cortical granules in human oocytes with reference to Golgi, nucleolar, and microfilament activity. In: In Vitro Fertilization and Embryo Transfer. M Seppala, RG Edwards (eds) *Ann N Y Acad Sci* 442: 251–264, 1985.
54. Suzuki S, Kitai H, Tojo R, Seki K, Oba M, Fujiwara T, Iizura R: Ultrastructural and some biologic properties of human oocytes and granulosa cells cultured in vitro. *Fertil Steril* 35: 142–148, 1981.
55. Szollosi W, Gerrard M: Cytoplasmic changes in the mammalian oocytes during the preovulatory period. In: Fertilization of the Human Egg in Vitro. HM Beier, HR Lindner (eds), Berlin: Springer Verlag, p 35–55, 1983.
56. Sathananthan AH: Maturation of the human oocyte in vitro: nuclear events during meiosis (an ultrastructural study). *Gamete Res* 12: 237–254, 1985.
57. Fawcett DW: The Cell. Philadelphia: WB Saunders, 1981.
58. Tesarik J, Dvorak M: Human cumulus oophorus preovulatory development. *J Ultrastruct Res* 78: 60–72, 1982.
59. Cho WK, Stern S, Biggers JD: Inhibitory effect of dibutyryl cAMP on mouse oocyte maturation in vitro. *J Exp Zool* 187: 383–386, 1974.
60. Schultz RM, Wassarman PM: Specific changes in the pattern of protein sythesis during meitoic maturation of mammalian oocytes in vitro. *Proc Natl Acad Sci USA* 74: 538-541, 1977.
61. Albertini DF, Anderson E: The appearance and structure of intercellular connections during the ontogeny of the rabbit ovarian follicle with particular reference to gap junctions. *J Cell Biol* 63: 234 –250, 1974.
62. Wassarman PM, Schultz RM, Letorneau GE: Protein synthesis during meiotic maturation of mouse oocytes in vitro. Synthesis and phosphorylation of a protein localized in the germinal vesicle. *Dev Biol* 69: 94–107, 1979.
63. Dekel N, Beers WH: Development of the rat oocyte in vitro: inhibition and induction of maturation in the presence or absence of the cumulus oophorus. *Dev Biol* 75: 247–254, 1980.
64. Schultz RM, Montgomery RR, Ward-Bailey PF, Eppig JJ: Regulation of oocyte maturation in the mouse: possible roles of intercellular communication, cAMP and testosterone. *Dev Biol* 95: 294–304, 1983.
65. Dekel N: Regulation of oocyte maturation: A paracrine role for cyclic AMP. Fifth World Congress on In Vitro Fertilization and Embryo Transfer. Norfolk, Virginia, ps-023, 1987.
66. Bornslaeger EA, Shultz RM: Regulation of mouse oocyte maturation: effect of elevating cumulus cell cAMP on

oocyte cAMP levels. *Biol Reprod* 33: 698–704, 1985.
67. De Felici M: Binding of flourescent lectins to the surface of germ cells from fetal early postnatal gonads. *Gamete Res* 10: 423–432, 1984.
68. Centola GM, Anderson LD, Channing CP: Oocyte maturation inhibitor (OMI) activity in porcine granulosa cells. *Gamete Res* 4: 451–461, 1981.
69. Iwamatsu T, Yanagimachi R: Maturation in vitro of ovarian oocytes of prepubertal and adult hamsters. *J Reprod Fertil* 45: 83–90, 1975.
70. Sorensen RA, Wassarman PM: Relationship between growth and meiotic maturation of the mouse oocyte. *Dev Biol* 50: 531–536, 1976.
71. Canipari R, Palombi F, Riminucci M, Mangia F: Early programming of maturation competence in mouse oogenesis. *Dev Biol* 102: 519–524, 1984.
72. Channing CP, Liu CQ, Jones GS, Jones H: Decline of follicular oocyte maturation inhibitor coincident with maturation and achievement of fertilizability of oocytes recovered at midcycle of gonadotropin treated women. *Proc Natl Acad Sci USA* 80: 4184–4188, 1983.
73. Hillensjo T, Brannstorm M, Chari S, Daume E, Magnusson C, Nilsson L, Sjogren A, Tornell J: Oocyte maturation as regulated by follicular factors. In: In Vitro Fertilization and Embryo Transfer. M Seppala, RG Edwards (eds), *Ann N Y Acad Sci* 442: 73–79, 1985.
74. McLachlan RI, Healy DL, Robertson DM, De Kretser DM, Burger HG: Inhibin: circulating levels in women during ovulation induction and detection in human placentae by specific radioimmunoassay. Fifth World Congress on In Vitro Fertilization and Embryo Transfer. Norfolk, Virginia, ps-019, 1987.
75. McNatty KP: Follicular fluid. In: Vertebrate Ovary, RE Jones (eds), New York: Plenum Press, p 215–259, 1978.
76. Yanagishita M, Hascall VC: Biosynthesis of proteoglycans by rat granulosa cells cultured in vitro. *J Biol Chem* 258: 12849–12856, 1979.
77. Yanagishita M, Hascall VC, Rodbard D: Biosynthesis of proteoglycans by rat granulosa cells cultured in vitro: Modulation by gonadotropins, steroid hormones, protaglandins and a cyclic nucleotide. *Endocrinology* 109: 1641–1649, 1981.
78. Grimek HJ, Bellin ME, Ax RL: Characteristics of proteoglycans isolated from small and large bovine ovarian follicles. *Biol Reprod* 30: 397–409, 1984.
79. Bellin ME, Ax RL: Chondroitin sulfate: an indicator of atresia in bovine follicles. *Endocrinology* 114: 428–434, 1984.
80. Lens RW, Ball GD, Lohse JK, First NL, Ax RL: Chondroitin sulfate facilitates an acrosome reaction in bovine spermatozoa as evidenced by light microscopy, electron microscopy and in vitro fertilization. *Biol Reprod* 28: 683–690, 1983.
81. Tesarik J, Kopecny V, Dvorak M: Selective binding of human cumulus cell-secreted glycoproteins to human spermatozoa during capacitation in vitro. *Fertil Steril* 41: 919–925, 1984.
82. Tesarik J, Kopency V: Late preovulatory synthesis of proteoglycans by the human oocyte and cumulus cells and their secretion into the oocyte–cumulus-complex extracellular matrices. *Histochemistry* 83: 523–528, 1986.
83. Motta PM: Sur l'ultrastructure des "corps de Call et Exner" dans l'ovaire du lapin. *Z Zellforsh Mikrosk Anat* 68: 308–319, 1965.
84. Zamboni L: Fine Morphology of Mammalian Fertilization. New York: Harper and Row, 1971.
85. Motta PM, Nesci E: The Call and Exner bodies of mammalian ovaries with reference to the problem of rosette formation. *Arch Anat Micr Morph Exp* 58: 283–290, 1969.
86. Anderson E, Wilkinson RF, See G, Meller S: A correlative microscopical analysis of differentiating Ovarian follicles of mammals *J Morphol* 156: 339–365, 1978.
87. Motta PM, Van Blerkom J: A scanning electron microscopic study of the luteo-follicular complex. I. Follicle and oocyte. *J Submicr Cytol* 6: 297–310, 1974.
88. Motta PM, Van Blerkom J: Structure and ultrastructure of ovarian follicles. In: Human Ovulation. ESE Hafez (eds), Amsterdam: Elsevier, North-Holland, p 17–38, 1979.
89. Motta PM, Makabe S: Morphodynamic changes of the mammalian ovary in normal and some pathological conditions. A scanning electron microscopic study. *Biomedl Res* 2s: 325–339, 1981.
90. Motta PM: Ovulation: A three-dimensional correlative analysis by scanning and transmission electron microscopy. In: Psychoneuroendocrinology in Reproduction. L Zichella, P Pancheri (eds), Amsterdam: Elsevier North-Holland Biomedical Press, p 145–155, 1979.
91. Makabe S: Scanning electron microscopy of normal and anovulatory human ovaries. Advances in the Morphology of Cell Tissues. EA Vidrio, MA Galina (eds), New York: Alan Liss Inc, p 321–330, 1981.
92. Amsterdam A, Koch Y, Siebermann E, Lindner HR: Distribution of binding sites for human chorionic gonadotrophin in the preovulatory follicle of the rat. *J Cell Biol* 67: 894–899, 1975.
93. Motta PM, Di Dio LJA: Microfilaments in granulosa cells during the development of the follicle and its transformation in corpus luteum in the rabbit ovary. J *Submicrosc* Cytol 6: 15–27, 1974.
94. Cavallotti C, DiDio LJA, Familiari G, Fumagalli G, Motta PM: Microfilaments in granulosa cells of rabbit ovary: immunological and ultrastructural observations. *Acta Histochem* 5: 253–256, 1975.
95. Batten EB, Anderson E: The distribution of actin in cultured ovarian granulosa cells. *Am J Anat* 167: 395–404, 1983.
96. Cran DG, Musk L: The distribution of actin in sheep ovaries. *J Exp Zool* 235: 375–380, 1985.
97. Ben Ze'ev A, Amsterdam H: Regulation of cytoskeletral proteins involved in cell contact formation during differentiation of granulosa cells on extracellular matrix. *Proc Natl Acad Sci USA* 83: 2894–2898, 1986.
98. Albertini DF, Fawcett DW, Olds PJ: Morphological variations in gap junctions of ovarian granulosa cells. *Tissue & Cell* 7: 389–405, 1975.
99. Fuksuhima M: Intercellular junctions in the human developing preovulatory follicle and corpus luteum. *Int J Fertil* 22: 206–216, 1977.

100. Familiari G, Toscano V, Motta PM: Morphological studies of polycystic mouse ovaries induced by dehydroepiandrosterone. *Cell Tissue Res* 240: 519–528, 1985.
101. Toshimori K, Oura C: Cellular interconnections in the young mouse ovary. *Cell Tissue Res* 224: 383–395, 1982.
102. Familiari G, Nottola SA, Motta PM: Focal cell contacts detected by ruthenium red, triton X 100 and Saponin in the granulosa cells of mouse ovary. *Tissue & Cell* 19: 207–215, 1987.
103. Albertini DF: Structural modifications of the granulosa cell plasma membrane during folliculogenesis. In: Biology of the Ovary. PM Motta ESE Hafez (eds), The Hague Boston: Martinus Nijhoff, p 138–149, 1980.
104. Hiura M, Fujita H: Electron microscopy of the cytodifferentiation of the theca cell in the mouse ovary. *Arch Istol Jap* 40: 95–105, 1977.
105. Hart DM, Baillie AH, Calman KC, Ferguson MM: Hydroxysteroid dehydrogenase development in the mouse adrenals and gonads. In: Developments in Steroid Histochemistry. New York: Academic Press, Chap 5, 1966.
106. Christensen AK, Gillim SW: The correlation of fine structure and function in steroidsecreting cells, with emphasis on those of the gonads. In: The Gonads. McKerns (ed), New York: Appleton-Century Crofts. p 415–488, 1969.
107. Channing CP, Coudert SP: Contribution of granulosa cells and follicular fluid to ovarian estrogen secretion in the rhesus monkey in vivo. *Endocrinology* 98: 590–597, 1976.
108. Moor RM: Sites of steroid production in ovine Graafian follicles in culture. *J Endocr* 73: 143–150, 1977.
109. Makris A, Ryan HJ: Progesterone, androstenedione, testosterone, estrone and estradiol synthesis in hamster ovarian follicle cells. *Endocrinology* 96: 694–701, 1975.
110. Fortune JE, Armstrong DT: Androgen production by isolated components of rat ovarian follicles. In: Ovarian Follicular Development and Function. AR Migdley, A Sadler (eds), New York: Raven Press, p 193–198, 1979.
111. Sporrong B, Kannisto P, Owman CH, Sjoberg NO, Walles B: Histochemistry and ultrastructure of adrenergic and acetylcholinesterase-containing nerves supplying follicles and endocrine cells in the guinea-pig ovary. *Cell Tissue Res* 240: 505–511, 1985.
112. Kanzaki O, Man I, Morikawa O, Nishimura T: Scanning Electron Microscopic observation of microcorrosion casts for rabbit ovarian vasculature; a preliminary study. *Acta Obst Gynaec JPN* 32: 422–426, 1980.
113. Fumagalli Z, Motta PM: Sulla presenza al Microscopio Elettronico di cellule muscolari lisce nell'ovaio di alcuni mammiferi. *Arch Ital Anat Embriol* 57: 1969.
114. Burden HW: Ultrastructural observations on ovarian perifollicular smooth muscle in the cat, guinea pig and rabbit. *Am J Anat* 133: 125–142, 1972.
115. Amsterdam A, Lindner HR, Groschal-Stewart U: Localization of actin and myosin in the rat oocyte and follicular wall by immunofluorescence. *Anat Rec* 187: 311 –328, 1977.
116. Amenta F, Allen DJ, Di Dio LJA, Motta PM: A transmission electron microscopic study of smooth muscle cells in the ovary of rabbits, cats, rats and mice. *J Submicrosc Cytol* 11: 39–51, 1979.
117. Pandergrass P, Talbot P: The distribution of contractile cells in the apex of the preovulatory hamster follicle. *Biol Reprod* 20: 205–213, 1979.
118. Motta PM, Familiari G: Occurrence of a contractile tissue in the theca externa of atretic follicles in the mouse ovary. *Acta Anat* 109: 103 –114, 1981.
119. Amenta F, Cavallotti C, Correr S, Familiari G: Innervazione adrenergica della componente endocrina dell'ovaio del topo. *Prog Med* 34: 175–179, 1978.
120. Walles B, Edvisson L, Falck B, Owman CH, Sjoberg UN, Svensson KG: Evidence for a neuromuscular mechanism involved in the contractility of the ovarian follicular wall: Fluorescence and Electron Microscopy and effects of tyramine on follicle strips. *Biol Reprod* 12: 239–248, 1975.
121. Moor RM, Warnes GM: Regulation of meiosis in mammalian oocytes. *Br Med Bull* 35: 99–100, 1979.
122. Bomsel-Helmreich O, Gougeon A, Thibault A, Saltarelli D, Milgrom E, Frydman R, Papiernik E: Healthy and atretic human follicles in the preovulatory phase: differences in evolution of follicular fluid. *J Clin Endocr Metab* 48: 686–694, 1979.
123. Ireland JJ, Roche JF: Development of nonovulatory antral follicles in heifers: changes in steroids in follicular fluid and receptors for gonadotropins. *Endocrinology* 112: 150–156, 1983.
124. Callesen H, Greene T, Hyttel P: Follicular steroids and ovarian development in superovulated dairy cattle. A preliminary report. J In Vitro Fert Em Transf 1: 101 a, 1984.
125. Liu HC, Sandow BA, Rosenwaks Z: Follicular fluid and oocyte maturation. In: In Vitro Fertilization. HW Jones Jr, G Seegar-Jones, GD Hodgen, Z Rosenwaks (eds), Norfolk: Pb Williams and Wilkins p 106–125, 1986.
126. Le Maire WJ, Janson PO, Kallfelt BJ, Holmes PV, Cajander S, Bjersing L, Ahren KEB: The preovulatory decline in follicular oestradiol is not required for ovulation in the rabbit. *Acta Endocrinol* 101: 452–458, 1982.
127. Tsafriri A, Bar-Ami S, Lindner HR: Control of the development of meiotic competence and of oocyte maturation in mammals. In: Fertilization of the Human Egg in Vitro. HM Beier, HR Lindner (eds), Berlin: Springer-Verlag, p 3–17, 1983.
128. Dvorak M, Tesarik J: Ultrastructure of human ovarian follicles. In; Biology of the Ovary. PM Motta, ESE Hafez (eds), The Hague/Boston: Martinus Nijhoff, p 121–137, 1980.
129. Okada A, Yanagimachi R, Yanagimachi H: Development of a cortical granule-free area of cortex and the perivitelline space in the hamster oocyte during maturation and following ovulation. *J Submicrosc Cytol* 18: 233–247, 1986.
130. Szollosi D, Mandelbaum J, Plachot M, Salat-Baroux J, Cohen, J: Ultrastructure of the human preovulatory oocyte. *J in Vitro Fert Emb Transf* 3: 232–242, 1986.
131. Sathananthan AH. Trounson AO: Ultrastructural observation on cortical granules in human follicular oocytes cultured in vitro. *Gamete Res* 5: 191–198, 1982.
132. Motta PM, Nottola SA, Familiari G, Micara G: Ultrastructure of human unfertilized oocytes and embryos in

an in vitro fertilization program. *Ann NY Acad Sci*, in press.

133. Phillips DM, Shalgi R: Surface architecture of the mouse and hamster zona pellucida and oocyte. *J Ultrastruct Res* 72: 1–12, 1980.
134. Ebensperger C, Barros C: Changes at the hamster oocyte surface from the germinal vesicle stage to ovulation. *Gamete Res* 9: 387–397, 1984.
135. Longo FJ, Chen DY: Development of surface polarity in mouse eggs. *Scanning Electron Microscopy* II: 23–36, 1984.
136. Wolf DE, Ziomek CA: Regionalization and lateral diffusion of membrane proteins in unfertilized and fertilized mouse eggs. *J Cell Biol* 96: 1786–1790, 1983.
137. Longo FJ, Chen DY: Development of cortical polarity in mouse eggs. Involvement of the meiotic apparatus. *Dev Biol* 107: 382–394, 1984.
138. Longo FJ: Fine structure of the mammalian egg cortex. *Am J Anat* 174: 303–315, 1985.
139. Suzuki F, Yanagimachi R: Freeze-fracture observations of ovulated hamster oocytes with their cumulus cells. *Cell Tissue Res* 231: 265–274, 1983.
140. Koehler JK, De Curtis I, Stenchever MA, Smith D: Interaction of human sperm with zona-free hamster eggs. A freeze-fracture study. *Gamete Res* 6: 371–386, 1982.
141. Koehler JK, Clark JM, Smith D: Freeze-fracture observations on mammalian oocytes. *Am J Anat* 174: 317–329, 1985.
142. Clark JM, Koehler JK, Smith WD: Freeze-fracture observations of unfertilized and fertilized hamster oocytes with special reference to the use of lipid probes. *Gamete Res* 14:129–147, 1986.
143. Albertini DF: Novel morphological approaches for the study of oocyte maturation. *Biol Reprod* 30: 13–28, 1984.
144. Hyttel P, Callesen H, Greene T: Ultrastructural features of preovulatory oocyte maturation in superovulated cattle. *J Reprod Fertil* 76: 645–656, 1986.
145. Hyttel P, Xu KP, Smith S, Greeve T: Ultrastructure of in vitro oocyte maturation in cattle. *J Reprod Fertil* 78: 615–625, 1986.
146. Zamboni L, Thompson PS, Moore-Smith D: Fine morphology of human oocyte maturation in vitro. *Biol Reprod* 7: 425–457, 1972.
147. Motta PM, Van Blerkom J: A scanning electron microscopic study of the luteo-follicular complex, II: Events leading to ovulation. *Am J Anat* 143: 241–264, 1975.
148. Sundstrom P: Interaction between spermatozoa and ovum "in vitro". In: Atlas of Human Reproduction by Scanning Electron Microscopy. ESE Hafez, P Kenemans (eds), Lancaster, Boston, The Hague: MTP Press Ltd, p 225 –230, 1982.
149. Nottola SA, Familiari G, Petrillo S, Micara G, Aragona L, Carenza L, Motta PM: La zona pellucida di ovociti umani sottoposti a fecondazione in vitro e la sua interazione con gli spermatozoi. Studio al microscopio elettronico a scansione. LXIV Congr Naz Soc It Ginec Ostet A cura di A Bompiani, L Carenza, B Salvadori, A Pachi (eds) Monduzzi, Bologna, p 2091–2096, 1986.
150. Familiari G, Nottola SA, Petrillo S, Micara G, Aragona L, Motta PM: The application of electron microscopy in the evaluation of in vitro unfertilized human oocytes. In: Morphological Basis of Human Reproductive Function. G Spera, DM de Kretser (eds), New York, London: Plenum Press, and Rome: Acta Meidca, p 99–104, 1987.
151. Bjersing L, Cajander S: Ovulation and the mechanism of follicle rupture. IV. Ultrastructure of membrane granulosa of rabbit Graafian follicles prior to induced ovulation. *Cell Tissue Res* 153: 1–14, 1974.
152. Dekel N, Kraicer PF: Induction in vitro of mucification of rat cumulus oophorus by gonadotropins and cAMP. *Endocrinology* 102: 1797–1802, 1980.
153. Cajander S, Janson PO, Lemaire WJ, Kallfelt BJ, Holmes PV, Ahrén K, Bjersing L: Studies on the morphology of the isolated perfused rabbit ovary. I° Effect of long-term perfusion. *Cell Tissue Res* 235: 59–63, 1984.
154. Cajander S, Janson PO, Le Maire WJ, Kallfelt BJ, Holmes PV, Ahrén K, Bjersing L: Studies on the morphology of the isolated perfused rabbit ovary. II. Ovulation in vitro after HCG treatment in vivo. *Cell Tissue Res* 235: 565–573, 1984.
155. Wallach EE, Okuda Y, Kanzaky Y, Okamura H, Santulli R, Wright KH: Ultrastructure of ovarian follicles "in vitro" perfused rabbit ovaries: response to human chorionic gonadotropin and comparison with in vivo observations. *Fertil Steril* 42: 127–133, 1984.
156. Mestwerdt W: Die Follikel. Granulosazellen in Beziehung zur SteroidBiosynthese in der Periovulationsphase. *Fortschr Med* 95: 361–368, 1977.
157. Makabe S, Kaneko Y, Kojima E, Omura G, Momose K. The human granulosa cell changes during luteogenesis viewed by scanning and transmission electron microscopy. In: In Vitro Fertilization, Embryo Transfer and Early Pregnancy. RF Harrison, J Bonnar, W Thompson (eds), Lancaster, Boston, The Hague: MTP Press, p 33 –37, 1983.
158. Nagai N: Electron microscopic study on the structure and the function of the granulosa cell in the human ovary. Ultrastructure and 3-beta hydroxysterod dehydrogenase activity. *Nippon Sanka Fujinka Gakkai Zasshi* 37: 871 –878, 1985.
159. Rotmesh S, Dor J, Furman A, Rudak E, Mashiach S, Amsterdam A: Ultrastructural characterization of human granulosa cells in stimulated cycles: correlation with oocyte fertilizability. *Fertil Steril* 45: 671–679, 1986.
160. Delforge JP, Thomas K, Roux F, Carneiro de Siqueira J, Ferin J: Time relationships between granulosa cell growth and luteinization, and plasma luteinizing hormone discharge in humans. I. A Morphometric analysis. *Fertil Steril* 23: 1–7, 1972.
161. Di Zerega GS, Marrs RP, Campeau JD, and Kling OR: Human granulosa cell secretion of proteins which suppress follicular response to gonadotropins. *J Chem* 56: 147–153, 1983.
162. Beers WH, Strickland S, Reich E: Ovarian plasminogen activator: relationship to ovulation and hormonal regulation. *Cell* 6: 387–394, 1975.
163. Beers WH: Follicular plasminogen and plasminogen activator and the effect of plasmin on ovarian follicle wall. *Cell* 6: 379–386, 1975.
164. Shimada H, Mori T, Takada A, Takada Y, Noda Y,

Takai I, Kohda H, Nishimura T: Use of chromogenic substrate S-2251 for determination of plasminogen activator in rat ovaries. *Thromb Haemostasis* 46: 507–510, 1981.
165. Espey LL, Shimada H, Okamura H, Mori T: Effect of various agents on ovarian plasminogen activator activity during ovulation in pregnant mare's serum gonadotropin-primed immature rats. *Biol Reprod* 32: 1087–1094, 1985.
166. Solomkin JS, Simmons RL: Cellular and subcellular mediators of acute inflammation. *Surg Clin North Amer* 63: 225–243, 1983.
167. Valinsky JE, Reich E, Le Dourain NM: Plasminogen activator in cell migration and morphogenesis in the bursa of fabricius. In: Embryonic Development, Part A: Genetic Aspects. New York: Alan Liss Inc, p 473–480, 1982.
168. Espey LL: Cycloheximide inhibition of ovulation, prostaglandin biosynthesis and steroidogenesis in rabbit ovarian follicles. *J Reprod Fert* 78: 679–683, 1986.
169. Byskov AGS: Ultrastructural studies on the prevoulatory follicle in the mouse ovary. *Z Zellforsh* 100: 285 –299, 1969.
170. El-Fouly MA, Cook B, Nekola M, Nalbandov AV: Role of the ovum in follicular luteinization. *Endocrinol* 87: 288–297, 1970.
171. Dekel N, Phillips DM: Maturation of the rat cumulus oophorus. A Scanning Electron Microscopic study. *Biol Reprod* 21: 9–18, 1979.
172. Nicosia SV, Mikhail G: Cumuli oophori in tissue ultrastructure: hormone production, ultrastructure and morphometry of early luteinization. *Fertil Steril* 26: 427–448. (1975)
173. Talbot P, DiCarlantonio G: Architecture of the hamster oocyte cumulus complex. *Gamete Research* 9: 261–272, 1984.
174. Talbot P, Di Carlantonio G: The oocyte cumulus complex: Ultrastructure of the extracellular components in hamsters and mice. *Gamete Research* 10: 127–142, 1984.
175. Hyttel PJ: Bovine cumulus–oocyte disconnection in vitro. *Anat Embryol* 176: 41–44, 1987.
176. Salustri A, Siracusa G: Metabolic coupling, cumulus expansion, and meiotic resumption in mouse cumuli oophori cultured in vitro in the presence of FSH, of cAMP, or stimulated in vivo by HGC. *J Reprod Fertil* 68: 335–341, 1983.
177. Larsen J, Wert E, and Brunner D: A dramatic loss of cumulus cell gap junctions is correlated with germinal vesicle breakdown in rat oocytes. *Dev Biol* 113: 517–521, 1986.
178. Tsafriri A, Channing CP: An inhibitory influence of granulosa cells and follicular fluid upon porcine oocyte meiosis in vitro. *Endocrinology* 96: 922–927, 1975.
179. Tsafriri A, Channing CP: Influence of follicular maturation and culture condition on the meiosis of pig oocytes in vitro. *J Reprod Fertil* 43: 49–152, 1975.
180. Leibfried L, First NL: Follicular control of meiosis in the porcine oocyte. *Biol Reprod* 23: 705–709, 1980.
181. Bjersing L, Cajander S: Ovulation and the mechanism of follicle rupture. IV. Ultrastructure of theca interna and the inner vascular network surrounding rabbit Graafian follicles prior to induced ovulation. *Cell Tissue Res* 153, 31–77, 1974.
182. Otsuki Y, Magari S, Sugimoto O: Lymphatic capillaries in rabbit ovaries during ovulation: an ultrastructural study. *Lymphology* 19: 55–64, 1986.
183. Parr EL: Histological examination of the rat ovarian follicle wall prior to ovulation. *Biol Reprod* 11: 483–503, 1974.
184. Cherney DD, Di Dio LJA, Motta PM: The development of the rabbit ovarian follicles following copulation. *Fertil Steril* 26: 257–270, 1975.
185. Espey LL: Ovulation, In: The Vertebrate Ovary. RE Jones (eds), New York: Plenum Press, p 503–526, 1978.
186. Reed M, Burton FA, Van Diest PA: Ovulation in the guinea-pig. I. The ruptured follicle. *J Anat* 128: 195–206, 1979.
187. Kitai H, Yoshimura Y, Wright KH, Santulli R, Wallach EE: Microvasculature of preovulatory follicles. Thirty-First Annual Meeting of Society for Gynecologic Investigation, San Francisco, California, p 127, 1984.
188. Bjersing L, Cajander S: Ovulation and the mechanism of follicle rupture. V. Ultrastructure of tunica albuginea and theca externa of rabbit Graafian follicles prior to induced ovulation. *Cell Tissue Res* 153: 15–30, 1974.
189. Bell C: Autonomic nervous control of reproduction: Circulatory and other factors. *Pharm Rev* 24: 657–736, 1972.
190. Zuckerman S, Weir BJ: The Ovary 2nd edition. Vol 1. New York: Academic Press, 1977.
191. Espey LL: Ovarian contractility and its relationship to ovulation. A review. *Biol Reprod* 19: 540–551, 1978.
192. Di Dio LJA, Allen DJ, Correr S, Motta PM: Smooth musculature in the ovary. In: Biology of the Ovary. PM Motta, ESE Hafez (eds), The Hague/Boston: Martinus Nijhoff, p 106–118, 1980.
193. Talbot P, Chacon RS: In vitro ovulation of hamster oocytes depends on contraction of follicular smooth muscle cells. *J Exp Zool* 224: 409–415, 1982.
194. Martin G, Van Steenwyk G, Miller Walker C: The fate of thecal smooth muscle cells in postovulatory hamster follicles. *Anat Rec* 207: 267–277, 1983.
195. Owman C, Sjobert NO, Swesson KG, Walles B: Autonomic nerves mediating contractility in the human Graafian follicle. *J Reprod Fertil* 45: 553–556, 1975.
196. Motta PM, Charney DD, Di Dio LJA: Scanning and transmission electron microscopy of the ovarian surface in mammals with special reference to ovulation. *J Submicr Cytol* 3: 85–100, 1971.
197. Martin GG, Miller-Walker C: Visualization of three-dimensional distribution of collage fibrils over preovulatory follicles in the hamsters. *J Exp Zool* 225: 311–319, 1983.
198. Cajander S, Bjersing L: Fine structural demonstration of acid phosphatase in rabbit germinal epithelium prior to induced ovulation. *Cell Tissue Res* 164: 279–289, 1975.
199. Cajander S, Bjersing L: Further studies of the surface epithelium covering preovulatory rabbit follicles with special reference to lysosomal alterations. *Cell Tissue Res* 169: 129–141, 1976.

200. Strickland S, Beers W: Studies of the enzymatic basis and hormonal control of ovulation. In: Ovarian Follicular Development and Function. AR Migdley, WA Sadler (eds), New York: Raven Press, p 143–153, 1979.
201. Espey LL: Ovarian proteolytic enzymes and ovulation. *Biol Reprod* 10: 216–235, 1974.
202. Okamura H, Takenaka A, Yajima Y, Morikawa H, Nishimura T: An ultrastructural study of ovulatory changes in stromal tissue of the follicular apex in the human. *Acta Obst Gynaec Jap* 32: 53–60, 1980.
203. Espey LL: Ultrastructure of the apex of the rabbit Graafian follicle during the ovulatory process. *Endocrinology* 81: 267–276, 1967.
204. Bjersing L, Cajander S: Ovulation and the mechanism of follicle rupture. III. Transmission electron microscopy of rabbit germinal epithelium prior to induced ovulation. *Cell Tissue Res* 149: 313–327, 1974.
205. Espey LL: Ovarian proteolytic enzymes and ovulation. *Biol Reprod* 10: 216–235, 1974.
206. Espey LL: Multivesicular structures in proliferating fibroblasts of rabbit ovarian follicles during ovulation. *J Cell Biol* 48: 437–442, 1971.
207. Okamura H, Takenaka A, Yajima Y, Nishimura T: Ovulatory changes in the wall at the apex of the human Graafian follicle. *J Reprod Ferti* 58: 153–155, 1980.
208. Willoughby DA, Sedgwick A, Edwards J: The inflammatory response. In: Marker Proteins in Inflammation. RC Allen, J Bienvenu, P Laurent, RM Suskind (eds), New York: Walter de Gruyter Press, p 45–48, 1982.
209. Morales TI, Woessner JF Jr, Marsh JM, Le Maire WJ: Collagen, collagenase and collagenolytic activity in rat Graafian follicles during follicular growth and ovulation. *Biochim Biophys Acta* 756: 119–122, 1983.
210. Reich R, Tsafriri A, Mechanic GL: Studies on the involvement of collagenase and of plasmin in ovulation. *Biol Reprod* 30 (suppl 1): 93, 1983.
211. Ichikawa S, Ohta M, Morioka H, Murao S: Blockage of ovulation in the explanted hamster ovary by a collagenase inhibitor. *J Reprod Fertil* 68: 17–19, 1983.
212. Reich R, Tsrafriri A, Mechanic GL: The involvement of collagenolysis in ovulation in the rat. *Endocrinology* 116: 522–527, 1985.
213. Curry TE Jr., Dean DD, Woessner JF Jr, Le Maire WJ: The extraction of a tissue collagenase associated with ovulation in the rat. *Biol Repord* 33: 981–991, 1985.
214. Espey LL: Ovulation as an inflammatory reaction. A hypothesis. *Biol Reprod,* 22: 73–106, 1980.
215. Cajander S: Structural alterations of rabbit ovarian follicles after mating with special reference to the overlying surface epithelium. *Cell Tissue Res* 173: 437–449, 1976.
216. Blandau RJ: Anatomy of ovulation. *Clin Obstet Gyecol* 10: 347–360, 1967.
217. Pendergrass B, Reber M: Scanning electron microscopy of the Graafian follicle during ovulation in the golden hamster. *J Reprod Fertil* 59: 21–24, 1980.
218. Van Blerkom J, Motta PM: A scanning electron microscopic study of the luteo-follicular complex. III. Formation of the corpus luteum and repair of the ovulated follicle. *Cell Tissue Res* 189: 131–154, 1978.
219. Osterholzer HO, Johnson JH, Nicosia SV: An autoradiographic study of rabbit ovarian surface epithelium before and after ovulation. *Biol Reprod* 33: 729 –738, 1985.
220. Makabe S, Hefez ESE, Motta PM: The ovary and ovulation. In: Atlas of Human Reproduction by Scanning Electron Microscopy. ESE Hafez, P Kenemans (eds), Lancaster: MTP Press, p 135–144, 1982.
221. Nicosia SV: Cytological analysis of in vivo and in vitro luteinization. In: Endocrine Physiopathology of the Ovary. R Tozzin, G Reeves, RL Pineda (eds), Amsterdam: Elsevier North-Holland, p 101–119, 1980.
222. Yuh KC, Possley RM, Brabec RK, Keyes PL: Steroidogenic and morphological characteristics of granulosa and thecal compartments of the differentiating rabbit corpus luteum in culture. *J Reprod Fertil* 76: 267–277, 1986.
223. Soto EA, Kliman HJ, Strauss JF, Paavola LG: Gonadotropins and cyclic adenosine 3′, 5′ — monophosphate (cAMP) alter the morphology of cultured human granulosa cells. *Biol Reprod* 34: 559–569, 1986.
224. Theodosis DT, Wooding FB, Sheldrick EL, Flint AP: Ultrastructural localization of oxytocin and neurophysin in the bovine corpus luteum. *Cell Tissue Res* 243: 129–135, 1986.
225. Wathes DC, Swann RW: Is oxytocin an ovarian hormone? *Nature* (London) 297: 225–227, 1982.
226. Wakins WB:Immunohistochemical localization of neurophysin and oxytocin in sheep corpora lutea. *Neuropeptides* 4: 51–54, 1983.
227. Guldenaar SEF, Wathes DC, Pickering BT: Immunocytochemical evidence for the presence of oxytocin and neurophysin in the large cells of the bovine corpus luteum. *Cell Tissue Res* 237: 349–352, 1984.
228. Sheldrick EL, Flint APF: Circulating concentrations of oxytocin during the estrous cycle and early pregnancy in sheep. *Prostaglandins* 22: 631–636, 1981.
229. Shams D, Prokopp S, Barth D: The effect of active and passive immunization against oxytocin on ovarian cyclicity in ewes. *Acta Endocr Copenh* 103: 337–344, 1983.
230. Vaupel MR, Sherwood OD, Anderson MB: Immunocytochemical studies of relaxin in ovaries of pregnant and cycling mice. *J Histochem Cytochem* 33: 303–308, 1985.
231. Fields PA, Fields MJ: Ultrastructure localization of relaxin in the corpus luteum of the nonpregnant, pseudopregnant and pregnant pig. *Biol Reprod* 32: 1169–1179, 1985.
232. Henderson KM: Gonadotrophic regulation of ovarian activity. *Brit Med Bull* 35: 161–166, 1979.
233. Schmidt G, Owman CH, Sjoberg NO: Histamine induces ovulation in the isolated perfused rat ovary. *J Reprod Fertil* 78: 159–166, 1986.
234. Le Maire WJ, Yang NST, Behrman HH, Marsh JM: Preovulatory changes in the concentration of prostaglandins in rabbit Graafian follicles. *Prostaglandins* 3: 367 –376, 1973.
235. Marsh JM, Le Maire WJ: The role of cyclic AMP and prostaglandins in the actions of luteinizing hormone. In: Gonadotropins and Gonadal Function. NR Moudgal (eds), New York: Academic Press, p 376–390, 1973.
236. Rigler GL, Peake GT, Rainer A: Effects of follicle sti-

mulating hormone and luteinizing hormone on ovarian cyclic AMP and prostaglandin E. In vivo in rats treated with indomethacin. *J Endocrinol* 70: 285–291, 1976.

237. Zor U, Strulovici B, Nimrod A, Lindt HR: Stimulation by cyclic nuceleotids of prostaglandin E production in isolated Graafian follicles. *Prostaglandins* 14: 947–959, 1977.
238. Yang NST, Marsh JM, Le Maire WJ: Post ovulatory changes in the concentration of prostaglandis in rabbit Graafian follicles. *Prostaglandins* 6: 37–44, 1974.
239. Zor U, Lamprecht SA: Mechanisms of prostaglandin action in endocrine glands. In: Biochemical Actions of Hormones, Vol 4. G Litwack (ed), New York: Academic Press, p 85–133, 1977.
240. Espey LL, Coons P, Marsh J, Le Maire WJ: Effect of indomethacin on preovulatory changes in the ultrastructure of rabbit Graafian follicles. *Endocrinology* 108: 1040 –1048, 1981.
241. Martin GG, Talbot P: Drugs that block smooth muscle contraction inhibit in vivo ovulation in hamsters. *J Exp Zool* 216: 483–491, 1981.
242. Downs SM, Longo FJ: An ultrastructural study of preovulatory apical development in mouse ovarian follicles: effects of indomethacin. *Anat Rec* 205: 159–168, 1983.
243. Ischikawa SE, Saito T, Yoshida S: The effect of prostaglandins on the release of arginine vasopressin from the guinea pig hypothalamo-neurohyphophyseal complex in organ culture. *Endocrinology* 108: 193–198, 1981.
244. Cheesman DW, Schlegel R, Sagasay AM, Forsham PH: Anovulatory effect of synthetic analogs of arginine vasotocin in the rat. *Endocrinology* 112: 267–269, 1983.
245. Punnonen RH, Viinamaki O, Kujansun ET, Heinonen PK, Pystynen PP: A possible role of vasopressin in the control of ovarian activity. In: In Vitro Fertilization and Embryo Transfer. M Seppala, RG Edwards (eds), *Ann N Y Acad Sci* 442: 236–239, 1985.
246. Wathes DC, Guldenaar SEF, Swann RW, Webb R, Porter DG, Pickering BT: A combined radioimmunoassay and immunocytochemical study of ovarian oxytocin production during the periovulatory period in the ewe. *J Reprod Fertil* 78: 167–183, 1986.
247. Flint APF, Sheldrick EL: Ovarian secretion of oxytocin is stimulated by prostaglandin. *Nature* 297: 587–588, 1982.
248. Balboni GC, Vannelli GB: La relaxine dans les follicules de l'ovaire. *Bull Ass Anat* 67: 149–162, 1983.
249. Bryant-Greenwood GD, Too CK, Koai E, Greenwood FC: Relaxin as a local hormone in ovulation and parturition. 1st Int Conference on Human Relaxin, Florence, 1982.
250. Westergaard L, Sinosich MJ, Grudzinskas JG, Boeton T, McNatty KP, Saunders DM, Teisner B, Westergaard J: Pregnancy-associated plasma protein-A in preovulatory and nonovulatory healthy and atretic human ovarian follicles during the natural cycle. In: In Vitro Fertilization and Embryo Transfer. M Seppala, RG Edwards (eds), *Ann NY Acad Sci* 442: 205–211, 1985.
251. Sinosich MJ, Saunders DM, Grudzinskas JG: Pregnancy-associated Plasma Protein A and placental Protein 5 in human ovarian follicular fluid. In: In Vitro Fertilization and Embryo Transfer. M Seppala, RG Edwards (eds), *Ann NY Acad Sci* 442: 269–275, 1985.
252. Imoedemhe D, Shaw RW: Follicular fluid alpha, 1 antitrypysin — correlation with fertilizing capacity of oocytes. *Br J Obstet Gynaeco* 93: 863–868, 1986.
253. Sinosich MJ, Westergaard LW, Teismer B, Westergaard JD, Grundizinskas JG, Saunders DM: Pregnancy associated plasma protein A in human ovarian follicular fluid. In: Proceedings of XXXII Colloquium on Protides of Biological Fluids. Oxford: Pergamon Press, p 259–262, 1984.
254. Bagdasarian A, Wheeler J, Stewart GJ, Ahmed SS, Colman RW: Isolation of alpha 1 — protease inhibitor from human normal malignant ovarian tissue. *J Clin Invest* 67: 281–291, 1981.
255. Sinosich MJ, Porter R, Sloss P, Bonifacio MD, Saunders DM: Pregnancy associated plasma protein A in human follicular fluid. *J Clin Endocrino Metab* 58: 500–504, 1984.
256. Jewelewicz R: Management of infertility resulting from anovulation. Am J Obstet Gynecol 122: 309–316, 1975.
257. Temmerman M, Devroey P, Naaktgeboren N, Amy JJ, Van Steirteghem AC: Incidence, recurrence and treatment of the luteinized unruptured follicle syndrome. *Acta Eur Fertil* 15: 179–316, 1984.
258. Hamilton CJCM, Wetzels LCG, Evers JLH, Hoogland HJ, Mujtjens A, De Haan J: Follicle growth curves and hormonal patterns in patients with the luteinized unruptured follicle syndrome. *Fertil Steril* 43: 541–548, 1985.
259. Schenken RS, Werlin LB, Williams RF, Prihoda TJ, Hodgen GD: Histologic and hormonal documentation of the luteinized unruptured follicle syndrome. *Am J Obstet Gynecol* 154: 839–847, 1986.
260. Coulam CB, Hill LM, Breckle R: Ultrasonic evidence for luteinization of unruptured preovulatory follicles. *Fertil Steril* 37: 524–531, 1982.
261. Gibbons WE, Buttram VC Jr, Rossavik IK: The observed incidence of luteinized unruptured follicles in a population of infertile women undergoing ovulation monitoring by ultrasound. *Fertil Steril* 41: 19–27, 1984.
262. Nottola SA, Micara G, Morgia F, Familiari G: Osservazioni ultrastrutturali sulle cellule della granulosa che circondano ovociti umani non fecondati dopo inseminazione "in vitro". 42° Congresso Nazionale Società Italiana di Anatomia, Siena, 27 Settembre–1 Ottobre 1987. Abstract book, p 145, 1987.
263. Nottola SA, Familiari G, Micara G, Aragona C, Motta PM: On the structure and function of the cumulus-corona cells surrounding human oocytes and polypronuclear ova. VIII International Symposium on Morphological Sciences, Rome, 10–15 July 1988. Quaderni di Anatomia Pratica, Serie XLIV: 288, 1988.

Author's Address:
Professor G. Familiari
Department of Anatomy
University of Rome "La Sapienza"
Via Borelli 50
00161 Rome, Italy

CHAPTER 5

Developmental failure in human reproduction associated with preovulatory oogenesis and preimplantation embryogenesis

JONATHAN VAN BLERKOM

1. Early developmental failure — an overview

The rapid and worldwide application of in vitro fertilization (IVF) for the treatment of human infertility has presented the opportunity to examine, by noninvasive (1) and invasive protocols (2,3), molecular and cellular aspects of normal and abnormal oocyte and early embryonic development in large and clinically varied patient populations. The collective findings from thousands of attempts at IVF demonstrate that the vast majority of fertilized human eggs do not result in a pregnancy and that most embryos appear to arrest development during the pre- and early postimplantation stages of gestation (4,5).

The extent to which maternal factors account for developmental failure during early embryogenesis is difficult to determine precisely. Clearly, the uterine milieu may be incapable of supporting the initiation or maintenance of an early pregnancy because of disease or inadequate hormonal preparation. However, after failure to achieve pregnancy during a stimulated (IVF) cycle, pregnancies have been initiated at a relatively high frequency with cryopreserved embryos transferred to the uterus during subsequent unstimulated cycles (Chapter 6, this volume; 6). This observation suggests that for some women, morphophysiological and/or biochemical conditions within the female reproductive tract during stimulated cycles may be inconsistent with the establishment of pregnancy.

On the other hand, there are also inherent defects in the oocyte that lead to the failure of meiotically mature oocytes to fertilize and to the failure of grossly normal-appearing eggs to progress through the preimplantation stages. Determinations of the developmental ability of human oocytes from hyperstimulated ovaries in which fertilization and early embryogenesis occurred entirely in vitro (4,5,7,8) or in vivo (9) demonstrate an increased developmental failure between the early cleavage and hatched blastocyst stages (see Chapter 7, this volume).

One explanation for in vitro developmental failure is that our present understanding of the physical and nutritional requirements necessary to support the normal development of the preimplanation stage human embryo is inadequate. Consequently, current conditions of culture and formulations of media may allow only limited development. In the same sense, ovarian hyperstimulation by the administration of drugs (clomiphene citrate) and gonadotropic hormones (FSH, LH) may alter the normal pattern of biochemical and morphological changes (10) such that the milieu of the reproductive system is incapable of supporting embryonic development.

One approach to the study of early developmental failure in the human is to consider the ability of an oocyte to be fertilized or of an embryo to progress through the preimplantation stages to be inherently limited. Developmental failure or aberration may be a consequence of a prior perturbation in the normal sequence of chromosomal maturation and molecular and cytoplasmic differentiation of the oocyte or newly fertilized egg (11–22). This notion is difficult to test directly because to obtain unambiguous results often requires invasive analyses (e.g., fixation for microscopy, homogenization for biochemistry, etc.). While such an approach may not be lethal to the oocyte or embryo (e.g., isotopic pulse-labeling, embryo dissection), it may preclude or compromise continued development.

The selection of oocytes and embryos for analysis is critical in a clinical situation. Each oocyte and embryo must have sufficient time to express its full developmental potential. The element of time is especially important in those cases where mature oocytes do not fertilize at the same approximate time, or where fertilized eggs do not develop in synchrony. Embryos dividing more slowly are not necessarily developmentally impaired and therefore not necessarily the most appropriate candidates for analysis. Culturing oocytes and embryos under suboptimal conditions may lead to defects unrelated to and possibly obscuring the original cause(s) of developmental failure.

The intent of this chapter is twofold: (1) to discuss deviations from the normal pattern of preovulatory maturation and early human embryogenesis that may provide clues as to the causes of developmental failure that may be inherent to the oocyte and early embryo, as contrasted with those that may originate from external influences (i.e., culture conditions and media, maternal factors); and (2) to indicate and discuss specific experimental approaches to the analysis of early developmental failure in the human. With the aforementioned caveats taken into account concerning the nature of experimentation that can be systematically undertaken in a clinical setting, this chapter focuses on the following phenomena related to the reproductive process: (1) developmental failure of human oocytes and early preimplantation-stage embryos of a nonchromosomal origin; (2) the developmental potential of chromosomally aberrant oocytes and embryos; and (3) the developmental potential of morphologically anomolous oocytes and embryos. Collectively, the findings presented in this chapter lead to the conclusion that even under controlled and defined conditions, developmentally viable human embryos may not be produced in the majority of occasions when sperm and oocyte meet.

1.1. Developmental failure at fertilization

Frequencies of fertilization of mature preovulatory human oocytes in excess of 80% are currently reported by most experienced IVF programs (4, 5). This baseline rate of fertilization appears to be generally independent of culture media formulation or of the hormonal protocols employed for ovarian hyperstimulation (23–25). The baseline frequency of fertilization observed at Reproductive Genetics In Vitro between 1982 and early 1987 has been 94% (2). Although this figure is relatively high, it does not necessarily indicate that the oocytes obtained from each patient have an equally high probability of being fertilized. The frequency of fertilization for any IVF program is a value based on aggregate results of many attempts at fertilization. However, the results for any individual patient can range from 0 to 100%. While this distribution is obvious, it underscores the basic observation that situations occur where none of the apparently normal preovulatory oocytes are fertilized. The failure to fertilize may involve sperm dysfunction due to an identifiable physiological condition (oligospermia, inadequate motility, etc.) or morphophysiological state (structural abnormalities; 26). Likewise, gross aberrations in the process of meiotic (chromosomal) maturation of the preovulatory oocyte (5,27–29) may be associated with subtle perturbations in the normal progression of molecular and cellular differentiation that determines the fertilizability of an oocyte (15–18,22,30). By contrast, the causes of fertilization failure are more obscure when mature oocytes that appear morphologically and chromosomally normal fail to fertilize in the presence of sperm of known fertility. The analyses presented in the following paragraphs are derived from observations of oocytes and newly fertilized eggs of patients treated at Reproductive Genetics In Vitro between September 1982 and January 1987. The discussions of oocyte-related causes of fertilization failure involve only those cases where a sperm factor was excluded and where oocytes appeared chromosomally normal after staining with DNA-specific probes.

1.2. *Cortical and plasma membrane dysfunction*

The exocytosis of cortical granules (cortical reaction) is the primary mechanism by which the penetration of the oocyte by more than one spermatozoon is prevented. The dispersal of cortical granule contents into the perivitelline space is followed by a chemical alteration of the zona pellucida (zona reaction) that entraps spermatozoa in the zona matrix and therefore inhibits the continued passage of accessory sperm to the surface of the oocyte (10,31–34). As discussed in greater detail later in this chapter, a slow or incomplete cortical/zona reaction may be the most common cause of polyspermic penetration in otherwise normal oocytes (33–35). A premature cortical granule discharge can also be the basis for the failure of fertilization owing to the incomplete passage of spermatozoa though the zona pellucida.

The inability of an oocyte to respond to the presence of a spermatozoon by initiating a cortical reaction is another potential cause of the failure of fertilization. Electron microscopic observations of 18 mature but unpenetrated oocytes (from four patients) demonstrated the presence of at least one spermatozoa in proximity to the oocyte plasma membrane (Fig. 5-1). Indeed, in some of these oocytes, multiple spermatozoa were present in the perivitelline space. Two features common to all 18 oocytes were the absence of any indication of (1) cortical granule discharge or (2) membranous interactions between oocyte and spermatozoon (Fig. 5-1). The four patients from which these oocytes were obtained all had a history of normal ovulatory cycles, patent fallopian tubes, multiple attempts at intrauterine insemination on normal and stimulated cycles, and diagnosis of infertility attributed to idiopathic causes.

The failure of fertilization to occur with these oocytes cannot be attributed to the absence of a spermatozoon in the perivitelline space. Rather, a defect in the cortical cytoplasm or plasma membrane of the oocyte, or both, may preclude the binding of the sperm to the oocyte surface and therefore the juxtaposition of the gametes necessary for sperm penetration. The presumed inability of the oocyte plasma membrane to interact with a spermatozoon may be the result of absent or nonfunctional cell surface glycoproteins or microvilli that normally recognize and interact with sperm-associated molecules (36). The cortical cytoplasm may be defective in its ability to respond to the presence of sperm by failing to initiate the normal progression of morphodynamic changes associated with the initial stages of fertilization — namely, the exocytosis of cortical granules and the interaction between sperm and oocyte plasma membranes that promotes penetration of the sperm head into the oocyte cytoplasm (35).

Although the possible causes of fertilization failure in patients exhibiting perivitelline spermatozoa but no sperm penetration are relatively easy to envisage, direct proof is more difficult to obtain because analytical methods would be invasive and might preclude fertilization. The hypothesis that dysfunction at the cortical cytoplasmic or plasma membrane levels, or both, is the direct and immediate cause of the failure to fertilize may not be unreasonable to test in the mature oocytes of specific patients whose oocytes consistently fail to fertilize in vitro, but where perivitelline spermatozoa can be recognized by high resolution light microscopy. A fundamental issue in such an approach is the question of what cortical and plasma membrane characteristics should be examined.

Findings from the analysis of fully grown, in-vitro-matured mouse oocytes and newly fertilized eggs may be of significant relevance to the human situation, insofar as they indicate a sequence of stage-related changes in membrane fluidity (37) and in the distribution of cortical microfilaments, microvilli and cell surface glycoproteins (19,20, 38–41). Collectively, these changes describe a series of differentiative events that are related to developmental viability. For example, the region of the mouse oocyte plasma membrane that overlies the first and second meiotic metaphase spindles is characterized by a depopulation of microvilli and certain classes of glycoproteins (19, 20,38). The cortical cytoplasm associated with this region exhibits a highly localized and marked thickening of microfilaments (19,20,40). The observed differentiation of the cortical cytoplasm and plasma membrane specifically localized to the first and second meiotic metaphase spindles probably establishes conditions permissive for the

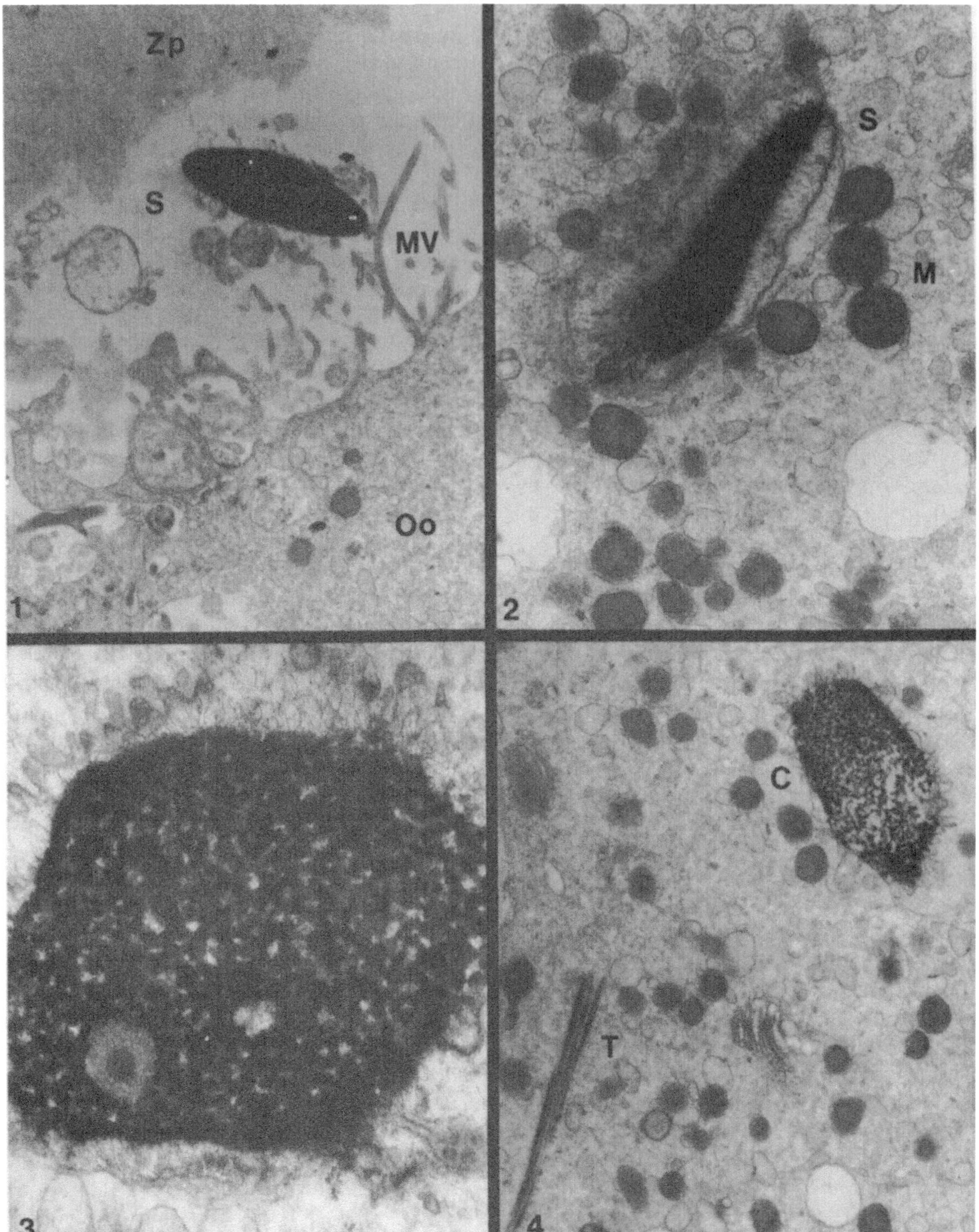
Zp
S
MV
Oo
1
S
M
2
3
C
T
4

abstriction of the first and second polar bodies (19,20).

In the presence of the drug nocodozole, which destabilizes the microtubular component of the metaphase spindle, individual chromosomes, no longer confined to an organized spindle, disperse throughout the cortical cytoplasm. Under these circumstances, the same sequence of cortical and plasma membrane differentiation normally associated with a subjacent metaphase spindle occurs in the presence of individual chromosomes at multiple sites throughout the circumference of the oocyte (19,20). The transfer of individual bivalent chromosomes to oocytes anucleated at specific times prior to germinal vesicle breakdown (GVB) suggested that the capacity of the cortical cytoplasm and plasma membrane to respond to the presence of a chromosome develops approximately 45–60 minutes after the resumption of meiosis in vitro (20).

With respect to the preovulatory differentiation of the mouse oocyte, these findings suggest the following. (1) A sequence of molecular and cytoplasmic changes occurs in the fully grown oocyte prior to GVB and the appearance of condensed bivalent chromosomes. These changes are associated with the development of the ability of the cytoplasm to respond at a later time to the presence of chromosomes by initiating specific alterations in the structure, organization, and molecular composition of the cortical cytoplasm and oolemma (15,18,19,22). (2) The differentiation of the cortical cytoplasm and overlying plasma membrane occurs in response to the physical presence of a chromosome (19,20). The expression of genomic information does not appear to have a role in these early differentiative events because mouse meiotic chromosomes appear to be transcriptionally inactive (42,43). Collectively, these observations indicate that intrinsic differentiative processes beginning prior to GVB may initiate a series of structural and organizational changes in the cytoplasm and cytocortex (18). These changes may also be of direct consequence in the development of the capacity of the mature oocyte to promote the cytoplasmic internalization of the fertilizing spermatozoon and, as discussed below, may be relevant to the question of early developmental failure in the human.

Observations and experimental protocols derived from animal models, such as the mouse, provide both a conceptual basis and an analytical approach to the question of whether or not fertilization failure of the human oocyte may be the result of perturbations in cortical cytoplasmic or plasma membrane differentiation in the preovulatory oocyte. Molecular probes that differentially resolve plasma membrane components (37), cellular organelles (44–47), chromosomes (20,44), and cytoarchitectural elements (19,20,46,48) can be applied to the study of preovulatory differentiation of the human oocyte. With these probes it should be possible to determine definitively whether a pattern of chromosome-mediated cytocortical and plasma membrane differentiation, such as that observed in the mouse, is associated with the preovulatory maturation of the human oocyte. Such analyses could be particularly relevant for those patients whose oocytes permit the passage of spermatozoa through the zona pellucida but are not penetrated. In these cases, experimental observations should focus on the cytocortical and plasma membrane processes normally associated with abstriction of the polar bodies and internalization of the fertilizing spermatozoon.

The molecular differentiation of the human oocyte plasma membrane during preovulatory maturation is another area of experimentation that could provide direct evidence to explain longstanding fertilization failure (in vivo and in

←

Figures 5-1–5-4.
1: Transmission electron micrograph of a meiotically mature human oocyte (Oo) in which a single acrosome-reacted spermatozoon (S) had traversed the zona pellucida (Zp) but failed to attach to the oolemma. Mv: microvilli. X 15,000.
2: Transmission electron micrograph of a penetrated human egg at 36 hours postinsemination in which decondensation of sperm DNA failed to occur to any significant degree. S: sperm head; M: mitochondria. X 28,000.
3 and 4: Penetrated human eggs that arrested development at the one-cell stage and displayed only partially decondensed sperm DNA (C). T: tail segment. Figure 5-3, X 44,000; Figure 5-4, X 16,000.

vitro) in a segment of the patient population whose infertility is attributed to idiopathic causes. Typically, the oolemma of a mammalian oocyte develops the capacity to initiate molecular interactions with spermatozoa during the latter stages of preovulatory maturation. Early interactions of developmental consequence during the perifertilization stage include (1) binding of the spermatozoon to the oocyte cell surface and (2) initiation of a physiochemical or electrochemical signal at the site of sperm binding that is propagated throughout the circumference of the oocyte and is associated with the discharge of cortical granules. It is generally assumed that the initial interaction between sperm and oocyte is mediated by glycoproteins (35,36,38). The identification of those specific cell surface glycoproteins involved in sperm–oocyte interaction is currently in the early stages of analysis (14,16,17,35). However, information concerning the temporal, spatial and biochemical aspects of stage-related glycoprotein expression may be of direct relevance in understanding the causes of fertilization failure in the human. For example, do fertilization-associated cell surface glycoproteins preexist in the (cortical) cytoplasm or are they posttranslationally modified in situ? Aberrations in the process by which glycoproteins are inserted into the plasma membrane or the absence of an effective enzymatic system to produce carbohydrate chains of the appropriate composition and order could render the oocyte refractory to the presence of spermatozoa, even when the gametes are in direct physical contact. The identification of specific glycoproteins involved in sperm recognition and binding, as well as information concerning the mechanism by which these proteins become functional, will contribute significantly to an appreciation of the extent to which subtle metabolic disorders may be responsible for the failure of sperm penetration to occur in otherwise normal-appearing, mature oocytes.

2. Incomplete male pronucleus formation and decondensation of DNA

In an attempt to determine whether a specific cause(s) of fertilization failure can be identified, standard practice at Reproductive Genetics In Vitro has been to examine unfertilized ova by high resolution differential interference contrast and epifluorescence microscopy after staining with DNA-specific probes. These analyses have often been followed by transmission electron microscopy. One of the more unexpected findings derived from the examination of apparently unfertilized ova was that approximately 2% of the oocytes had in fact been penetrated by a spermatozoon. Routine light microscopic observation had indicated that fertilization had not occurred owing to the absence of pronuclei and a second polar body after as many as 36 hours postinsemination.

For eggs subsequently identified as having been penetrated, electron microscopic examination of a representative sample revealed spermatozoa in varying states of decondensation (Figs. 5-2–5-4) and male pronuclei arrested at various stages of early development (Figs. 5-5–5-8). Many eggs contained spermatozoa in which no significant decondensation was evident (Fig. 5-2). These findings demonstrate the existence of a class of human oocytes that permit sperm penetration but are unable to initiate or maintain either the process of sperm DNA decondensation or the formation of the male pronucleus, or both. The possibility of sperm dysfunction in these situations was reduced by two observations: (1) the sperm were derived from men of proven fertility, and (2) other oocytes, derived from the same patients during retrieval cycles and inseminated concomitantly, fertilized and developed normally in vitro.

Several factors may contribute to the failure of penetrated ova to promote the decondensation of sperm DNA or the complete formation of the male pronucleus. In some patients, ovarian hyperstimulation produces a developmentally heterogeneous or maturationally asynchronous population of oocytes (49). If the biochemical maturation of the human oocyte (3) follows the pattern observed in other mammals (11–18,30, 50–55), the development of the ability of the cytoplasm to decondense sperm DNA probably develops during the final stages of the preovulatory period. Asynchrony in the maturational process among a group of oocytes obtained from a cohort of stimulated follicles may produce oocytes that are chromosomally mature (meiosis

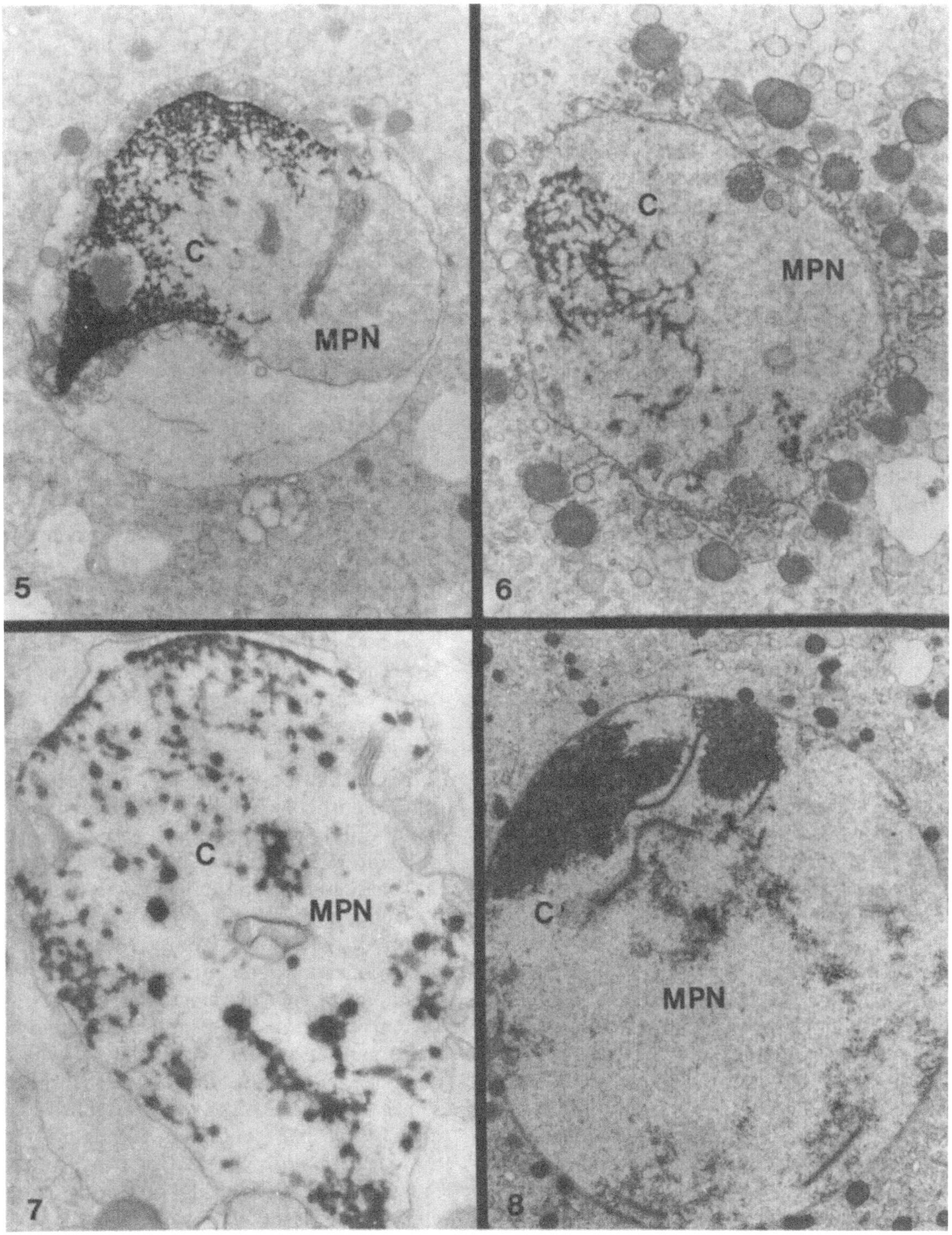

Figures 5-5–5-8. Four examples of developmentally arrested human eggs that displayed male pronuclei (MPN) in varying stages of formation. C: chromatin. Figure 5-5, X 15,000; Figure 5-6, X 24,000; Figure 5-7, X 36,000; Figure 5-8, X 9,000.

arrested at metaphase II) but cytoplasmically immature such that molecular changes and cellular reorganizations are incomplete or inadequate. This notion is supported by animal studies that demonstrate that some aspects of chromosomal and cytoplasmic maturation (e.g., stage-related protein synthesis) in the preovulatory oocyte may not necessarily involve developmentally coupled or mutually dependent processes at all stages of maturation (30,55–58). Consequently, at the time of insemination, a chromosomally mature human oocyte may not have undergone or completed the molecular and cytoplasmic maturation necessary for sperm DNA decondensation.

The number of hours after the completion of meiotic maturation at which time sperm penetration occurs may also play a critical role in determining the success of fertilization. Mammalian oocytes have a finite temporal capacity for fertilization such that an oocyte penetrated during the latter stages of the fertilizable period may have a diminished ability to fully decondense sperm DNA. This reduced capacity may be reflected in the description of an oocyte as *postmature*, an ill-defined cause of developmental failure frequently suggested as being responsible for the failure of eggs to develop.

Our understanding of the biochemistry of sperm DNA decondensation and male pronuclear formation is limited at present and is mostly derived from in vitro experiments (see 59 for review). The ability of the oocyte cytoplasm to permit the formation of the male pronucleus has been traditionally thought to involve a so-called Male Pronuclear Growth Factor (60,61). The precise nature of this factor is undetermined. However, it may be speculated that the ability of the cytoplasm to promote sperm DNA decondensation and male pronuclear formation involves the addition of histones (62), removal of protaminelike proteins (63), and the appearance of a series of relatively short-lived enzymes, some of which may be involved in the reduction of disulfide bonds within the DNA (64). These enzymes would be expected initially to be expressed in the mature oocyte, and the subsequent ability of the oocyte to decondense sperm DNA and form a male pronucleus may correlate with quantitative changes in the levels of these putative proteins. Alternatively, stage-related posttranslational modifications of existing proteins may be involved in establishing cytoplasmic conditions favorable for postpenetration development. In this regard, stage-related quantitative and qualitative changes in the pattern of protein synthesis (13,14,16–18, 65) and modification (12,13,17,30,66,67) associated with preovulatory oocyte maturation have been described in detail for several mammalian species.

Although no specific oocyte proteins have been identified as causative agents in the decondensation/pronculear formation process, those proteins whose expression can be specifically correlated with this period of oogenesis/postpenetration development are candidates for experimental analysis. It is also reasonable to speculate that similar molecules may be involved in the cycle of chromosomal condensation and decondensation that occurs in the embryo with mitotic divisions. Therefore, the identification of proteins whose expression is transient in the mature oocyte and limited to specific phases of the blastomere cell cycle may provide a defined population of proteins for detailed examination. Experimental analyses in which quantitative and qualitative aspects of specific protein(s) expression are measured with respect to time/stage of appearance and functional capacity of the cytoplasm to decondense sperm DNA and promote pronuclear formation may be particularly relevant in understanding the finite ability of the oocyte to develop after penetration.

The organization of the oocyte ooplasm may be another factor involved in establishing the ability of the cytoplasm to promote sperm DNA decondensation and pronuclear formation. In addition to specific molecular changes, stage-related alterations in cytoarchitecture that involve the organization and distribution of cytoskeletal components, such as microfilaments and microtubules (19,20,22,41,48) may establish localized intracytoplasmic conditions permissive for pronuclear formation. The extent to which developmental processes during oogenesis and early embryogenesis involve stage-related changes in the cytoarchitecture have only recently become evident. For example, stage-related redistributions of mitochondria in the preovulatory mouse oocyte (20,22,44) appear to be mediated by microtubules that radiate into the cytoplasm from perinuclear microtubule organizing centers (22,68, 69). A similar cytostructural mechanism appears to be involved in the perinuclear translocation of lysosomelike elements during the resumption of meiosis in the rat oocyte (47,70). Highly regional changes in the density of subcortical microfilaments are associated with abstriction of the first and second polar bodies (19,20,40), as well as in the generation and maintenance of polarity in the early morula-stage mouse embryo (15,21,46). The localization of mitochondria to the cortical cytoplasm at regions of intercellular association in the morula-stage mouse embryo is another example of the involvement of cytostructural components in the stage-related cytoplasmic distribution of organelles during early development (71).

Whether a specific relationship (cause or effect?) exists between the cytoarchitectural state of an oocyte and the initiation of a sequence of developmental events has only recently been considered for analysis in the mammal. Van Blerkom (18) proposed that enzymes involved in protein modification during preovulatory maturation and early postfertilization development could be associated with cytoarchitectural components such that changes in cytoarchitectural configurations could lead to the differential activation, inactivation, or changes in the activity of these putative enzymes. This notion suggests that enzymatic activities (e.g., protein phosphorylation or dephosphorylation; see 30) that may be required for, or involved in, specific developmental processes (e.g., cortical granule discharge, sperm DNA decondensation) are already present and differentially distributed or compartmentalized in the cytoplasm. One potential consequence of differentially positioning such proteins (i.e., cortex vs. perinuclear region) is that the occurrence of developmental events, such as formation of a meiotic metaphase spindle or attachment of a spermatozoon to the plasma membrane, is anticipated. Therefore, highly localized changes in cytoarchitecture that are initiated or mediated by, for example, the presence of chromosomes or by the attachment of a spermatozoon could be associated with local rather than global changes in molecular expression or enzymatic activity. One aspect of this hypothesis is that prepositioning and compartmentalization of molecular function could permit the oocyte or early embryo to initiate rapid and differential developmental changes that do not require concomitant genomic participation (i.e., new transcription). With respect to the preovulatory mouse oocyte, major differentiative changes occur throughout the cytoplasm, cytocortex and plasma membrane in the absence of any detectable concurrent genomic contribution (16–18). The development of molecular probes (tagged antibodies, labeled mRNA) to identify proteins whose expression or activity (i.e., function) is stage-related will go a long way in providing the analytical ability to test this hypothesis by allowing the following considerations to be addressed experimentally: (1) localized vs. ubiquitous distribution; (2) association with cytoplasmic organelles and cytoarchitectural elements; and (3) temporal and spatial characteristics of their synthesis and distribution during oogenesis.

2.1. Developmental failure at the pronuclear stage

One of the more perplexing aspects of early developmental arrest in the human is the failure of normally fertilized eggs to progress beyond the pronuclear stage. Observations made at Reproductive Genetics In Vitro demonstrate that eggs remaining at the pronuclear stage usually terminate development with pronuclei in juxtaposition rather than separated by any significant distance. One approach to the question of developmental arrest at this stage is to examine the normal morphodynamic and molecular processes that accompany pronuclear changes which lead to syngamy and the first mitotic cell division. The findings discussed below were derived from brief, timed observations of newly fertilized eggs examined under differential interference contrast optics. The images were electronically enhanced to increase resolution and recorded both photographically and on video tape. The results of this study were obtained during a three-year period, and details as to number and type of observations are described in detail by Van Blerkom et al. (2).

Approximately 14 hours after insemination, pronuclei are typically in close contact near the center of the oocyte (Fig. 5-9). Nucleoli are dispersed within the nucleoplasm, although some

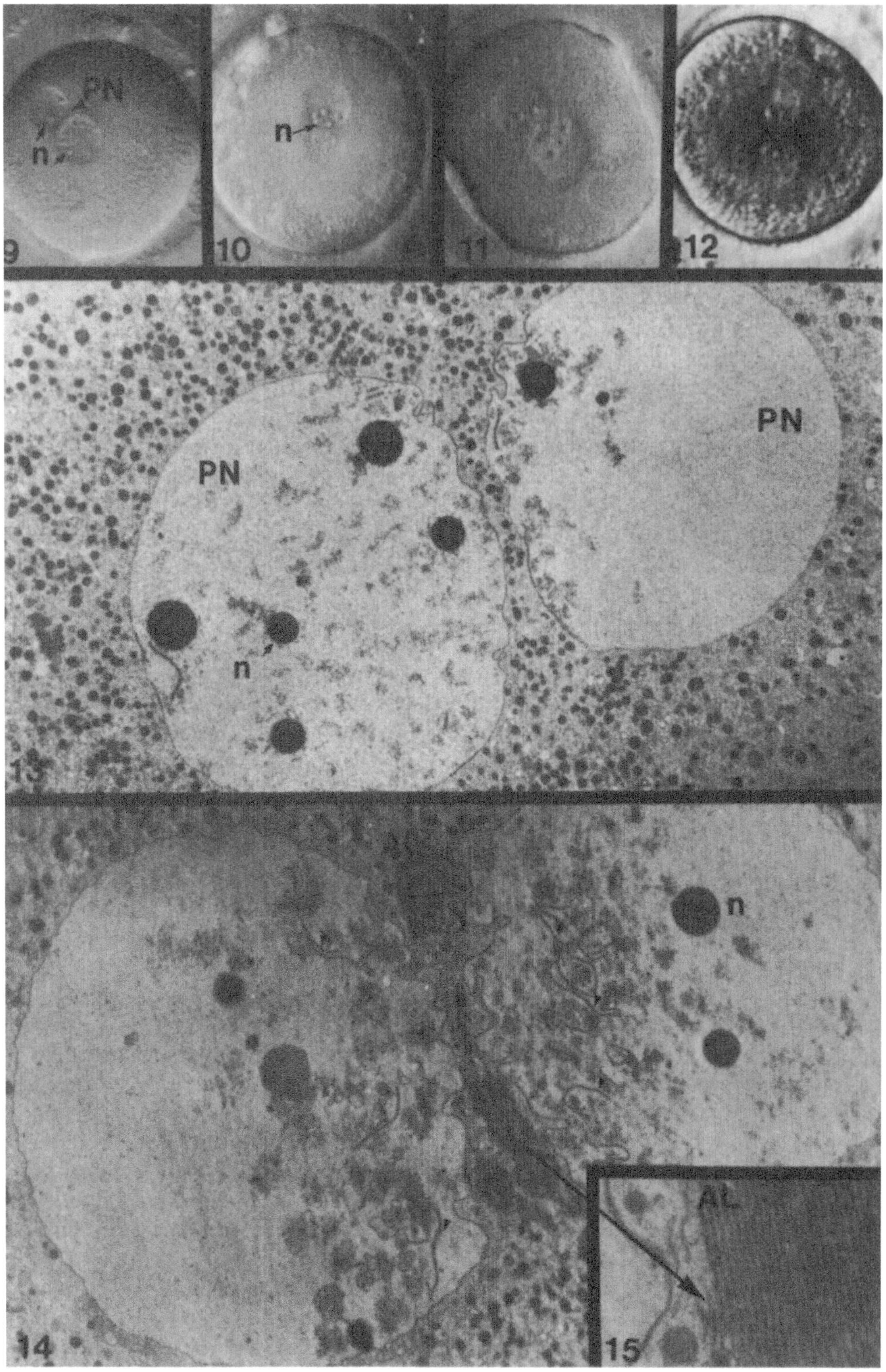
PN
n
9
n
10
11
12
PN
PN
n
13
n
14
AL
15

nucleoli are located in proximity to the nuclear membrane (arrows, Fig. 5-9). Between 16 and 18 hours, pronuclei are clearly juxtaposed and are usually located in the center of the egg. Nucleoli begin to show a pronounced localization to the region of juxtaposed nuclear membranes at approximately 15–16 hours after insemination (Fig. 5-10). Morphodynamic changes in the shape of the pronuclei are clearly evident at this time in regions of juxtaposition. At the light microscopic level, the pronuclear membranes appear invaginated such that the overall shape of the two pronuclei is that of an hourglass (Fig. 5-10). As development progresses to syngamy, the invaginations in regions of juxtaposition become more pronounced (Fig. 5-11) and the membranous interface between pronuclei is no longer apparent (Fig. 5-12). At the level of the transmission electron microscope, the invaginations of the pronuclear membranes are observed initially to result from a process of interdigitation and subsequently include fragmentation of the pronuclear membranes at regions of juxtaposition (Figs. 5-13 and 5-14).

The normal progression of dynamic changes associated with pronuclear juxtaposition and syngamy provide specific morphological determinants that may be useful in assessing the normality of development in the living human egg (2). Although developmental arrest at the pronuclear stage occurs relatively infrequently, two basic patterns of pronuclear morphodynamics appear to correlate with and be predictive of a nonprogressing embryo. In the first situation, pronuclei develop in an apparently normal manner but either fail to migrate or do not become juxtaposed (Fig. 5-16). Most of the nucleoli do not become associated with the pronuclear membrane and no change from the spherical shape occurs, even after prolonged culture in vitro (two to three days). The more common situation involves the failure of juxtaposed pronuclei to progress to syngamy (Fig. 5-17). In some of these situations, the sequence of pronuclear development appears normal; nucleoli become associated with the nuclear membrane at regions of pronuclear juxtaposition; changes in the shape of the pronuclear membranes visible at the light microscopic level and normally associated with this stage of development take place. However, electron microscopic images reveal that although the pronuclear membranes are extensively interdigitated, they do not undergo dissolution (Fig. 5-14). In some pronuclear-stage eggs that fail to progress to syngamy, enlarged stacks of annulate lamellae are observed interposed between the interdigitated membranes (Figs. 5-14 and 5-15). Although static images frequently cannot provide a definitive indication of the dynamics of formation or site of origin of a particular cytoplasmic component, serial section analysis of human eggs that failed to progress to syngamy indicated that stacks of annulate lamellae arose by delamination and aggregation of the pronuclear membranes (Fig. 5-15). The presence of such extensive stacks of annulate lamellae between interposed pronuclear membranes is probably an abnormal feature of the perisyngamic phase and may be indicative of a developmentally nonprogressing embryo.

An understanding of the factors associated with, or responsible for, arrest at the pronuclear stage has the potential of providing not only information of clinical importance, but also information relevant to the normal developmental processes that occur immediately after fertilization. The current state of knowledge related to developmental arrest at the pronuclear stage is of a descriptive nature. For example, humans eggs

←

Figures 5-9–5-14. Changes in pronuclear (PN) shape and nucleolar (n) location in in vitro fertilized human eggs at 14 hours. Figure 5-9, 15-16 hours; Figure 5-10, 18 hours; Figure 5-11 and 22 hours postinsemination (Figure 5-12). Progressive developmental changes include deformation from a spherical shape and aggregation of nucleoli at regions of pronuclear juxtaposition. At the level of the transmission electron microscope, the invaginations of the pronuclear membranes are observed initially to result from a process of interdigitation and subsequently to include fragmentation (arrowheads, Figure 5-14) of the pronuclear membranes at regions of juxtaposition. Figure 5-13, early stage of interaction between juxtaposed pronuclear membranes. X 4,000. Figure 5-14, more advanced stage of membrane interdigitation and fragmentation. X 4,500. Some of the human eggs that arrest development at the pronuclear stage contain dense aggregates of annulate lamellae (AL, Figures 5-14 and 5-15) deposited between the juxtaposed pronuclei. Figure 5-15, X 20,000 (From 2).

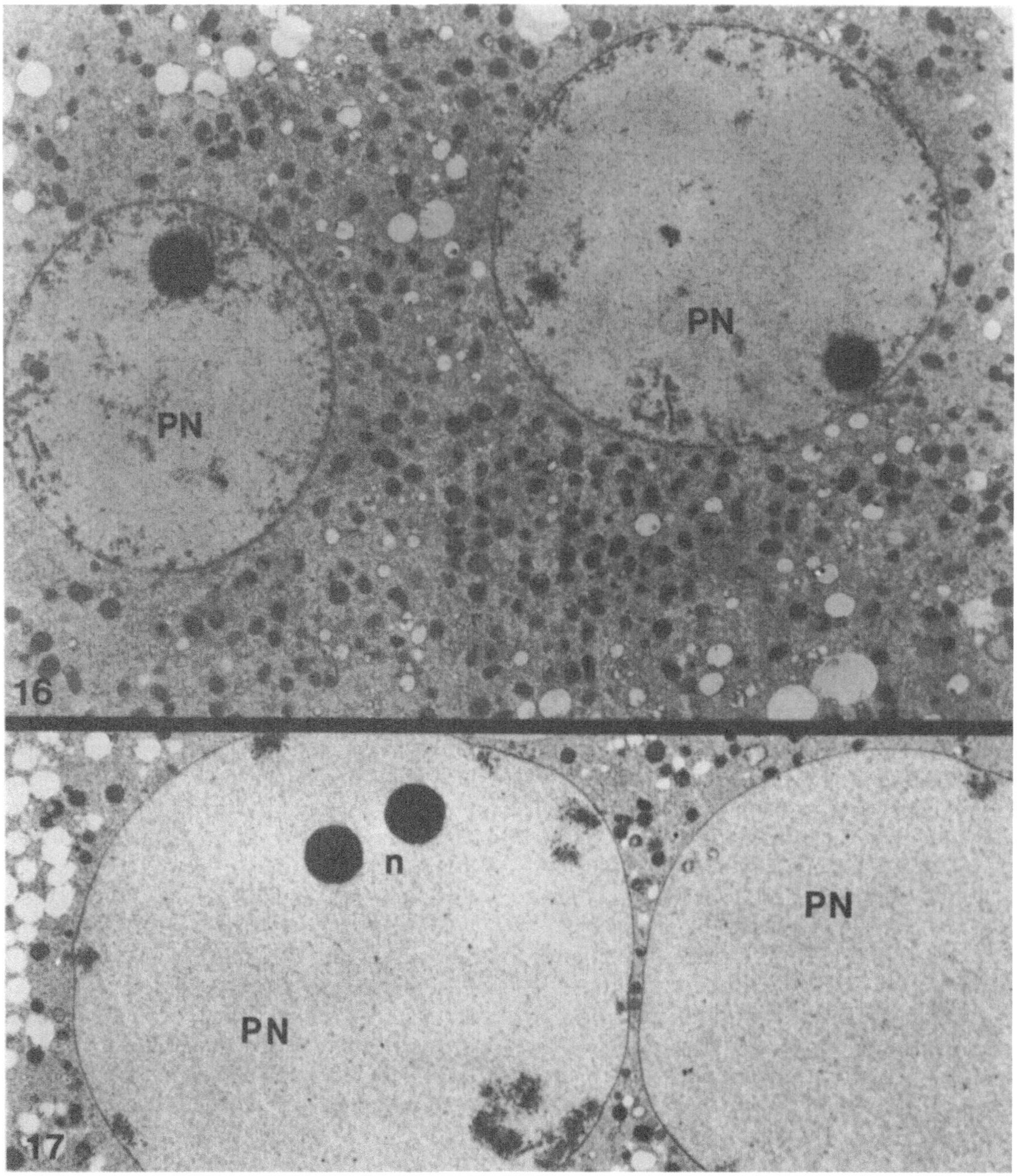

Figures 5-16 and 5-17. Frequently, human eggs that arrest development at the one-cell stage display pronuclei (PN) that are either in close proximity (Figure 5-16) or juxtaposed (Figure 5-17). Of particular interest is that for many of these eggs, pronuclei remain spherical and nucleoli (n) do not become located in regions of pronuclear membrane juxtapostion. Figure 5-16, X 5,000; Figure 5-17, X 9,000.

exhibiting apparently normal pronuclear formation but failure of pronuclei to move, or become juxtaposed, or undergo membrane fragmentation, have been described. However, the underlying cause(s) of this type of developmental failure is obscure.

Observations of pronuclear behavior in human eggs and the results of experimental studies with newly fertilized mouse eggs offer important insights that may explain developmental failure at the pronuclear stage. Nocodozole is a rapidly reversible inhibitor of microtubule polymerization. Mouse oocytes inseminated and cultured in the continuous presence of nocodozole do not form pronuclei (19). By contrast, in mouse eggs exposed to nocodozole at two hours or later after insemination, pronuclei form but their migration is suppressed completely (19). In a similar series of experiments, Maro et al. (41) examined the role of microfilaments in the process of pronuclear formation and migration. With anti-actin antibody staining of newly fertilized eggs, these investigators were able to observe the progressive development of peripronuclear microfilaments. Cytochalasin D is a drug that induces microfilament depolymerization. Mouse oocytes inseminated in the presence of cytochalasin D formed normal-appearing pronuclei but the subsequent migration of the pronuclei was completely inhibited. Results from nocodozole and cytochalasin D experiments with mouse eggs strongly suggest that (1) the failure of pronuclei to form may in part be related to the absence of an adequate population of organized microtubules and (2) once formed, the ability of pronuclei to migrate and become juxtaposed may rely on a system of peripronuclear microtubules or microfilaments. Similar findings were reported for the preovulatory mouse (2,20,22) and rat oocyte (47) in which the perinuclear translocation of mitochondria and acidic organelles, respectively, was totally inhibited during culture in the presence of drugs that disrupt microtubule and microfilament organization.

The role of microtubules in pronuclear formation, and of both microtubules and microfilaments in pronuclear migration in the mammalian egg, needs to be investigated in more detail. For example, to what extent, if any, are quantitative variations in microtubule density or in the spatial organization of microtubules and microfilaments associated with the failure of pronuclei to develop completely or, once formed, to migrate? The availability of molecular probes to identify cytostructural elements, and of image enhancement/processing technologies (72) to determine the spatial distributions of these structures, will go a long way in identifying cytoplasmic factors that may be involved in the normal progression of development, as well as in the failure of development to progress in the newly fertilized human egg.

The developmental dysfunction responsible for the failure of juxtaposed pronuclei to undergo complete fragmentation is unknown. Without the complete breakdown of the pronuclear membranes, the joining of maternal and paternal chromosomes (syngamy) cannot occur, and consequently the formation of the first mitotic metaphase is precluded. Experimental studies from our laboratory with newly fertilized mouse eggs may provide a potential clue as to one possible cause of developmental arrest at the stage of pronuclear juxtaposition (73). The drug aphidicolin is a specific inhibitor of DNA synthesis. Newly fertilized eggs cultured in the presence of aphidicolin do not synthesize DNA. Our findings demonstrated that pronuclear formation, migration and normal development to the stage of pronuclear juxtaposition occurred in the absence of detectable DNA synthesis. However, breakdown of the pronuclear membranes, and therefore syngamy, was blocked. Because DNA synthesis is initiated during the pronuclear phase (74–76), these experimental observations suggest the possibility that replication is required for, or associated with, subsequent breakdown of the pronuclear membranes. Preliminary support for this notion has been derived from the quantitative analysis of the DNA content of pronuclei in human eggs (Van Blerkom, unpublished observations). Quantitative fluorescence analysis of three eggs that arrested development at the stage of pronuclear juxtaposition demonstrated that the DNA in each pronucleus of two eggs had not replicated. An accurate measurement could not be obtained for the third egg.

The establishment of a definitive correlation between the occurrence of DNA synthesis and the arrest of development at the stage of pronuclear juxtaposition will require the analysis of a substantially larger number of eggs than has been examined to date. The nature of the association of DNA with the pronuclear membranes in normally developing and developmentally arrested eggs may also be an important factor in determining the stability of the pronuclear membranes. For example, does the association of replicated DNA with the pronuclear membranes have a destabilizing influence on pronuclear membrane integrity? In the absence of pronuclear DNA replication, the pronuclear membranes may be less prone to breakdown.

The spatial redistribution of nucleoli may also be an important factor in determining the stability of the pronuclear membranes. We have observed that nucleoli migrate to the regions of pronuclear juxtaposition prior to the stage of extensive membranous interdigitation and fragmentation (2). By contrast, the spatial distribution of nucleoli appears to be nonspecific in eggs arrested at the stage of pronuclear juxtaposition. Perhaps the nature of the association of nucleoli with the pronuclear membrane is one determining factor in the temporal and spatial specificity of membrane stability. A somewhat similar correlation between nucleolar–nuclear membrane association has been observed during the stage of germinal vesicle breakdown in the mouse oocyte (22). In the case of the immature mouse oocyte, a change in the nature of the association between nucleoli and nuclear membrane precedes the dissolution of the germinal vesicle.

Support for the speculations described above clearly will require extensive ongoing analyses. However, these speculations do provide a conceptual framework for focused experimentation and analysis in an attempt to elucidate the mechanisms involved in normal development. With such efforts, the molecular and cellular conditions, states, or processes that may contribute to developmental failure should become increasingly apparent.

2.2. *The causes and developmental consequences of polyspermic penetration of human oocytes*

A variable percent of human eggs fertilized in vitro are multipronuclear. Reported frequencies of multipronuclear human eggs range from a few percent (77,78) to greater than 30% in oocytes morphologically classified as mature (first polar body) or immature (no first polar body), respectively, at the time of insemination (79). It would appear reasonable to assume that the frequency of polyspermy is related to the number of spermatozoa utilized for insemination. For most human IVF protocols, insemination with concentrations that range from 30,000 to several hundred thousand motile spermatozoa/ml are routine. This compares to perhaps a few hundred sperm that are in the immediate vicinity of an oocyte at the site of fertilization in vivo, the ampullary portion of the fallopian tube (80). From a probabilistic view, the higher number of potentially fertilizing sperm present in vitro might be expected to lead to a higher frequency of polyspermy owing to the increased chance that two or more sperm could arrive at the oolemma within the same approximate time. Van Der Ven et al. (79) described a marked dependence of the incidence of polyploid fertilizations on the number of spermatozoa added per oocyte. These authors reported that the frequency of multipronuclear eggs was 7% with 500,000–800,000 spermatozoa/ml, and 18% after insemination with 1.5 to 2.0 million spermatozoa/ml. Although a correlation may exist between sperm number and the frequency of polyspermy (81), results from inseminations with high numbers of spermatozoa should be interpreted with care for the following reasons. (1) The concentration of spermatozoa used in the study of Van Der Ven et al. (79), for example, was at least an order of magnitude higher than is necessary to achieve fertilization in vitro at high frequency (2,4,7). Consequently, the occurrence of polyspermy may be exaggerated by the presence of extraordinarily high numbers of spermatozoa. (2) It is difficult to preclude the possibility that the deleterious conditions of culture (change in pH) or alterations in culture medium composition (accumulation of products of metabolism from sperm) could develop during the relatively long-term exposure of oocytes (16–20 hours) to medium containing sperm concentrations as high as 2 million/ml. The high rate of sperm metabolism and the possibility of precocious release of acrosomal enzymes could have a negative influence on oocyte viability such that oocytes (perhaps depending upon the state of preovulatory maturation) may become increasingly predisposed to polyspermic penetration.

However, a correlation between polyspermy and sperm number was indicated by the study of Englert et al. (81) who reported a tripling of the frequency of polyspermy with relatively low sperm concentrations that increased from 20,000 to 120,000/ml.

In contrast to the situation described by Van Der Ven et al. (79), Diamond et al. (82) reported that the incidence of polyspermy after human IVF was unrelated to the concentration of spermatozoa present at insemination or to the state of maturity of the oocyte. Diamond et al. (82) also noted that the occurrence of polyspermy was not associated with the method of ovarian simulation. The frequency of polyspermy observed in their study was approximately 10%. The incidence of polyspermy was no different after insemination of oocytes with concentrations of progressively motile sperm ranging from 250,000 to 500,000/ml. In this respect, baseline frequencies of polyspermy at or significantly below 10% have been reported with sperm concentrations in the 30,000 to 50,000 range (2,78).

Several studies have addressed the question of whether specific biochemical and hormonal parameters associated with the follicular fluid of preovulatory follicles may predispose an oocyte to polyspermic penetration. In a retrospective analysis, Ben-Rafael et al. (83) related the concentrations of estradiol, progesterone, androstenedione, sex hormone-binding globulin, prolactin and insulin concentrations of follicular fluid obtained at the time of oocyte recovery (32–36 hours post hCG) to whether subsequent fertilization was normal or polyspermic. The only factor associated with an increased incidence of polyspermy was a level of follicular fluid progesterone that was several thousand ng/ml higher than that observed in follicles from which oocytes that fertilized normally were derived. In a very similar study of estradiol, progesterone and androstenedione levels in follicular fluids, Webster et al. (84) concluded that follicular fluid steroid measurements from follicles containing oocytes that became polyspermic were not statistically different from measurements in follicles containing oocytes that fertilized and cleaved normally. This was especially true for progesterone levels that were nearly identical in follicles from which polyspermic and normally fertilized eggs were obtained. With the same protocol of ovarian stimulation, the average preovulatory follicular fluid progesterone concentration measured by Ben-Rafael et al. (83) for follicles that produced oocytes that fertilized normally or abnormally was three- and four-fold higher, respectively, than the progesterone levels observed by Diamond et al. (82). The extreme variation in follicular fluid progesterone levels and, of greater importance, the opposite conclusions presented in these two major clinical studies of polyspermy, underscore the difficulty in attempting to identify definitively specific exogenous factors that may contribute to aberrant fertilization.

Observations made at Reproductive Genetics In Vitro may provide an additional insight into conditions that favor the occurrence of polyspermy. Of the first 1000 fertilized eggs produced at Reproductive Genetics In Vitro, 20 (2%) were unambiguously multipronuclear (2,78). Quantitative analysis of the estradiol and progesterone levels of follicles from which polyspermic oocytes were derived showed no significant difference from the levels observed in follicles from which normally fertilized eggs were obtained. These findings are in agreement with the results of Webster et al. (84).

The morphology and degree of expansion of the cumulus oophorus is one indicator of the extent of luteinization (and therefore follicular fluid progesterone level) that occurs prior to ovulation (85). Observations made at Reproductive Genetics In Vitro indicated that the morphology and degree of expansion of the cumulus oophorus varied significantly among oocytes that subsequently were penetrated by more than one spermatozoon. The appearance of the cumulus oophorus ranged from fully expanded to highly compacted. All oocytes were cultured in individual dishes and exposed to between 30,000 and 50,000 motile sperm per ml. Polyspermic penetration was not associated with a particular protocol of ovarian stimulation (2,78).

The only significant factor that correlated with the occurrence of polyspermy in our study was the motility of sperm during incubation prior to insemination. The most common protocol in current use in the prepartion of spermatozoa for human IVF is to dilute the ejaculate with culture medium followed by a series of low-speed centrifugations. The sperm are allowed to migrate (swim up) from the pellet in a test tube over a period of hours, and the most motile spermatozoa are collected for insemination. All 20 cases of definitive polyspermy that were observed at Reproductive Genetics In Vitro occurred when oocytes were inseminated with sperm that had migrated a minimum of 13 mm during a four-hour incubation prior to insemination. These findings, albeit based on a limited number of observations, do suggest that polyspermic penetration may be more likely to occur with exceptionally motile spermatozoa. In the case of sperm that exhibit high motility during the swim-up procedure, it may be useful to inseminate with a reduced number of spermatozoa, whereas in men with less motile sperm, concentrations can be increased without an attendant increase in the frequency of polyspermy.

An alternative explanation for polyspermy is that rather than being a random event based on the coincidental arrival of two or more spermatozoa at the oolemma, some oocytes may be predisposed to polyspermic penetration. This interpretation implies that although oocytes at the time of insemination may have completed chromosomal maturation (arrest at metaphase II), cytoplasmic maturation, and especially the development of cortical mechanisms involved in the prevention of polyspermy, may be incomplete (77,78,84). The swelling and subsequent exocytosis of cortical granule contents into the perivitelline space (the cortical reaction) after contact with the oolemma by a spermatozoon is considered the primary mechanism by which polyspermic penetration is prevented at the level of the cell surface (10,33–35,86). The interaction of the cortical granule contents with the zona pellucida initiates a physical change in the structure of the zona pellucida that inhibits the continued forward progression of spermatozoa present in the zona pellucida. This secondary consequence of the cortical reaction is termed the zona reaction (for review see 35). Soupart and Strong (31) suggested that the site of the polyspermic block in the human is at the level of the zona pellucida rather than the oolemma.

Incomplete cortical granule formation or the total or partial failure of the cortical reaction has widely been assumed to be the primary cause of polyspermic penetration (see 33,34,78). Cortical granules are formed by Golgi complexes during the period of reinitiated meiosis, and for some species it has been suggested that cortical granules continue to be produced even after ovulation (for review see 10). Detailed studies of cortical granule formation in the human oocyte by Sathananthan and Trounson (33,34) and Sathananthan et al. (87) demonstrate that cortical granule production is completed prior to ovulation. Sathananthan et al. (87) reported that although cortical granule production is a continuous process, there are two distinct waves of cortical granule formation from Golgi complexes, the first being more prolific than the second. The first wave of cortical granule formation occurs in small antral follicles, when there is a peak of Golgi activity, and the second at the germinal vesicle stage, particularly at the onset of the preovulatory resumption of meiosis. Two other important findings from the study of Sathananthan et al. (87) were that cortical granule production diminishes significantly after abstriction of the first polar body, and that the mature human oocyte is characterized by a paucity of Golgi complexes. With respect to the state of cortical maturation in the preovulatory human oocyte, these observations suggest the following: (1) cortical granule formation and deployment are complete prior to ovulation, and (2) cortical granules are most probably not synthesized after ovulation.

As observed by transmission electron microscopy, the appearance of fully grown, germinal vesicle-stage human oocytes obtained from very early antral follicles in unstimulated ovaries (Figs. 5-18 and 5-20) is shown in Figures 5-19, 5-21, and 5-22. In support of the findings of Sathananthan et al. (87), it is clear from these electron microscopic images that cortical granules are present in the subplasmalemma and cortical cytoplasm of fully grown, immature oocytes derived from very early antral follicles.

The premature release of cortical granules may be responsible for the failure of an oocyte to be fertilized, or a dysfunction in cortical granule exocytosis may predispose an oocyte to polyspermic penetration. Sathananthan et al. (87) described a well-defined band of microfilaments

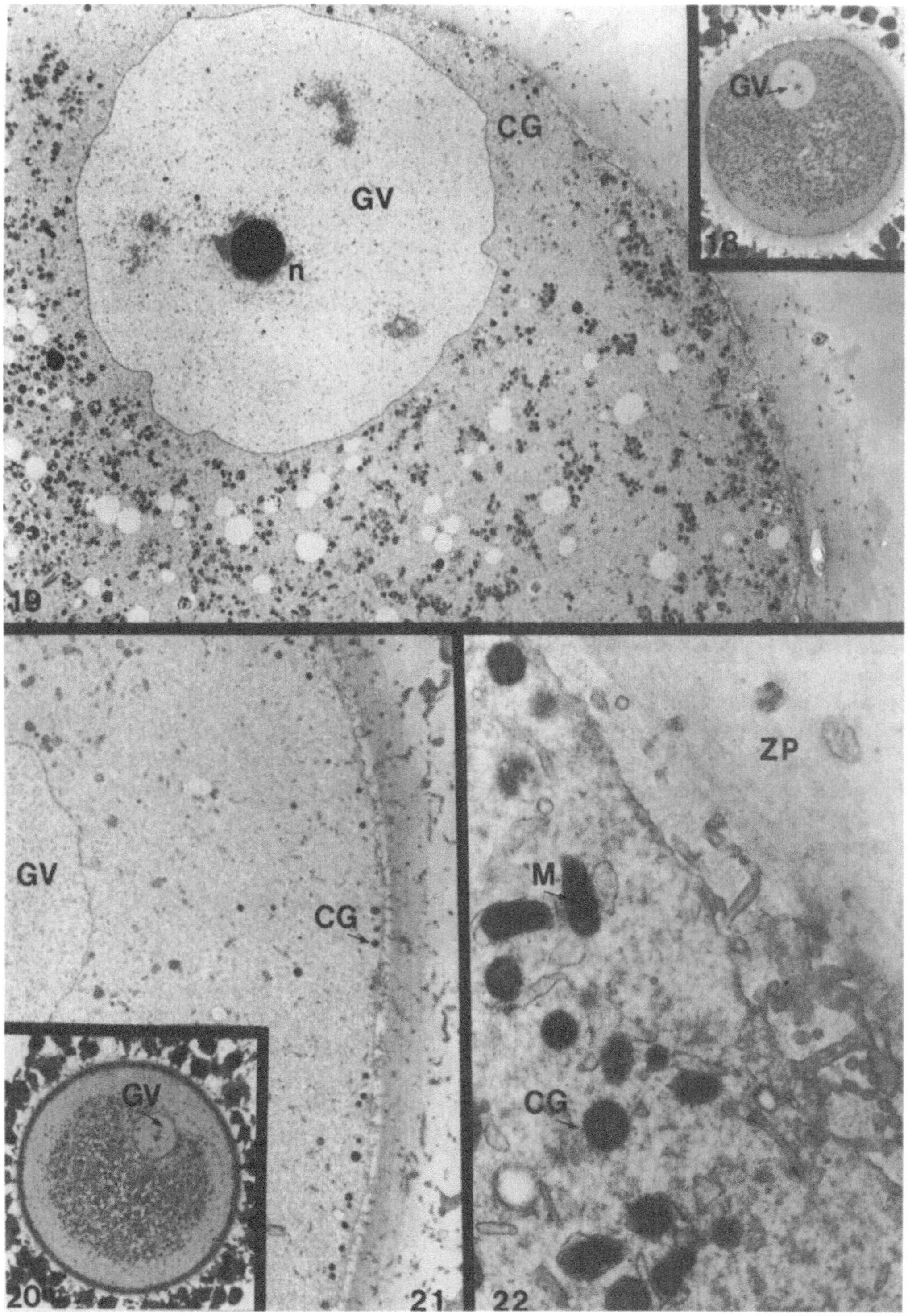

Figures 5-18–5-22. Light (Figures 5-18 and 5-20) and transmission electron (Figures 5-19, 5-21 and 5-22) microscopic images of fully grown germinal vesicle-stage human oocytes (GV) obtained from early antral follicles present in unstimulated ovaries (i.e., not exogeneously stimulated). Electron microscopy confirmed the presence of cortical granules (GC) in the cortical (Figure 5-19) and subplasmalemmal cytoplasm (Figures 5-22 and 5-23). M: mitochondria; ZP: zona pellucida. Figure 5-19, X 2,700; Figure 5-21, X 3,300; Figure 5-22, X 19,000.

in GV-stage oocytes retrieved from small antral follicles. This band gradually became disorganized at the germinal vesicle stage as oocytes resumed meiosis. The existence of a cortical band of microfilaments prior to the resumption of meiosis appears to present a physical barrier that prevents the juxtaposition of cortical granules with the oolemma. If it is assumed that such a juxtaposition is necessary for the occurrence of a cortical reaction, the presence of a cortical band of microfilaments may preclude the premature swelling and exocytosis of the cortical granule contents. The disorganization of the cortical band of microfilaments during the preovulatory resumption of meiosis is associated with the migration of cortical granules and their subsequent subplasmalemmal deposition. From this location, cortical granule exocytosis is induced as a consequence of an alteration in the physiochemical properties of the cortical cytoplasm and plasma membrane. The exocytosis is initiated at and propagated from the site of attachment of the fertilizing spermatozoon (88). It would appear reasonable to examine the structure of the cortical cytoplasm, with special attention to microfilaments in cases of polyspermic penetration, or where penetration has failed to occur in the presence of a perivitelline spermatozoon. Perhaps the structural basis of these developmentally abnormal situations is the failure of the cortical microfilament band to reorganize during preovulatory oocyte maturation.

Transmission electron microscopic observations of multipronuclear human eggs that resulted from polyspermic penetration have not shown retained cortical granules (2,77,78). This observation suggests that a dysfunction in the cortical reaction is probably not the primary underlying cause of polyspermic fertilization in human eggs described to date. Light microscopic analyses of living eggs that contained more than two definitive pronuclei revealed the presence of accessory spermatozoa contained within the matrix of the zona pellucida (78). These spermatozoa did not penetrate further into the zona pellucida during cleavage, strongly suggesting that for these particular eggs, cortical granule discharge produced a normal zona reaction. If the occurrence of polyspermy is the result of developmental abnormality during preovulatory maturation it appears that those factors predisposing to polyspermy (intrinsic or extrinsic) may be more subtle than a dysfunction in the cortical or zona reactions.

It is also worthwhile to reconsider the notion that polyspermy may be a random event that, from a probabilistic point of view, may be more likely to occur in the presence of a population of spermatozoa that is both highly motile and, relative to the situation that exists in vivo, that represents a concentration that is not physiological. It is also possible that oocytes vary with respect to the rate at which the cortical reaction is propagated through the circumference of the oocyte. In the presence of several spermatozoa progressing through the zona pellucida, a difference of several minutes in the completion of the cortical reaction means the difference between a normal and abnormal fertilization. At present there is no definitive information concerning the time required to initiate and to complete the cortical reaction in the human oocyte. Experimental observations designed to determine both the time required to complete the cortical reaction and the extent to which such a time may vary among oocytes will go a long way in providing a more comprehensive understanding of oocyte-related factors that may contribute to polyspermic penetration.

3. The developmental potential of embryos derived from multipronuclear eggs

The presence of three or more definitive pronuclei in eggs with an abstricted second polar body is a clear indication of polyspermic penetration (Fig. 5-23). The ability of multipronuclear human eggs to develop through the preimplantation stages in a grossly normal manner is one of the more remarkable findings to come from detailed observations of human embryo development after in vitro fertilization (2,77,78,89). In the immediate postfertilization period, pronuclear formation, migration, juxtaposition, membrane interdigitation, and syngamy follow the same temporal and spatial pattern of morphodynamic change observed in normally fertilized eggs (Fig. 5-23). As is the situation for normally developing fertilized eggs, tripronuclear eggs exhibit progressive changes in membrane shape and

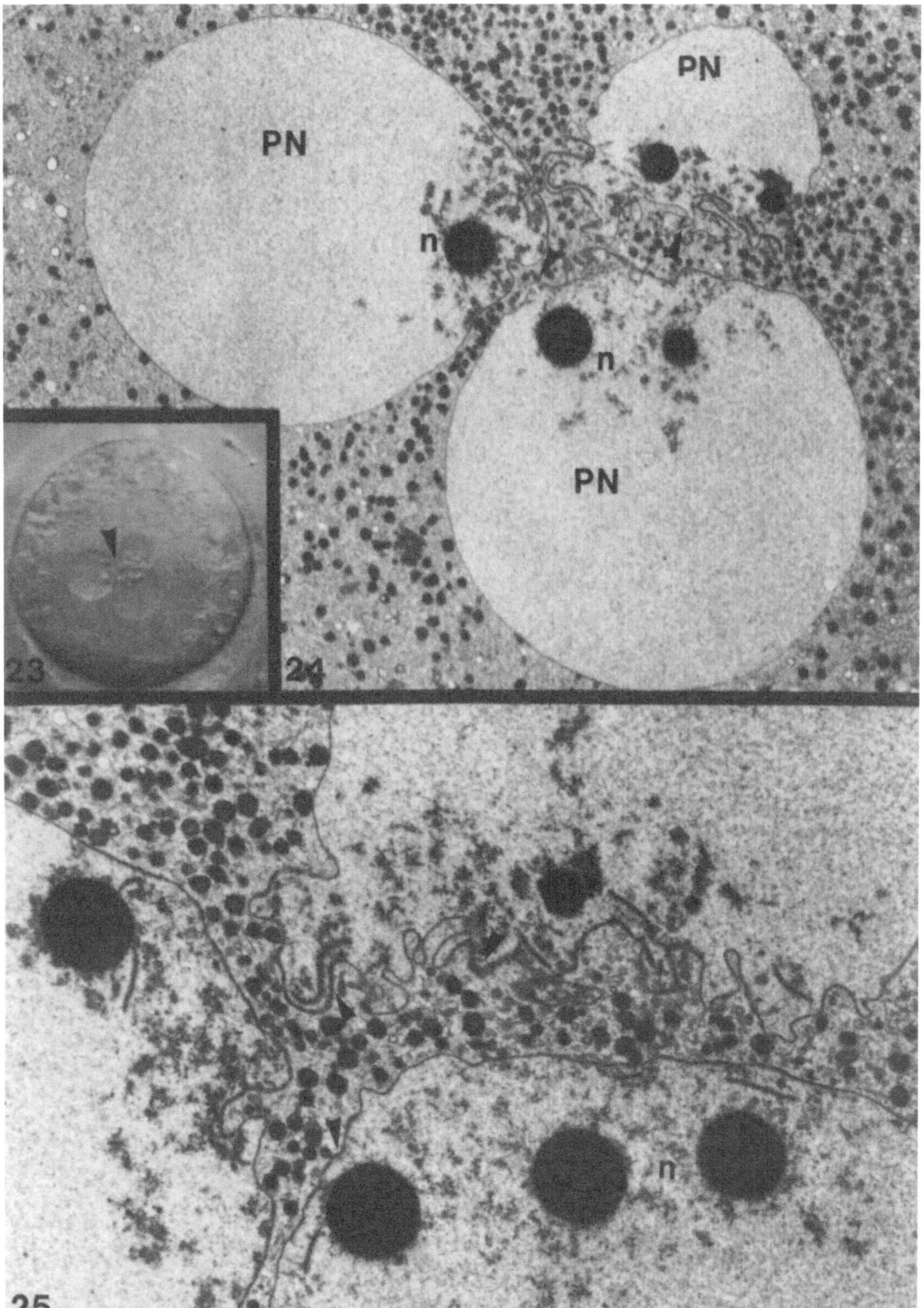

Figures 5-23–5-25. Light (Figure 5-23) and transmission electron (Figures 5-24 and 5-25) microscopic images of a tripronucleate human egg during the perisyngamic stage of development. Changes in pronuclear shape at regions of membrane juxtaposition (arrowhead, Figure 5-23) are the result of membrane interdigitation and fragmentation (arrowheads, Figures 5-24 and 5-25). As is the situation for normally fertilized human eggs, nucleoli (n) in multipronucleate eggs migrate to regions of pronuclear membrane association. Typically, changes in pronuclear shape and nucleolar location indicate that such an egg will complete at least the first cleavage division. Figure 5-24, X 3,300; Figure 5-25, X 7,000. Figure 5-24, from 2.

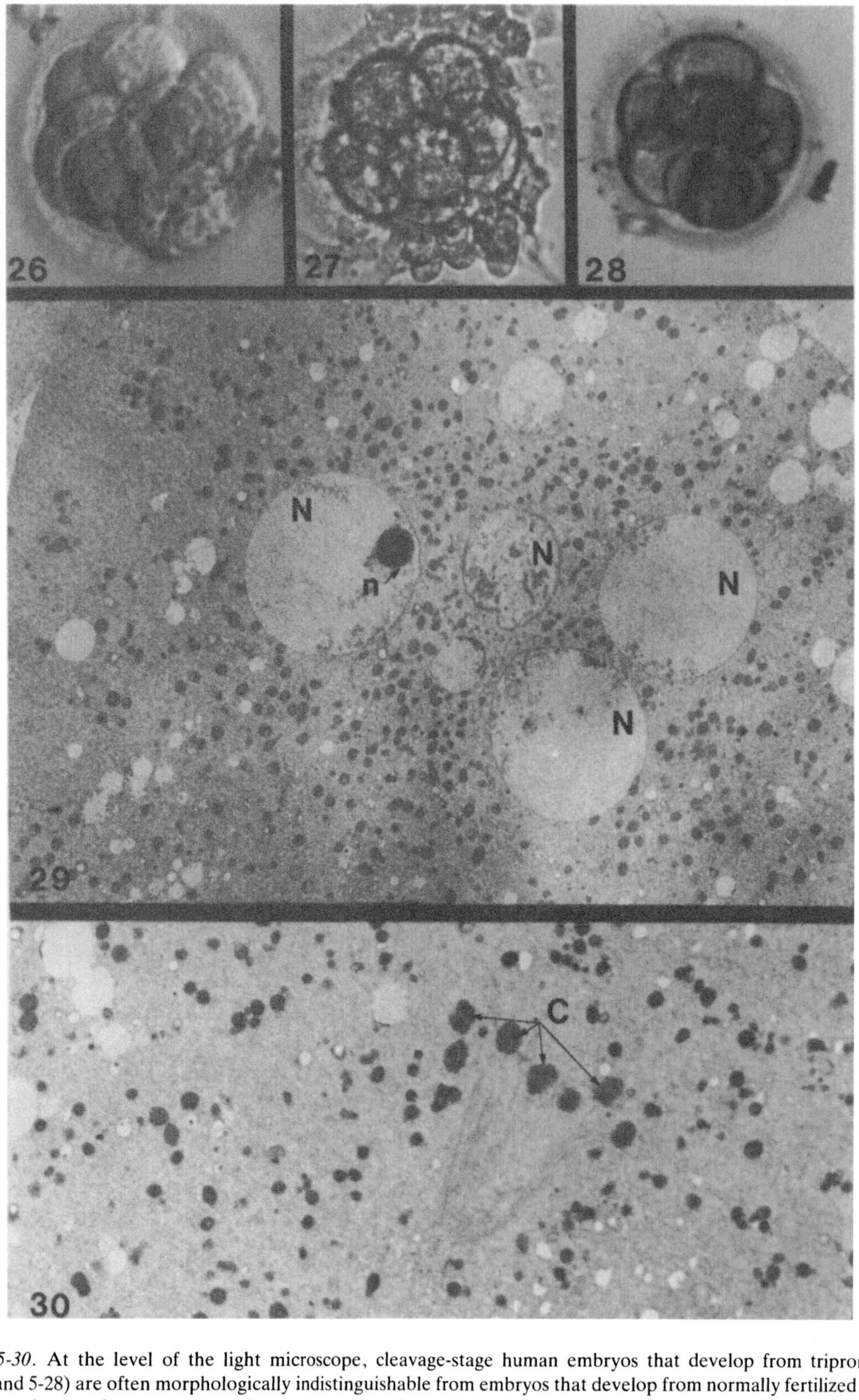

Figures 5-26–5-30. At the level of the light microscope, cleavage-stage human embryos that develop from tripronuclear eggs (Figures 5-27 and 5-28) are often morphologically indistinguishable from embryos that develop from normally fertilized eggs (Figure 5-26). Electron microscopic examination of morula-stage embryos derived from multipronuclear eggs often reveals nuclear and chromosomal anomalies. Two situations that are not uncommon in cleavage-stage embryos that developed from tripronuclear (triploid) eggs are the presence of (1) multinucleated blastomeres (N, Figure 5-29) and (2) blastomeres containing the normal diploid complement of chromosomes (C, Figure 5-30). Figure 29, X 3,600; Figure 30, X 6,000.

nucleolar position at regions of pronuclear juxtaposition (Figs. 5-24 and 5-25). These changes are highly predictive of progression to syngamy and completion of at least the first cell division (2).

Multipronuclear eggs that arrest development do so at the same stages and display the same pronuclear morphology as observed in eggs that fertilize normally but do not progress further than the pronuclear stage. For example, some eggs contain multiple pronuclei that failed to migrate. Other eggs exhibit multiple juxtaposed pronuclei in which none of the morphodynamic changes normally associated with the presyngamic phase are apparent. As was the case for nonprogressive, normally fertilized eggs, nucleolar position and pronuclear morphology remained unchanged for these eggs during an additional three days of culture (2).

Approximately 20%, or 32, of the multipronuclear eggs observed at Reproductive Genetics In Vitro remained at the one-cell stage. The rest underwent the first and subsequent cleavage divisions. In comparison to normally fertilized eggs, of which approximately 94% were at the four-cell stage at 48 hours post-insemination (2), embryos developed from tripronuclear eggs divided at a slightly faster rate, with most (70%) having attained the six-to-eight-cell stage by 48 hours. By routine light microscopic inspection, cleavage-stage embryos developed from normally fertilized (Fig. 5-26) and tripronuclear eggs (Figs. 5-27 and 5-28) cannot usually be distinguished by gross morphology (2,77,78). More detailed analysis by high-resolution phase contrast or differential interference contrast optics, followed by transmission electron microscopy (Fig. 5-29), often reveals the presence of multiple nuclei in some of the blastomeres of embryos developed from multipronuclear eggs. In this regard, fine structural analysis of some cleavage-stage embryos that developed from multipronuclear eggs has demonstrated the occurrence of chromosomally normal blastomeres (Fig. 5-30) adjacent to multinucleated blastomeres (Fig. 5-29).

The presence of blastomeres with multiple nuclei is not necessarily associated only with polyspermic fertilization. Multiple nuclei are observed in blastomeres of embryos derived from normally fertilized eggs (as described in Section 8). It might be anticipated that such embryos may also be chromosomally abnormal. In this regard, Plachot et al. (90) reported that 16% of normal-appearing eight-cell human embryos derived from monospermic fertilization showed chromosomal abnormalities, essentially diploid/haploid or triploid/haploid mosaicism. This finding indicates that a relatively high percent of normally fertilized human eggs may develop potentially lethal chromosomal abnormalities during subsequent cleavage divisions. By contrast, multipronuclear eggs are clearly chromosomally abnormal from the outset. The demonstrated ability of such eggs to develop through the early cleavage stages in a morphologically normal fashion emphasizes the need to determine pronuclear number prior to syngamy (77,78,89,90). However, as discussed later in this chapter (Section 7.1), a definitive pronuclear identification can be confounded by the development of cytoplasmic structures that resemble pronuclei (pseudopronuclei).

In spite of the observation that multipronuclear human eggs are able to undergo successive cleavage divisions, a major question of clinical relevance concerns the extent to which these embryos are developmentally viable. Approximately 60% of the embryos derived from tripronuclear eggs at Reproductive Genetics In Vitro arrested development between the eight-cell and early morula stage (Figs. 5-27 and 5-28). Less than 10% developed to the expanded blastocyst stage. However, the blastocyst-stage embryos that developed from tripronuclear eggs were morphologically indistinguishable from blastocysts derived from normally fertilized eggs (Van Blerkom, unpublished observations). These findings indicate that embryonic development to the stage of implantation can occur after dispermic fertilization. The ability of such embryos to implant and continue in gestation is demonstrated by the appearance of triploid fetuses in spontaneous abortions (5) and by the birth of babies with a triploid complement of chromosomes (91).

It might be expected that the termination of development observed during cleavage for the majority of embryos derived from multipronuclear eggs is a manifestation of the aberrant chromosomal situation. Perhaps this is a mechanism by which genetically abnormal embryos are selectively eliminated from futher development. However, the cleavage arrest exhibited by these embryos may not necessarily be related to the fact that they were derived from polyspermic fertilization and therefore were known to be genetically abnormal. Approximately 30% of normally fertilized and in vitro-cultured human embryos arrest cleavage by the 16-cell stage, with another 12% progressing no further than the morula stage (8).

One explanation for developmental failure during the early preimplantation stages is that current medium formulations or culture conditions are simply suboptimal for the support of postmorula development. However, the likelihood of this possibility is reduced by the observed ability of embryos to develop into normal blastocyst-stage embryos. It seems more probable that developmental variability may be embryo-specific. The recent finding that a significant percent of human embryos that develop from normally fertilized oocytes exhibit nuclear (92) and chromosomal abnormalities (90) suggests that premature developmental arrest in the human may be associated with an unbalanced chromosomal complement, or from defects in molecular or cellular differentiation. Such defects are most probably intrinsic to the embryo and manifest lethality during the cleavage stages (5,11).

Observations of human preimplantation development in vivo after ovarian hyperstimulation and timed intrauterine insemination strongly indicate that cleavage arrest is not necessarily a consequence of (1) fertilization and development in vitro, (2) nutritional/metabolic variation among embryos from the same patient, or (3) the result of delayed fertilization. Studies of human preimplantation development in vivo reported by Buster et al. (9) demonstrate that at the time blastocyst-stage embryos would be expected (100–130 hours postinsemination), only 31% of the embryos flushed from the uterus had attained this stage. Approximately 70% of the embryos had arrested cleavage between the two-cell and morula stage. Embryos in which development terminated during the preimplantation stages in vivo displayed the same types of gross morphology (i.e., normal-appearing to fragmented) observed after fertilization in vitro and the continued culture of embryos derived from normal and polyspermic eggs. It would appear that chromosomal excesses or deficits may be only one of several potential causes of early developmental failure.

4. Chromosomal correlates of developmental potential

The high percentage of developmental arrest during cleavage in vivo and in vitro suggests that human embryos may be developmentally compromised from the outset (for more detailed discussion, see Section 6.1). This notion is supported by the findings of several investigators who have examined the DNA and chromsosomal constitution of mature oocytes, fertilized eggs and cleavage-stage embryos. With a DNA-specific fluorescent probe, O'Rand et al. (93) made quantitative estimates of the DNA content of preovulatory human oocytes and of fertilized eggs that arrested development at the one-cell stage. The results of this study, albeit based on a very small sample population, indicated the occurrence of some oocytes in which DNA was not apportioned equally between the oocyte and first polar body. Some of the fertilized eggs that remained at the pronuclear stage had failed to undergo DNA replication.

Plachot et al. (94) reported that 22% of oocytes that failed to fertilize in vitro were chromosomally abnormal. Other investigators have suggested that greater than half of all mature human oocytes aspirated from hyperstimulated ovaries may be chromosomally abnormal (27,28). The observations of mature and unfertilized oocytes by Plachot et al. (94) and Wramsby et al. (28,29) are derived from a very limited number of observations in a highly selected population of patients treated for infertility by IVF. It should be cautioned that the apparently high percentages of chromosomally abnormal preovulatory oocytes described by some investigators may not be applicable to a large and diverse population of women. As discussed later in this chapter, the findings of Van Blerkom and Henry (72) suggest that a very small number of patients may account for a significant proportion of chromosomally abnormal metaphase II-stage oocytes.

With respect to the genetic status of preimplantation-stage human embryos, Plachot et al.

(90) reported that 11% of "healthy looking" eight-cell embryos and 19% of fragmented, cleavage-stage human embryos were chromosomally abnormal. Chromosomal perturbations may also be a major contributing factor to embryonic demise during the preimplantation stages even when fertilization and development occur entirely in vivo (5,9). Although the extent to which the chromosomal abnormalities observed in normal-appearing cleavage-stage embryos (as opposed to oocytes that do not fertilize) originate in the preovulatory oocyte remains to be determined (see Section 6.1), it is clear that a variable degree of preimplantation development can occur in the presence of such aberrations. This is obviously the situation for tripronuclear eggs that are triploid at fertilization and capable of at least several cleavage divisions (77,78,95). As noted previously, embryos derived from tripronuclear eggs can develop to the blastocyst stage and are capable of implantation and progressive post-implantation development (91). Considering the normal appearance of many cleavage-stage embryos that develop from tripronuclear eggs, the importance of recognizing apparent polyspermic penetration prior to the first cell division cannot be overemphasized.

The underlying factors related to the variable developmental potential of tripronuclear human eggs have yet to be established definitively. However, karyotypic, electron microscopic, and biochemical analyses are beginning to provide information that is of fundamental importance in understanding the molecular and cellular processes involved in both normal and abnormal early human development. Van Blerkom et al. (78) demonstrated that partial diploidy could be restored in triploid embryos. Several eight-cell embryos developed from tripronuclear eggs were shown to be a mosaic of chromosomally normal and abnormal cells, as some of the blastomeres examined by serial thin-section analysis had a diploid complement of chromosomes (Fig. 5-29). Similar findings were derived from cytogenetic analyses of two- to eight-cell embryos that developed from tripronuclear eggs (89). One explanation for the presence of diploid blastomeres after a clearly polyspermic fertilization is that accessory pronuclei do not participate in syngamy and are partitioned to one or both blastomeres at the first cleavage division. Presumably, late chromosomal condensation from one accessory pronucleus could generate a diploid/triploid mosaicism. An embryo containing normal diploid cells would occur if an accessory pronucleus failed to participate in syngamy and, during subsequent cell divisions, remained unchanged or degenerated.

Triploidy appears to be a significant cause of spontaneous abortion in human pregnancy. Some estimates suggest that triploidy may occur in 1.5% of all conceptions (96–98). However, even this figure may be an underestimate if certain forms of triplody, such as those carrying XYY sex chromosomes, are lethal at the preimplantation stages (5). Indeed, such triploids occur in preimplantation embryos after fertilization in vitro (99,100) but are rarely observed in abortuses (101). While triploidy is not an exceptionally rare event associated with fertilization in vivo, its occurrence in vitro appears to be significantly higher. Most IVF programs report a consistent incidence of tripronuclear eggs of approximately 5% (5,95). However, as discussed previously in this chapter, values in excess of 10–15% have also been reported.

Although some forms of triploidy (e.g., 69XYY) are almost certainly lethal at the preimplantation stages, the presence of three pronuclei does not necessarily preclude continued development, especially if some eggs revert to the diploid condition by excluding an accessory male pronucleus from syngamy. However, based on available cytogenetic analyses of embryos derived from tripronucleate eggs (89), reversion to the diploid state is relatively rare and many embryos develop as triploids. An embryo that is a mosaic of chromosomally normal and abnormal blastomeres may have an increased developmental potential that is related to the percentage of diploid cells (78,89). Van Blerkom et al. (78) suggested that even if abnormal blastomeres persist, the presence of such cells in the preimplantation embryo may not be inconsistent with the establishment of a normal pregnancy if these cells enter the trophoblast lineage and participate in the formation of extraembryonic structures. In this regard, Robinson and Puck (102) and Kalousek and Dill (103) reported

in the mouse blastocyst. By contrast, measurements of pyruvate uptake by preimplantation-stage human embryos follows a pattern that is similar to that identified in the mouse (118,123). The mature human oocyte has a high level of pyruvate uptake, but after fertilization this value falls and then slowly increases to reach a peak just prior to the morula stage, after which it decreases (124). An increase in pyruvate uptake prior to the morula stage is not observed in the mouse.

A determination of glucose-based metabolism by preimplantation human embryos cultured in the presence of radiolabeled substrates indicates a substantial increase in carbon dioxide and lactic acid production between days 2 and 3 following fertilization (4–12-cell stage), or approximately 24 hours earlier than for the mouse (124). Although glucose utilization appears to increase early in cleavage, it is apparently not accompanied by a significant rise in the level of oxygen consumption. Indeed, measurements of oxygen consumption during early human development indicate a higher level of oxygen consumption in the preovulatory oocyte (0.53 nl/hr) than in the blastocyst (0.34 nl/hr) (120). This observation is somewhat surprising considering the correlation between increased glucose utilization and increased oxygen consumption observed in the preimplantation-stage embryos of other (nonprimate) mammals.

It should be noted that the study of human embryo metabolism has been initiated very recently and the described findings, which are based on a limited number of observations, may need to be reinterpreted as more data become available. Such information will be especially important to obtain for stages of development between the one-cell and blastocyst. However, current evidence does suggest that the overall level of metabolism of the preimplantation human embryo is lower than expected when compared to preimplantation-stage embryos of other species. This low level may be a reflection of suboptimal culture conditions or an indication of an underlying developmental impairment that is not manifested by gross embryo appearance. Alternatively, measurements of embryo metabolism in highly inbred strains of laboratory mice may not be relevant to the human. Human embryos, even those present in the same woman, may be metabolically heterogeneous. This factor may even be more pronounced when embryos from different women are analyzed. Therefore, determinations of metabolic parameters may be more relevant to a particular embryo than a generalized statement of viability or developmental potential.

The apparently low level of metabolism observed in the early human embryo is consistent with the presence of a morphologically heterogeneous population of mitochondria during the preimplantation period (Fig. 5-31). As described previously, the studies of Dvorak and Tesarik (117) indicate that mitochondrial differentiation in the human occurs in a gradual fashion during the preimplantation stages. These changes involve the shape, configuration, and internal arrangement of mitochondria. Round or oval and elongated mitochondria occur together in all stages, but the number of elongated forms increases as cleavage proceeds to the blastocyst stage. The presence of morphologically undifferentiated mitochondria (round/oval) at all stages of preimplantation development may be reflected in the relatively low level of metabolism exhibited by the early human embryo. The increased uptake and utilization of glucose as a substrate for metabolism appears to coincide with a decrease in the density of the mitochondrial matrix and the concomitant increase in the number of cristae (Fig. 5-32); both biochemical and fine structural changes are first observed at the early cleavage stages in the human.

Developmental arrest at the pre- and peri-implantation stages that may be associated with mitochondrial dysfunction would appear to be very difficult to determine unambiguously. The biochemical nature and the origin(s) (nucleus, cytoplasm) of the signals that regulate the transformation of mammalian embryonic mitochondria from a form that is comparatively inactive in metabolism to one that is associated with higher metabolic levels have not been identified. These signals are presumably generated as part of an intrinsic developmental program. The timing of mitochondrial differentiation does not appear to be regulated by factors extrinsic to the preimplantation embryo. For example, culture of early cleavage-stage embryos in medium containing glucose is not associated with a precocious shift either in mitochondrial structure or to higher levels of metabolism normally associated with

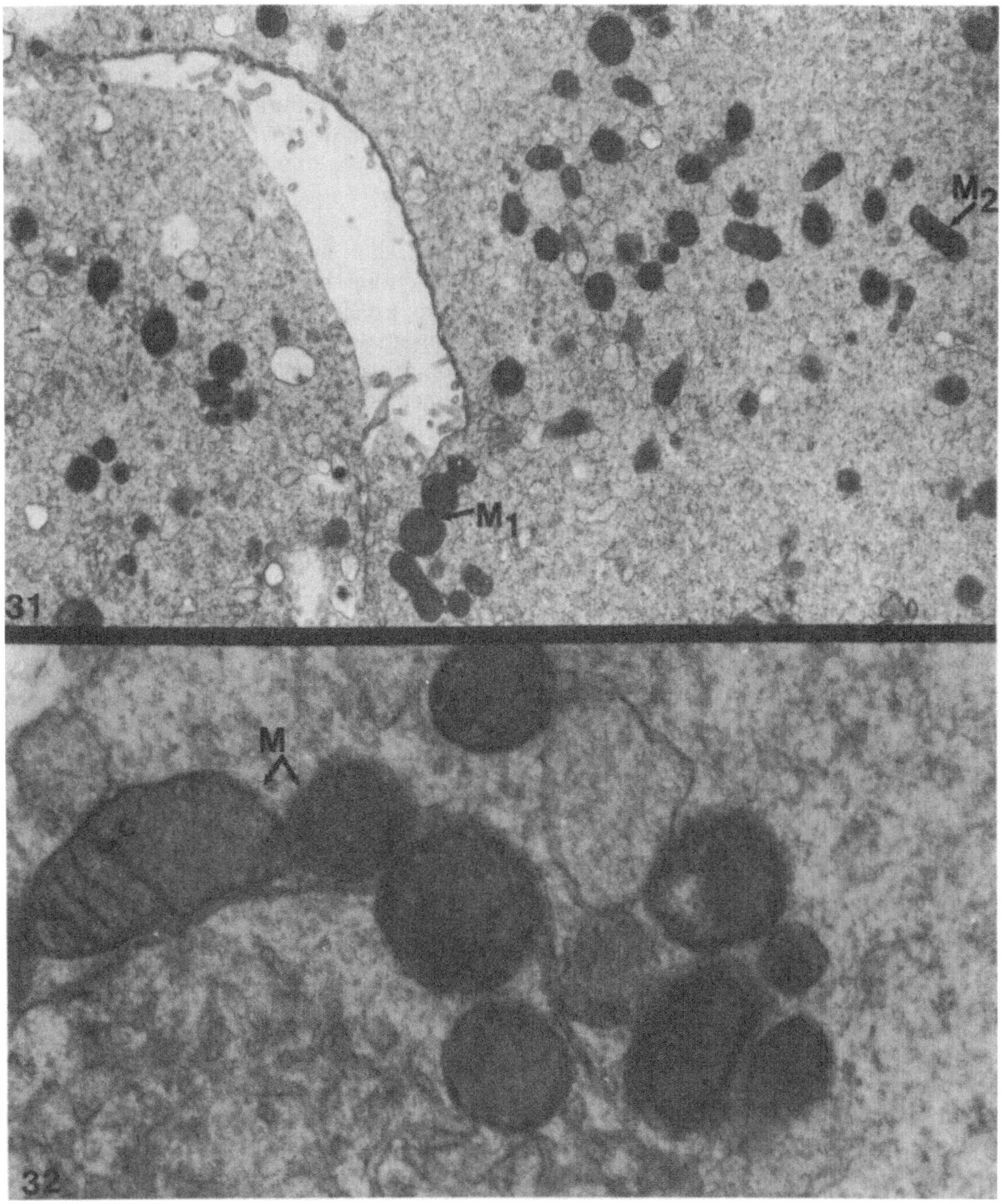

Figures 5-31 and 5-32. Mitochondria in early human embryos are pleomorphic with both round (M1) and elongated (M2) forms present during the preimplantation stages (eight-cell embryo, Figure 5-31). As development proceeds through the cleavage/morula stages (e.g., 16-cell embryo, Figure 5-32), a progressive increase occurs in the number of elongated mitochondria in which cristae (c, Figure 5-32) penetrate completely the mitochondrial matrix. This gradual shift in mitochondrial morphology is associated with an increased level of metabolism and glucose utilization. Figure 5-31, X 12,000; Figure 5-32, X 50,000.

oxidative phosphorylation in later-stage embryos.

The possibility of mitochondrial dysfunction in the human embryo that could result from quantitative or temporal variability in the provision of a transforming signal(s) is unrealistic to approach experimentally at present. A more productive approach may be the application of morphometric analyses in normally developing human embryos in order to correlate quantitative shifts in mitochondrial forms at specific stages of preimplantation development. This type of analysis may be especially relevant for the cells of the trophectoderm and inner cell mass (112). Perhaps the failure of mitochondrial replication or the inability of these organelles to support requirements for higher levels of metabolism in blastocyst-stage embryos could contribute to developmental failure during the peri-implantation stages. In this regard, preliminary findings from a fine structural examination of two human embryos that developed in culture to the mid-to-late blastocyst stage after ferilization in vitro may be of relevance (Van Blerkom, unpublished observations). Both embryos arrested development at the blastocyst stage and upon inspection at the electron microscopic level exhibited metabolically active-appearing mitochondria only in inner cell mass cells. These results, if confirmed by additional analysis of embryos that fail to progress in vitro, may provide an early indication of incipient developmental failure during the peri-implantation period.

The extent to which (if any) mitochondrial dysfunction, whether global or cell type-specific, impairs or compromises embryonic development is unknown. Indeed, it could be reasonably argued that failure of mitochondria to differentiate is a consequence or symptom of other aberrant processes that lead to a premorbid state. However, stage-related transitions in mitochondrial morphology are of functional significance because they relate to the state of embryonic metabolism (125), and therefore may provide a useful indicator of the normality of development at specific stages of early embryogenesis. Nonivasive techniques which can assess the state of metabolism in living embryos should provide a clear indication of the state of mitochondrial function as development progresses to implantation (126). In combination, noninvasive biochemical/physiological measurements and invasive morphometric analyses (127) for both normal and abnormal (e.g., multipronuclear, chromosomally aberrant) embryos could provide a direct indication of states of cytoplasmic differentiation associated with progressive and nonprogressive development.

5.2. *Nucleoli*

Changes in nucleolar composition and organization during the preimplantation stages of development are clearly associated with the formation of ribosomal RNA and a progressive increase in the population of cytoplasmic ribosomes and polysomes (for reviews see 10,59,112,128). Nucleolar differentiation and the resultant increase in the population of cytoplasmic ribosomes is also accompanied by increased levels of protein synthesis (107,128, for review see 10). During the early postfertilization stages of development, protein synthesis is supported by maternal (oocyte) ribosomes. The stage of development at which embryonic ribosomes/polysomes are first detectable in the cytoplasm and, by molecular analysis, are shown to contribute to the support of translation, varies among mammalian species. However, for all species studied to date, an increased rate of nucleolar-derived ribosomal RNA synthesis and a significant increase in the number of cytoplasmic ribosomes and polysomes are observed between the late morula and early blastocyst stages (128, 129). The clear correlation between nucleolar function and organization during preimplantation embryogenesis provides an opportunity to assess the extent to which nucleolar differentiation and early developmental failure or arrest in the human may be associated.

The report by Tesarik et al. (130) is the most comprehensive description of nucleologenesis presently available for the preimplantation human embryo. With a combination of autoradiographic and ultrastructural analyses, these investigators noted that the two- to four-cell embryo lacked functionally active nucleoli. Nucleoluslike bodies in early cleavage-stage human embryos are composed of a homogeneous mass of densely packed fine fibrils (Figs. 5-33 and 5-34), contain no recently replicated DNA, and do not show any detectable RNA synthesis. The first indication of nucleolar function is observed between the six- and eight-cell stage of human development. The

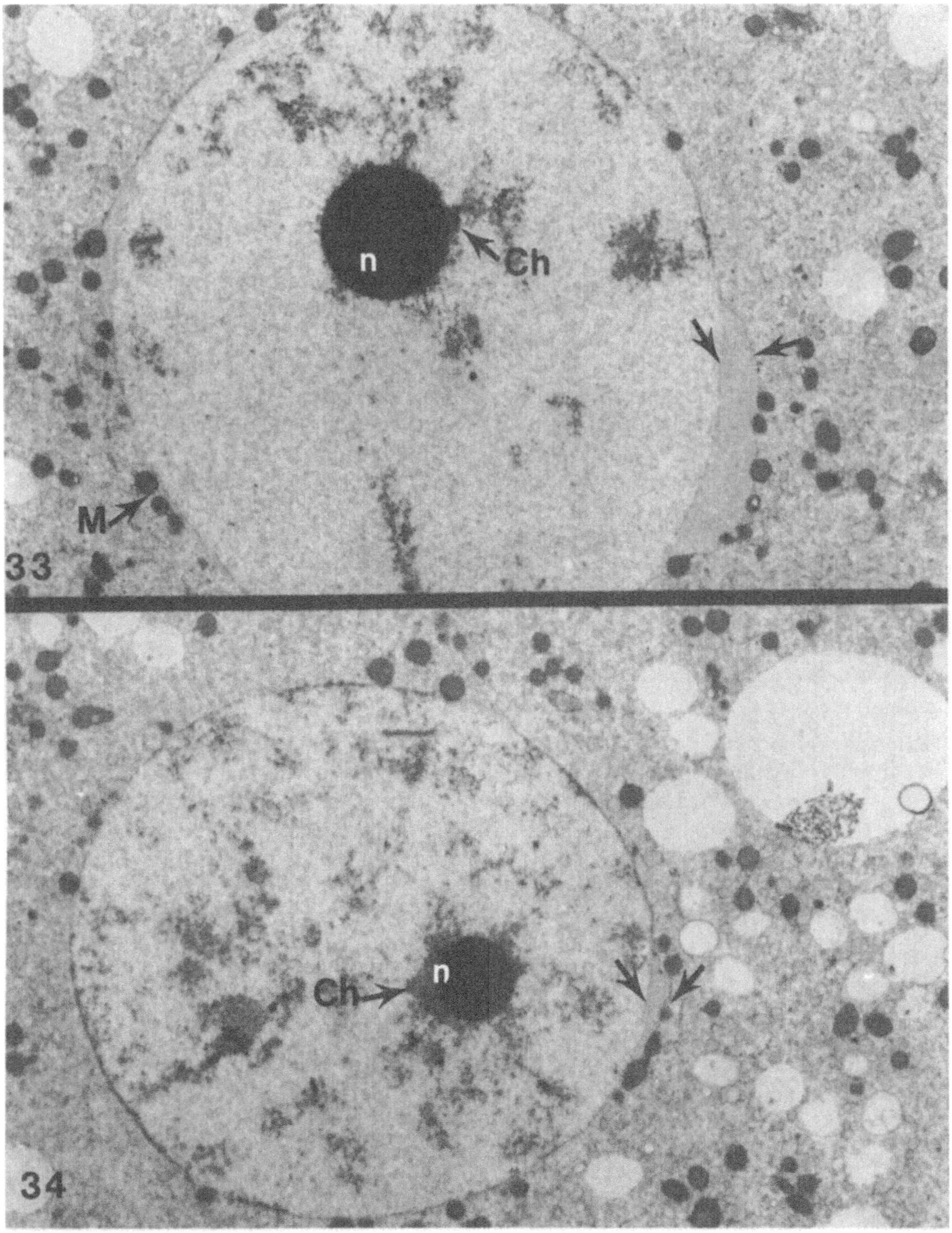

Figures 5-33 and 5-34. Transmission electron micrographs of nucleoli (n) in two- (Figure 5-33) and Four-cell (Figure 5-34) human embryos. Nucleoli at these stages of development are composed of a homogeneous network of densely packed fibrils. Aggregates of chromatin (Ch) are often observed at the periphery of these structures. This type of nucleolus is inactive in transcription. Nuclei in early cleavage-stage human embryos frequently display regions of separation between the inner and outer nuclear membranes (arrows). M: mitochondria. Figure 5-33, X 8,500; Figure 5-34, X 8,600.

transformation of nucleoluslike bodies into true nucleoli is a relatively rapid process that coincides with the beginning of RNA synthesis and processing as reflected by the appearance of the first nucleolar granules. Nucleolar differentiation is characterized by the progressive infiltration of the dense fibrillar mass by adjacent chromatin. By the 12-cell stage of human embryogenesis, the typical morphology associated with functioning nucleoli is established as a result of intranucleolar differentiation. This process gives rise to distinct fibrillar and granular components as well as to nucleolar interstices. The study of Tesarik et al. (130) indicates that in the human, the nucleologenetic process, once started, appears to be completed within one cell cycle.

Structural and temporal aspects of nucleolar differentiation observed by Tesarik et al. (130) are not unique to the human, but rather are very similar to those described for the rabbit (116). Nucleolar differentiation in the rabbit is virtually complete by the 16-cell stage. The appearance of a granular component by the eight-cell stage coincides with the first detectable synthesis of embryonic ribosomal RNA (131). The development of extensively reticulated nucleoli with large zones of granular material in the 12–16-cell embryo is associated with a major increase in rRNA synthesis (132) and with the appearance of a large population of cytoplasmic ribosomes and polysomes (116). Nucleolar differentiation in the rabbit and the mouse morula-stage embryo also coincides rather closely with progressive increases in the rate of protein synthesis (106,107,116, 131).

In contrast to the situation that prevails in the early rabbit and human embryo, nucleologenesis in the mouse begins as early as the two-cell stage and progresses in a gradual fashion during subsequent cleavage divisions (133,134). Sites of ribosomal DNA transcription identified by high-resolution autoradiography after pulse labeling with tritiated uridine demonstrate the initiation of rRNA synthesis between the late two- and early four-cell stages (134). The intensity of rRNA synthesis detected by autoradiographic analysis increases progressively from the four-cell stage to the blastocyst. These molecular findings correlate very closely with the fine structural differentiation of the mouse nucleolus that occurs between the late two-cell and early blastocyst stages (10,133, 134).

The timing of nucleolar activation appears to be closely related to the stage of development at which the embryonic genome becomes active in transcription. Prior to this period, embryonic protein synthesis utilizes ribosomes and is directed by messenger RNA inherited from the oocyte (13,15). With the activation of and expression from the embryonic genome, the regulation of development shifts from a maternally directed to an embryonically directed developmental program (15–18). In the mouse, this shift in developmental regulation occurs as early as the two-cell stage (135), but in the rabbit (57) and human (136), expression of the embryonic genome and embryonic participation in the regulation of development does not appear to occur until the eight-cell stage, if not somewhat later.

The state of nucleolar fine structure and organization in cleavage- and blastocyst-stage human embryos may provide an important, indirect indication of the extent to which the activation/ expression of the embryonic genome has occurred in a particular blastomere. The analysis of human embryos that developed from triploid embryos may be of particular interest, especially for those embryos that can be shown to be or are suspected of being composed of chromosomally normal and abnormal blastomeres. Likewise, embryos developed from normally fertilized eggs but that manifest cellular aberrations during cleavage represent suitable candidates for nucleolar analysis.

Observations of nucleolar fine structure in cleavage-stage embryos indicate that a correlation may exist between the state of nucleolar differentiation and the developmental status of a particular blastomere. At the 12–16-cell stage, some of the blastomeres of embryos that developed after a dispermic fertilization show no signs of nucleolar differentiation. This is especially evident in multinucleated cells and is in contrast to the situation that exists in adjoining blastomeres where nucleolar differentiation appears to be on schedule for the stage of development. This finding is subject to several interpretations: (1) The failure of nucleolar differentiation may be one visible symptom of a cell that has ceased to develop or is developmentally impaired; (2) A particular cell may be normal but developmentally delayed; (3) The appearance of the nucleolus may be related to the stage of the cell cycle at which time the embryo was prepared for electron microscopy.

Fine structural analyses alone cannot establish a definitive cause-and-effect association between nucleolar organization and the developmental normality or potential of a given blastomere. However, the nonlethal determination of the chromosomal constitution of an individual blastomere prior to fine structural analysis would go a long way in indicating whether or not the failure of nucleolar differentiation was associated with a particular chromosomal state. This type of determination, when combined with molecular analyses, could provide a comprehensive indication of the normality of development at the single-cell level. For example, Van Blerkom (11) demonstrated that newly translated polypeptides could be resolved from the individual blastomeres of cleavage-stage rabbit embryos by high-resolution, two-dimensional polyacrylamide gel electrophoresis. Analyses of quantitative and qualitative aspects of protein synthesis by normal preimplantation human embryos should provide a detailed view of stage-related changes in the pattern of molecular expression. A correlation between the pattern of polypeptide synthesis, the gross morphological appearance of a blastomere, and (potentially) chromosomal status could provide a specific indication of the extent to which cytogenetic, cytoplasmic and molecular differentiation in early human embryogenesis are mutually dependent or independent processes (11).

6. The developmental ability of oocytes and preimplantation-stage embryos that are chromosomally or morphologically variant

6.1. Chromosomal abnormalities in the preovulatory oocyte

The results of several recent studies of preovulatory and unfertilized human oocytes indicate that a significant percent of mature human oocytes obtained from hyperstimulated ovaries may be chromosomally abnormal (28,29). Wramsby and Fredga (27) reported that when ovarian stimulation involved clomiphene citrate and human menopausal gonadotropin, 65% of the karyotypes obtained from oocytes that failed to fertilize in vitro were abnormal. One of the more surprising findings of this study was that for 22% of the oocytes, chromosomes could not be detected after fixation and preparation for karyotyping. In a similar series of analyses, Wramsby et al. (28) reported that abnormal karyotypes were observed in nearly 50% of the preovulatory oocytes obtained from women stimulated only with clomiphene citrate and examined within five hours of follicular aspiration. Chromosomes were detected in all of the oocytes examined in this study. Wramsby and Fredga (27) interpreted this observation by suggesting the time difference between recovery and fixation for karyotpic analysis (five hours for uninseminated oocytes versus 48 hours for inseminated oocytes) could be associated with chromosomal loss or degeneration during the continued culture of unfertilized oocytes. Alternatively, differences in the method of ovarian hyperstimulation could be associated with the generation of oocytes with varying developmental potentials, a generation that is also reflected at the chromosomal level. As noted by Wramsby and Fredga (27), the addition of human menopausal gonadotropin, which gives a higher mean number of oocytes recovered, might also induce the development of a larger proportion of chromosomally abnormal oocytes.

In a recent comparison of superovulation protocols from two major in vitro fertilization centers, the frequency of fertilization was 69% (1097/2023) and and 54% (1185/2171) respectively (25). Both stimulation protocols used a combination of clomiphene citrate and human menopausal gonadotropin, but differed in the dose and timing of hormonal administration. With a database derived from several thousand mature oocytes, the high degree of fertilization failure suggests that current hormonal protocols for ovarian hyperstimulation may be associated with the preovulatory development of oocytes that have compromised or impaired abilities.

Chromosomal abnormalities seem to occur in a significant proportion of mature human oocytes that fail to fertilize in vitro (7,23,28,49,137). The extent to which, if any, chromosomal aberrations can be correlated with fertilization failure is fundamental not only to an understanding of the

Each chromosome was pycnotic (Figs. 5-38 and 5-39) and was enclosed by a nuclear membrane (Fig. 5-39). None of the oocytes in this group displayed any indication of an organized metaphase spindle. Degenerate, membrane-enclosed chromosomes were observed in both preovulatory and unfertilized oocytes (for details, see 72). The fine structure of the chromosomes is consistent with degenerative processes. The weak fluorescent response to DNA-specific staining indicates that the DNA may have been partially digested with only the protein scaffold remaining visible with the electron microscope. It is probable that these chromosomes do not survive processing for karyotypic analysis (139) and as a consequence, give the impression that chromosomes are absent in some human oocytes (27).

While the exact origin of this cytogenetic condition is unknown, the ultrastructure of the chromosomes and their enclosure by a nuclear membrane present a potential explanation. The abstriction of the first polar body and the presence of DNA-fluorescence in this structure demonstrates that the first meiotic division had been completed. The absence of an organized spindle and associated chromosomes indicates that the metaphase II spindle did not persist, or that it was transiently stable. The retention of chromosomal fluorescence in the first polar body suggests that degenerative processes may not have been as pronounced in the polar body as in the ooplasm. Enclosure of each chromosome by a nuclear membrane indicates that the cytoplasm retained the developmental ability to promote the formation of a pronuclear membrane. For these oocytes, a nuclear membrane developed around each chromosome, rather than in association with an aggregate of chromosomes normally present after fertilization.

One interpretation of these findings is that the resumption of meiosis occurred spontaneously and precociously (i.e., within the follicle and prior to the appearance of an ovulatory stimulus). The enclosure of dispersed chromosomes by a nuclear envelope may represent the expression of a developmental program or capacity that is initiated during preovulatory maturation and that continues to function in the oocyte even in the absence of fertilization (11,15,18,57). The formation of a nuclear membrane around individual chromosomes could also indicate that the cytoplasm activated spontaneously or parthenogenetically. In this regard, if an oocyte spontaneously resumed meiosis in vivo, it is not unreasonable to suspect that the intrafollicular conditions that promoted precocious meiotic maturation could also be associated with parthenogenetic activation in the mature oocyte. Enclosure of dispersed chromosomes by a pronuclear membrane may be one manifestation of cytoplasmic activation (133).

The vast majority of unfertilized oocytes examined at Reproductive Genetics In Vitro (Fig. 5-40) displayed an apparently normal metaphase II spindle located in proximity to the first polar body (Figs. 5-36 and 5-41). This situation prevailed even in those oocytes that exhibited obvious abnormalities and distortions in cytoplasmic structure and organization (see following section). Meiotic chromosomes could be visualized by fluorescence microscopy in living oocytes (Figs. 5-36 and 5-41), and their arrangement with respect to the spindle could be observed and determined unambiguously (Fig. 5-42). Because of the compact arrangement of chromosomes on the metaphase spindle, the number of chromosomes could not always be quantified definitively. However, as described by Van Blerkom and Henry (72), this approach to a cytogenetic analysis of living oocytes was of sufficient sensitivity to detect major deviations from the normal haploid complement, such as those described by Wramsby et al. (28). With the exception of oocytes in which chromosomes were not detected by routine fluorescence microscopy, or oocytes in which chromosomes that were not associated with the metaphase spindle were identified (see below), virtually all of the mature oocytes appeared to have a normal spindle and, within the limits of resolution, a normal chromosome complement. However, karyotypic analysis of these apparently normal oocytes demonstrated that 7% were indeed chromosomally abnormal. Eighty percent of chromosomally aberrant oocytes were hypohaploid (one to three missing chromosomes).

Neither the protocol of follicular stimulation nor the duration of culture after insemination was associated with the incidence of aneuploid oocytes. By contrast, most of the oocytes with pycnotic, membrane-enclosed chromosomes were

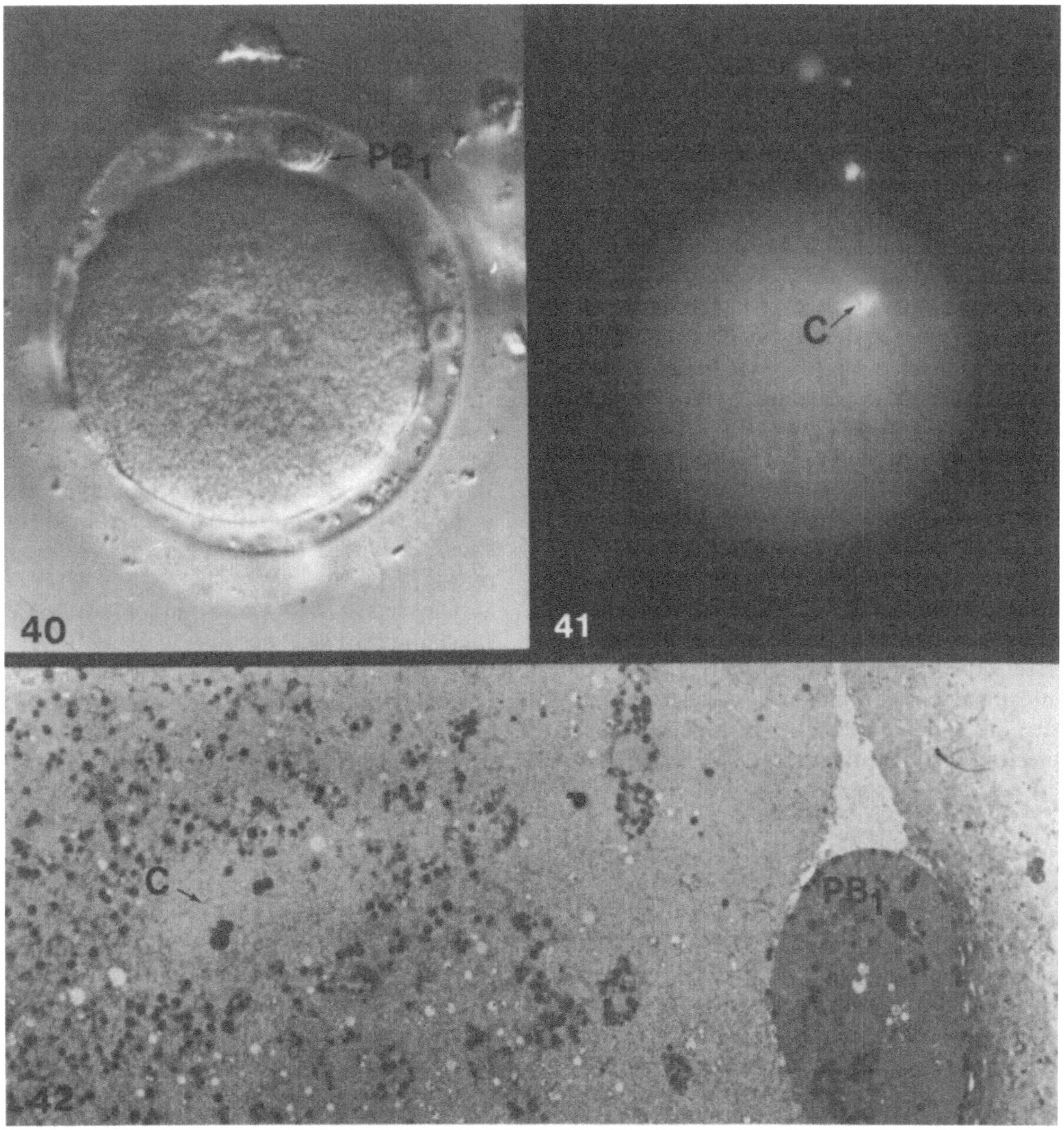

Figures 5-40–5-42. Light (Figure 5-40), DAPI-fluorescence (Figure 5-41) and transmission electron micrographs (Figure 5-42) of the typical mature but unfertilized oocyte observed at Reproductive Genetics In Vitro. DAPI fluorescence demonstrated the presence of chromosomes in the first polar body (Pb1) as well as in association with the metaphase II spindle (C). Figure 5-42, X 2,800.

observed after stimulation with clomiphene citrate and human menopausal gonadotropin. Of particular clinical relevance in this study was the finding that 27% of the aneuploid oocytes were obtained from two women. While these results must be considered tentative at present, they do present the intriguing notion that chromosomal perturbations in human oocytes may more likely be patient-specific than an obligate consequence of exogenous ovarian stimulation.

Van Blerkom and Henry (72) also identified living human oocytes in which some chromosomes were not associated with the metaphase II spindle. Subsequent karyotypes of these oocytes indicated a hypohaploid complement. The presence of chromosomes that are free in the cytoplasm may be a result of nonassociation at anaphase I (anaphase lag) or disassociation from the spindle at metaphase II (137). This observation demonstrates that the potential for chromosomal loss during destructive fixation for karyotyping can exist in human oocytes that contain an otherwise normal chromosomal constitution. It should be emphasized that while such oocytes are chromosomally normal, they would, if fertilized, most likely be developmentally compromised. As discussed previously in this chapter, the fertilization of such oocytes may contribute to the occurrence of preimplantation-stage human embryos that are morphologically normal but genetically abnormal.

The findings of Van Blerkom and Henry (72) and of Martin et al. (137) suggest that the frequency of chromosomally abnormal human oocytes may not be as high as reported by some investigators (e.g., 27,28). A preliminary conclusion based on an analysis of the preovulatory and unfertilized oocytes of 109 patients suggests that for most women, chromosomally abnormal oocytes may not be a frequent or expected consequence of hyperstimulation.

6.2. Genomic and extragenomic influences on preovulatory oocyte maturation

If a definitive relationship can be established between the chromosomal status of an oocyte and its subsequent fertilizability, an understanding of the molecular and cellular basis of this relationship may present one of the foremost experimental challenges in the study of early human development. For example, to what extent, if any, does the genome contribute to the regulation of chromosomal and cytoplasmic activites during oocyte maturation? The notion of genomic regulation must be reconciled with the fact that meiotic chromosomes are in a condensed state. However, transcriptional events taking place prior to germinal vesicle breakdown, when chromosomes are in a decondensed state, may be of fundamental importance for the normal progression of subsequent cytoplasmic and nuclear (chromosomal) maturation. In the mouse, transcription is detected up to the early stages of germinal vesicle breakdown (42,43). In the sheep, the addition of alpha-amanitin (an inhibitor of RNA polymerase II activity in vitro) to extrafollicular, cumulus-enclosed oocytes at explantation inhibits meiotic maturation and prevents many of the changes in polypeptide synthesis that normally accompany maturation in vitro (55). By contrast, these inhibitory effects are considerably reduced by either delaying the addition of alpha-amanitin for one to four hours or by denuding the oocyte of all associated cumulus cells at the outset of culture. Osborn and Moor (55) concluded that a transcriptional event at the onset of maturation in the sheep may be essential for the initiation of those changes in polypeptide synthesis required for the regulation of nuclear and cytoplasmic maturation. What could not be determined in this study was whether the putative transcriptional event(s) occurred within the cumulus cells or within the oocyte.

The report of Osborn and Moor (55) is of special importance because it suggests the possibility that molecules that may support, regulate or influence the course of preovulatory oocyte maturation could originate from a source external to the oocyte, such as the cells of the cumulus oophorus and corona radiata. Precedence for the endocytosis of proteins by oocytes exists not only in amphibians and insects (for reviews see 140, 141), but is also indicated by the analysis of absolute rates of protein synthesis during mouse oogenesis (14). Measurements of the rate of protein synthesis indicate that less than 60% of the protein present in fully grown mouse oocytes can be accounted for by protein synthesis during oocyte growth (142,143). This finding suggests that approximately 40% of the total protein present in the fully grown mouse oocyte may originate from an external source. However, this figure may be an overestimate. Errors in one or more of the measurements used in the calculation of the absolute rate of protein synthesis, the timing of growth, or total protein content could account for the discrepancy between the anticipated and the actual protein content of the fully grown mouse oocyte (14). Inadequate culture conditions for growth-stage oocytes may also adversely affect the observed rate of protein synthesis in vitro (144). Consequently, whether exogenous protein contributes to the total protein content of the fully grown mouse oocyte requires additional analysis. However, an unambiguous determination of whether or not mammalian oocytes require or receive exogenous protein may be a critical aspect of oogenesis as it relates to the developmental normality and potential of the fully grown oocyte.

Whether the human oocyte requires or receives exogenous protein remains to be determined. However, observations from animal studies, such as those of Osborn and Moor (55), provide some intriguing speculations concerning the process by which the human oocyte acquires maturational and developmental competence. The cells of the cumulus oophorous and corona radiata are the most obvious source of exogenous protein and other nutritional or regulatory molecules. The most obvious route by which such factors could be made available to the oocyte is by means of follicle cell processes which traverse the zona pellucida and become associated with the vitelline membrane through gap and tight junctions (30, 145–147). Therefore, transmission of follicle cell-derived molecules to the oocyte could occur directly, or indirectly by export from follicle cell processes to the perivitelline space and endocytosis by the oocyte.

If follicle cell-derived molecules are involved in human oogenesis, especially at the onset of preovulatory maturation, then the normality of oocyte development may be influenced by the timing or character of the transmitted/secreted molecules. In this regard, the hormonal protocols employed for ovarian stimulation may have a significant and differential effect on follicle cell activities that have consequences for oocyte development and maturation. For example, the degree of expansion or physical distensibility of the cumulus oophorus is one indication of the mucopolysaccharide content of the extracellular matrix and consequently the state of preovulatory development of the cumulus oophorus (85). The density of cytoplasmic lipid droplets visible at the light microscopic level in cumulus cells provides a useful indication of the degree of follicle cell luteinization (148).

The distensibility and degree of luteinization of the cumulus oophorus have been suggested to correlate with the state of meiotic maturation of the oocyte (85). However, detailed analyses of human oocytes have shown that a considerable asynchrony can exist between the stage of meiotic maturation and state of differentiation of the cumulus oophorus (149). Laufer et al. (150) have described the frequent occurrence of meiotically mature human oocytes enclosed by an unexpanded (compact) and unluteinized cumulus oophorus, as well as the reverse situation in which an apparently fully developed cumulus oophorus-enclosed immature oocytes. This finding would tend to suggest that preovulatory changes normally associated with the cells that surround the human oocyte may not significantly influence the process of meiotic maturation. However, the studies of Laufer et al. (150) were not designed to determine whether chromosomes were equally distributed between oocyte and polar body, or whether asynchronous development of oocyte and cumulus was related to fertilizability and normal postfertilization embryogenesis.

Comprehensive analyses of the degree of developmental synchrony between the cells of the corona radiata/cumulus oophorus and the oocyte could contribute significantly to an understanding of developmental failure or chromosomal aberrations in human reproduction that may originate with the oocyte. For example, if the granulosa cells that surround the oocyte influence the process of oocyte maturation, what is the nature and timing of such influence and to what extent is the function of these cells determined by biochemical conditions within the follicle? This may be of developmental significance if cellular activities of the cumulus oophorus and corona radiata are regulated or altered by the follicular milieu at the onset of resumed meiosis. If stage-related molecular activities of follicle cells are involved in preovulatory maturation, alterations in the nature, or perturbations in the timing of synthesis or transport of follicle cell-derived factors may have major consequences for the subsequent developmental potential of the oocyte. This particular notion may be of major clinical relevance in evaluating the mode of ovarian hyperstimulation with respect to the degree of developmental synchrony between the cumulus cells and the oocyte and (1) the normality of chromosomal maturation, (2) the fertilizability and normality of preimplantation development, and (3) the frequency of initiated and continuing pregnancies.

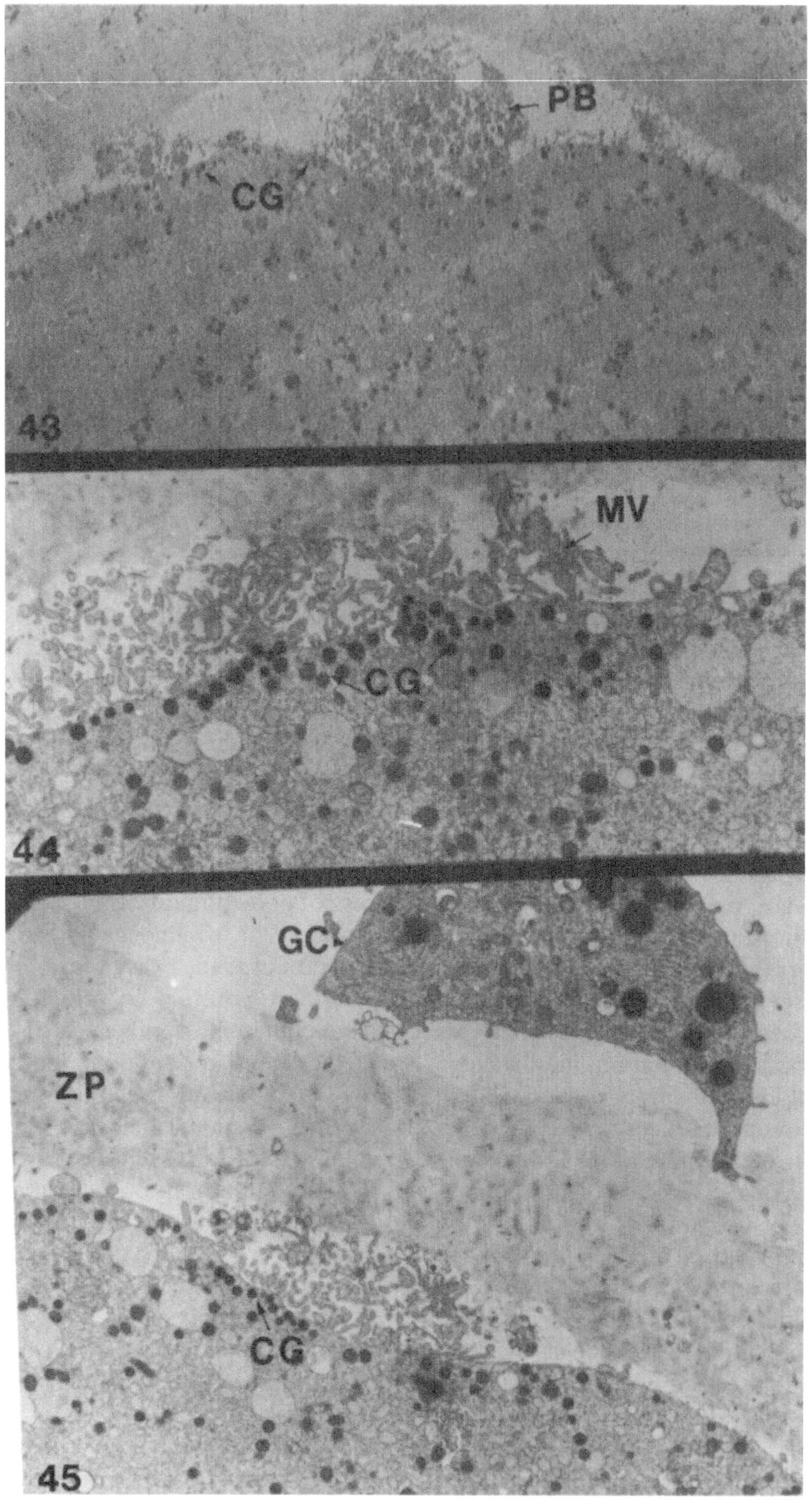
PB
CG
43
MV
CG
44
GC
ZP
CG
45

Although the role of follicle cells in the preovulatory maturation of the human oocyte is speculative at present, the experimental basis exists to determine if, when and what follicle cells contribute to the oocyte. For example, GV-stage human oocytes will spontaneously resume meiosis in vitro (151). Karyotypic analysis of denuded and intact human oocytes cultured from the GV stage should indicate whether the type and incidence of chromosomal abnormalities are related to the presence or absence of cumulus and coronal cells. In the same respect, radiolabeling of intact and denuded oocytes may indicate if and when follicle cell-specific proteins enter the ooplasm. Culture of oocytes in the presence of transcriptional inhibitors, such as alpha-amanitin, prior to or at the onset of meiotic maturation, could suggest whether or not transcription that occurs prior to chromosomal condensation may be required for normal cytoplasmic and meiotic maturation.

Because a high incidence of developmental failure and chromosomal abnormality appears to be associated with preovulatory maturation in the human, the type of experiments described above, although rather basic in design, may be quite informative with respect to the identification of molecular correlates of developmental failure in the human. A comparison of experimental findings from the human with those derived from the intact and denuded oocytes of laboratory (e.g., mouse, 152) and livestock species (e.g., sheep) that were matured and fertilized in vitro may indicate the extent, if any, to which genomic or follicle-cell participation, or both, is involved in the regulation of preovulatory oocyte maturation.

7. Fine structural aberrations in grossly normal-appearing mature oocytes

As discussed previously in this chapter, the causes of fertilization failure are difficult to identify unambiguously for mature oocytes that appear grossly normal. This is especially true for women with patent fallopian tubes, a history of idiopathic infertility and who, with different protocols of ovarian stimulation, consistently produce mature oocytes that fail to fertilize after multiple IVF cycles. For many of these patients, we have been able to examine cytoplasmic and chromosomal organization immediately after follicular retrieval. By light and fluorescence microscopic criteria, gross morphology, the state of meiotic maturation and the appearance of the cumulus oophorus in these oocytes were identical to the oocytes exposed to sperm. Twenty-seven oocytes from eight women in this category were examined by differential interference contrast, fluorescence and transmission electron microscopy.

A strikingly similar state of cellular organization was observed in all 10 oocytes obtained from three patients. DAPI fluorescence demonstrated a metaphase II spindle and associated chromosomes subjacent to the first polar body. Chromosomes dissociated from the spindle and free in the ooplasm were not observed. At the fine structural level, the organization of the cytoplasm and the appearance and distribution of cytoplasmic components were characteristic of normal, mature human oocytes. The only visible abnormality in these oocytes was a restriction of cortical granules to a highly localized region of the cytocortex. The particular region of the cytocortex to which cortical granules were restricted varied among oocytes. For some of these oocytes, serial section analysis demonstrated cortical granules restricted to the portion of the cytocortex associated with the first polar body (Figs. 5-43–5-45). Frequently, the first polar body in these oocytes was degenerate (Fig. 5-43) (72), suggesting that its stability may be compromised after abstriction. By contrast, no chromosomal, cytoplasmic, or cytocortical perturbations that could potentially be correlated with fertilization

←

Figures 5-43–5-45. The oocytes of some patients in which fertilization consistently failed to occur revealed a consistent pattern of restricted cortical granule distribution. A feature common to all of these oocytes is a restriction of cortical granules (GC) to the region of the cytocortex in proximity to the first polar body (PB, Figure 5-43; see text for discussion of the physical state of the polar body). This situation prevailed even in those oocytes prepared for electron microscopy immediately after retrieval from the ovary. Figures 5-43–5-45 show a series of thin sections taken through this specific region of cytocortex. A second structural feature common to this particular class of oocyte is the elaboration of a relatively dense population of microvilli (MV) from the region of plasma membrane that overlies the cortical granules. Figure 5-43, X 2,500; Figure 5-44, X 5,000; Figure 5-45, X 5,400.

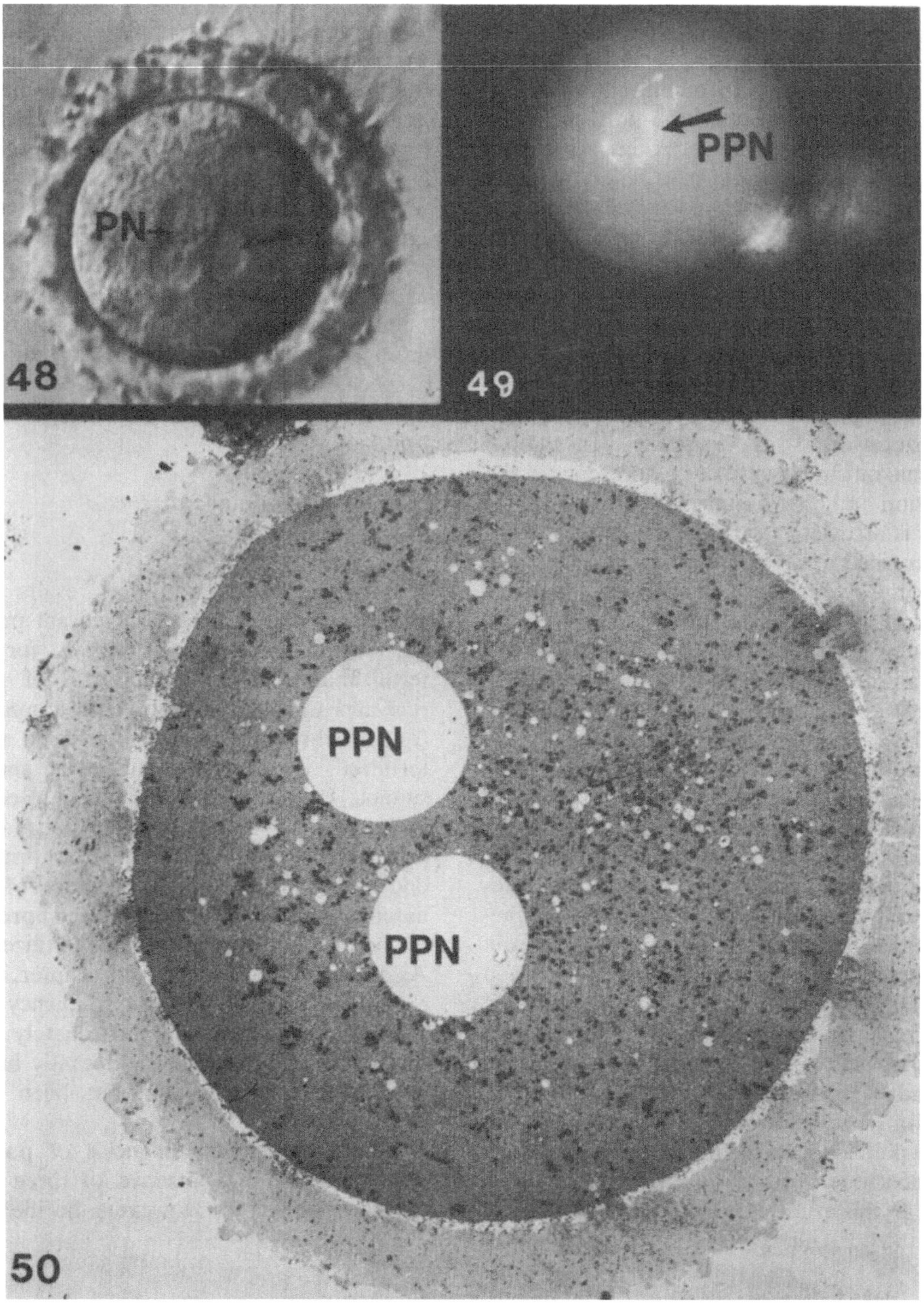

Figures 5-48–5-50. Differential interference contrast (Figure 5-48) and DAPI-fluorescence photomicrographs of a normally fertilized human egg containing two pronuclei (PN) and a pseudopronucleus (PPN). Pseudopronuclei are also observed in meiotically mature oocytes that are examined immediately after retrieval from the ovary (PPN, Figure 5-50). Figure 5-50, X 1,400. (From 2).

of polyspermic fertilization can be confounded by cytoplasmic structures that, by routine light microscopic examination, give the impression of accessory pronuclei and therefore suggest an aberrant fertilization. This type of structure, noted by Wentz et al. (77), has been termed pseudopronuclei by Van Blerkom et al. (2). Detailed analysis of the origin and occurrence of pseudopronuclei by Van Blerkom et al. (2) demonstrated that pseudopronuclei could develop either in the preovulatory oocyte or in the newly fertilized egg. Pseudopronuclei can be distinguished from true pronuclei (Fig. 5-48) at the light-microscopic level by the absence of nucleoli and by the lack of morphogenetic changes normally associated with pronuclei during the presyngamic stages of development. This characterization is supported by fluorescent and electron microscopic analyses (2). No fluorescence was associated with the pseudopronuclei after staining with the DNA-specific fluorescent probes (Fig. 5-49). During subsequent cleavage divisions of embryos displaying this morphology, the pseudopronuclei remained spherical and, by light-microscopic inspection, were associated with a single blastomere(s) at the 4–12-cell stage (2).

In a detailed examination of 27 presumed multipronuclear eggs, Van Blerkom et al. (2) reported that approximately 40% of these eggs were in fact normally fertilized, but contained one or more pseudopronuclei. Observations of putative multipronuclear eggs at two-hour intervals (from 14–24 hours postinsemination) clearly demonstrated that pseudopronuclei could move within the cytoplasm in the same manner as true pronuclei and could become juxtaposed with normal pronuclei. This finding indicates that intracellular processes involved in pronuclear migration and juxtaposition are not necessarily pronuclear-specific. The ability of pseudopronuclei to move within the cytoplasm and to become juxtaposed with true pronuclei demonstrates the importance of identifying the occurrence of these structures at the first inspection for fertilization.

Although based on a relatively small sample, the findings of Van Blerkom et al. (2) indicate that pseudopronuclei can develop both during preovulatory maturation and after fertilization. For example, inspection by differential-interference contrast microscopy immediately after retrieval from the ovary (34 hours post-hCG) of the the oocyte shown by transmission electron microscopy in Figure 5-50 indicated the presence of two pronuclearlike structures. Fluorescence and transmission electron microscopic examination demonstrated that the oocyte was meiotically mature (arrested at metaphase II) and contained two structures in close proximity and of pronuclear dimension. Approximately 1–2% of the 1400 meiotically mature oocytes obtained at Reproductive Genetics In Vitro between 1982 and 1986 exhibited pronuclearlike structures. By contrast, pseudopronuclei were not observed in any of the 112 GV-stage oocytes that were retrieved from antral follicles during stimulated cycles (2).

The above findings tentatively suggest that the formation of pronuclearlike vacuoles may be associated with the process of preovulatory maturation. Their occurrence in the MII-stage oocytes may be the result of aggregation and fusion of comparatively smaller cytoplasmic vacuoles that are characteristic of human oocytes (154,155). Pseudopronuclei can also develop de novo as a consequence of uncontrolled endocytosis initiated during the preovulatory resumption of arrest meiosis (2). Support for this notion is derived from the examination of human oocytes during maturation from the GV stage in vitro. As noted previously, pseudopronuclei have not been observed in the 112 GV-stage human oocytes examined to date at Reproductive Genetics In Vitro. When these eggs were subsequently cultured under conditions that promote the spontaneous resumption of meiosis (50,151), approximately 70% of the oocytes matured to metaphase II. Two oocytes were observed in the process of developing pseudopronuclei during maturation in vitro. In the first case, pronuclearlike structures appeared subsequent to the abstriction of the first polar body. The second oocyte is an extreme example of vacuolar formation de novo. In this particular oocyte, a pseudopronucleus was first observed to develop in proximity to the region of the cytocortex from which the first polar body would subsequently emerge. Within one hour, the first polar body had been abstricted and a single, relatively enormous vacuole had developed (Figs. 5-51,5-52 and 5-54). Serial section analysis by transmission electron microscopy de-

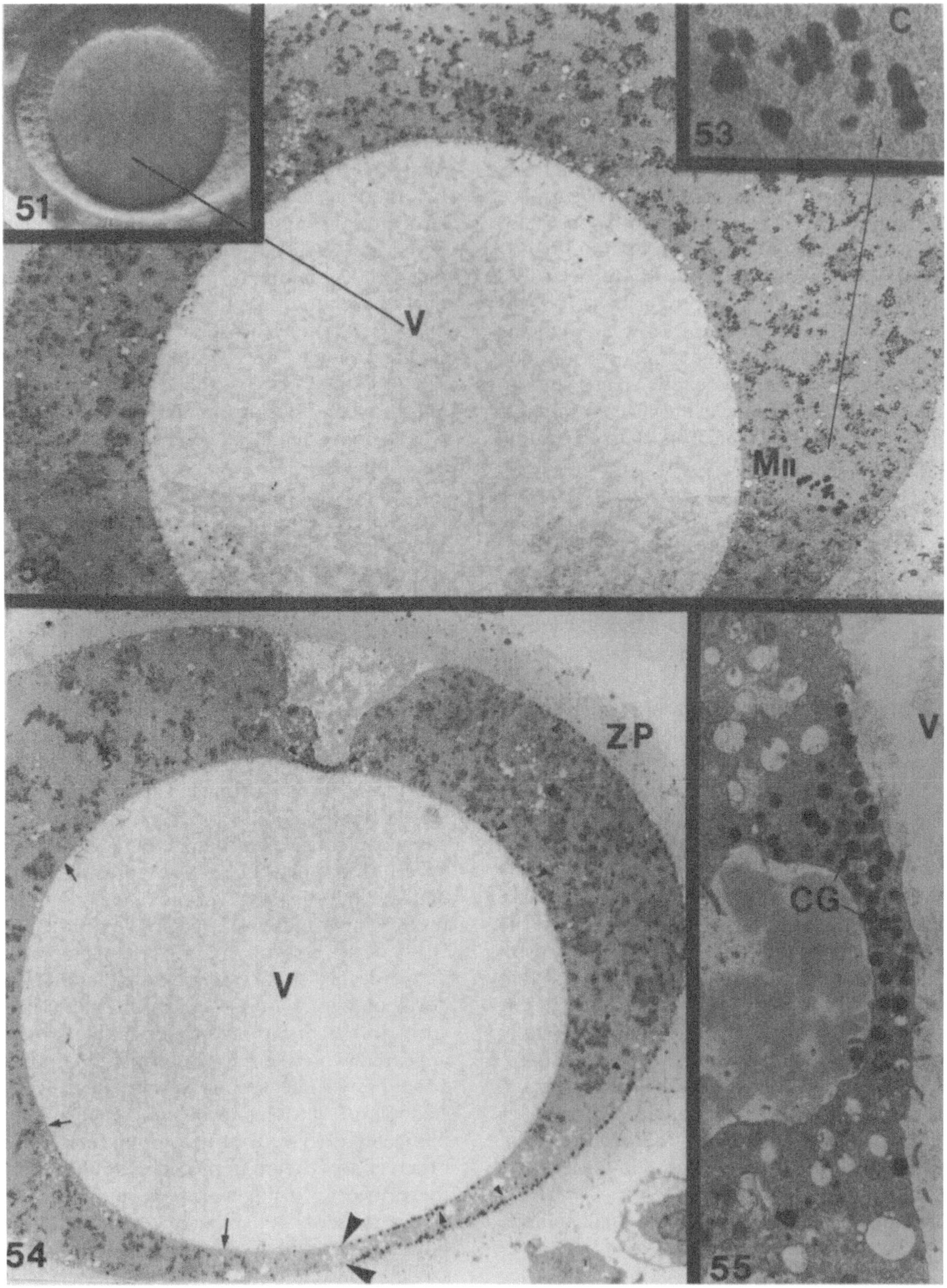
51
V
53
C
MII
52
ZP
V
54
V
CG
55

monstrated the vacuolar nature of this structure (Figs. 5-52 and 5-54), the presence of metaphase II chromosomes (Figs. 5-52 and 5-53), and the absence of any breaks in the continuity of oolemma (Figs. 5-54 and 5-55). Of particular interest is the observation of a polarized distribution of cortical granules in this oocyte. Analysis of serial sections demonstrated both the restriction of cortical granules to one-half of the oocyte (large arrows, Fig. 5-54) and the presence of these elements in the cortical and the perivacuolar cytoplasm (arrowheads, Fig. 5-54). This rather strange distribution indicates the occurrence of a major cytocortical reorganization or structural perturbation associated with the formation of this vacuole. The apparently rapid formation of cytoplasmic vacuoles in preovulatory oocytes appears to occur around the time of polar body abstriction. This observation suggests that transient and/or localized instabilities in the plasma membrane may occur in some oocytes during formation of the first polar body (2). A consequence of such instability may be aberrant or uncontrolled endocytosis that leads to the formation of intracellular vacuoles that have the approximate dimension and gross light-microscopic properties as true pronuclei.

Pseudopronuclei can also arise in fertilized eggs. The egg shown in Figure 5-56 is of particular relevance because it is representative of a class of normally fertilized eggs that displayed two definitive pronuclei when observed to confirm the occurrence of fertilization at 13 hours postinsemination. However, when this egg was reexamined at 16 hours, two additional pronuclearlike structures were encountered, a finding that initially suggested the possibility of an aberrant (polyspermic) or delayed fertilization by an accessory spermatozoon. Electron microscopic images of this egg not only demonstrated that these nascent structures were pseudopronuclei, but also provided an indication of how they could arise (2). Both the volume and the rapidity with which these structures developed (within three hours) reduced the likelihood of formation by aggregation and fusion of small endogenous vacuoles. High-magnification images of the pseudopronuclei demonstrated enclosure by an indistinct plasma (Fig. 5-57) rather than clearly defined pronuclear membrane (Fig. 5-58). Reconstructions of serial thin sections in the portion of the egg indicated by the large arrow in Figure 56 revealed that the interior of the pseudopronucleus was separated from the perivitelline space by an attenuated strand of cytoplasm (Fig. 5-59) and, in some regions, only by the oocyte plasma membrane (Fig. 5-60). The material contained within the pseudopronuclei had the same electron-microscopic characteristics as that observed in the perivitelline space (Figs. 5-59 and 5-60). The ruffled appearance of the plasma membrane (arrows, Figs. 5-56,5-59 and 5-60) suggest that this region of the oocyte plasma membrane may be unstable or highly active, or both.

Clearly, an aberrant process of endocytosis has occurred when the majority of the oocyte cytoplasm is occupied by a single vacuole. However the presence of pseudopronuclei is not inconsistent with normal embryogenesis and development after uterine placement. To date, three babies have been born after transfer of embryos containing one or more pseudopronuclei (2). The demonstrated normality of development indicates that eggs containing pseudopronuclei, whether formed by the aggregation of preexisting vacuoles, or by endocytosis, or both, may represent a class of morphological variants of fertilizable oocytes and developmentally competent eggs.

←

Figures 5-51–5-55. The development of an enormous cytoplasmic vacuole (V) subsequent to the abstriction of the first polar body and formation of the second metaphase spindle (MII, Figure 5-52; C, Figure 5-53) is shown in this series of light (Figure 5-51) and transmission electron micrographs (Figures 5-52–5-55). Of particular interest is the apparent perturbation of cytocortical organization as indicated by the highly polarized distribution of cortical granules (CG, arrows and arrowheads, Figures 5-54 and 5-55). Analysis of serial sections demonstrated both the restriction of cortical granules to one half of the oocyte (large arrowheads, Figure 5-54) and the presence of cortical granules in the cortical and perivacuolar cytoplasm (arrowheads, Figure 5-54). See text for discussion of these observations relevant to the formation of pseudopronuclei. Figure 5-52, X 1,700; Figure 5-53, X 5,000; Figure 5-54, X 1,400; Figure 5-55, X 7,500.

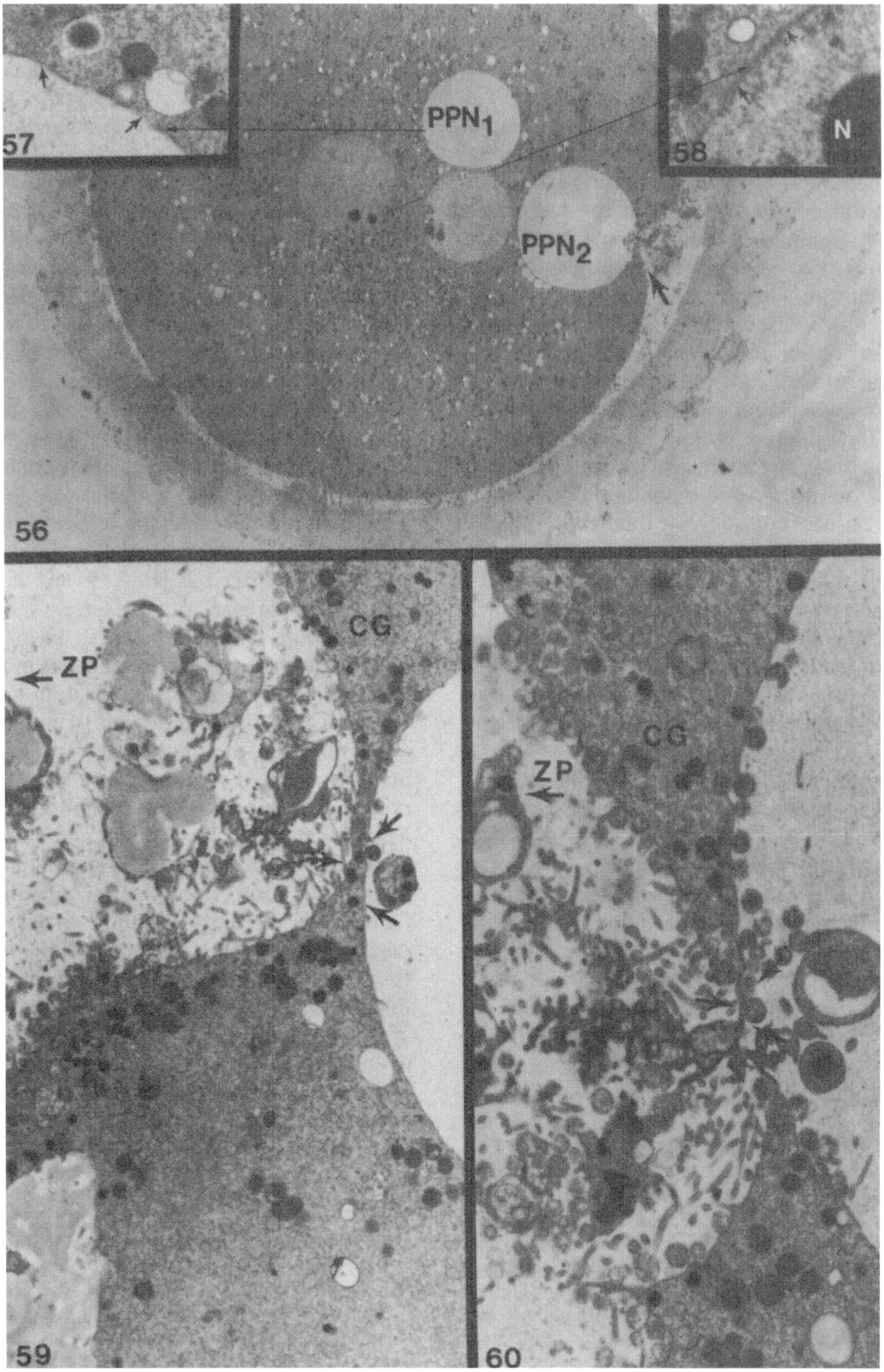
57
58
N
PPN1
PPN2
56
CG
ZP
59
CG
ZP
60

8. The occurrence and developmental consequences of multinucleated blastomeres in cleavage-stage human embryos

Blastomeres containing two or more nuclei are frequently observed in cleavage-stage embryos that develop from multipronuclear eggs (78). In situations where as many as six nuclei were present, staining with fluorescent probes specific for DNA and transmission electron microscopic examination of serial sections demonstrated that many but not all of the nuclei contained DNA and nucleoli (78). The intensity of DNA fluorescence, although not quantitated, suggested an unequal chromosomal distribution among the nuclei.

Plachot (156) reported that approximately 69% of four- to eight-cell human embryos developed from eggs that fertilized normally in vitro (two pronuclei) contained at least one multinucleated blastomere. As noted by Tesarik et al. (92), this high percent of abnormality indicates that blastomere multinucleation is a rather frequent phenomenon occurring in human embryos after normal monospermic fertilization. That multinucleation is not restricted to in vitro-fertilized eggs is demonstrated by a report showing four out of six early human embryos recovered from the reproductive tract contained multinucleated blastomeres (157). In the same respect, because multinucleation has been observed in a high percent of cleavage-stage human embryos that developed in different media and under different conditions of culture (92,149,156), this phenomenon does not appear to be associated with or mediated by environmental factors.

Multiple nuclei present in embryos that developed from normally fertilized eggs display the same fine structural features as observed in embryos derived from multipronuclear eggs (78). Most of the nuclei in multinuclear blastomeres are similar in appearance, dimension and composition to those present in the normal blastomeres of cleavage-stage embryos (Fig. 5-61), as well as a variable number of smaller nuclei (Figs. 5-62 and 5-63) (termed pseudonuclei by Tesarik et al. 92) are encountered in these abnormal cells from four- (Fig. 5-62) and 10-cell embryos (Fig. 5-63). Fluorescence microscopy with DNA-specific probes followed by transmission electron microscopic analysis of serial thin sections demonstrate the presence of DNA and nucleoli in most, but not all, of the multiple nuclei (78).

Ultrastructural and autoradiographic observations by Tesarik et al. (92) of multinucleated blastomeres in two- to 16-cell human embryos that developed in vitro from monospermic fertilization provide the most comprehensive description to date of DNA and RNA synthesis in these blastomeres. For 2–4-cell embryos, DNA labeling was detected in all multiple nuclei, although in comparison to the nuclei of normal blastomeres, irregularities in the intranuclear distribution of labeled DNA were often encountered. Normal-sized nuclei and pseudonuclei were both heavily labeled with ^{3}H-thymidine in the 8–16-cell embryo. In addition to nuclei and pseudonuclei, morula-stage embryos contained aggregates of condensed chromatinlike material that was free in the cytoplasm. Approximately 50% of these aggregates showed DNA labeling with the radioactive probe. With a ^{3}H-uridine probe for the detection of transcriptional activation, which begins in the human after the third cleavage division (130), nearly all of the nuclei and pseudonuclei in these cells could be classified as nonactivated. As noted by Tesarik et al. (92), the relatively low levels of RNA synthesis in multinucleated blastomeres, which also lack the ultrastructural markers

←

Figures 5-56–5-60. Pseudopronuclei can also form after fertilization. The egg shown in this series of transmission electron micrographs displayed two pronuclei when examined at 13 hours after insemination. However, at 16 hours, two pseudopronuclei (PPN) were observed, one of which appeared to be in the process of formation (Figure 5-56, PPN2). At high magnification, enclosure of the pseudopronuclei by an indistinct plasma (arrows, Figure 5-57) rather than distinct pronuclear membrane (arrows, Figure 5-58) is apparent. Reconstruction of serial sections demonstrated a continuous separation of the contents of the vacuole from that of the perivitelline space, although in some regions the two compartments were separated by an attenuated strand of cytoplasm (arrows, Figure 5-59) or by the oolemma (arrows, Figure 5-60). N: nucleolus; CG: cortical granule; ZP: zona pellucida. Figure 5-56, X 1,400; Figure 5-57, X 17,000; Figure 5-58, X 17,000; Figure 5-59, X 7,500; Figure 5-60, X 12,000. (From 2.)

targeted for degradation or cellular elimination, restoration to a diploid state might not result in a normal chromosomal constitution (e.g., occurrence of trisomies). At present, the extent to which chromosomally abnormal blastomeres exists in cleavage-stage embryos after monospermic fertilization is unknown because studies of the chromosomal status of cleavage-stage human embryos have not included an analysis of all of the blastomeres in each embryo (89,158). Perhaps those cells exhibiting multiple nuclei, pseudonuclei, or cytoplasmic aggregates of chromatin may not be amenable to definitive karyotypic analysis. This suggestion could explain the somewhat paradoxical observation that in spite of a high frequency of multinucleation in human cleavage-stage embryos, most of the analyzable blastomeres in embryos that develop from normally fertilized eggs are diploid. Therefore, it may be of some relevance to determine the degree of multinucleation in cleavage-stage human embryos prior to chromosomal analysis.

Collectively, available evidence suggests that blastomeres containing nuclear or chromosomal aberrations, or both, are not developmentally viable. Therefore, the developmental potential of such an embryo may be a function of the number of abnormal cells, and at what time during preimplantation development these cells arise and cease to continue to divide. The presence of diploid blastomeres in embryos that develop from polyspermic fertilization (78,89) demonstrates that the human embryo can on occasion restore the diploid state and presumably eliminate aberrant blastomeres from ongoing developmental and differentiative processes. One of the most intriguing questions in the study of early human development may well be whether restoration to chromosomal normality occurs by the demise of aberrant cells, or actually involves an intrinsic repair mechanism, or some combination of the two.

9. Conclusions and future experimental directions

Some of the most visible and understandable causes of early developmental failure in the human involve sperm dysfunction or an inadequate, if not hostile, environment in the female reproductive tract, or both. The observations discussed in this chapter deal with more subtle causes of reproductive failure. In this regard, detailed analysis of chromosomally or morphologically aberrant oocytes and embryos, rather than representing a collection of oddities, may in fact provide fundamental information in understanding the progression and patterns of normal and pathological early human development.

The findings discussed in this chapter are mostly derived from couples in which a male factor had been precluded and in which no obvious dysfunction in endocrine, ovarian, or uterine function was known or observed. The results indicate that chromosomal and morphological anomalies in oocytes and embryos are a frequent feature of human reproduction, and are often associated with early developmental failure. Some of the morphological aberrations are only identifiable after analysis with specific fluorescent molecular probes, or after fine structural or karyotypic examination. Clearly, further study is required to determine the extent to which they are indeed proximal causes of fertilization or postfertilization develomental failure. In the same respect, the accumulation of biochemical information will be invaluable in recognizing normal and abnormal patterns of molecular expression in preovulatory oocytes and preimplantation-stage embryos. The integration of results from cellular and molecular biological approaches to the analysis of early human development is of fundamental importance in the identification of specific biochemical, nuclear and cytoplasmic differentiative events that lead to fertilizability and developmental competence.

One of the more remarkable conclusions that can be derived from the study of early human reproduction is the ability of embryos to continue development in the presence of obvious genetic abnormalities and morphological perturbations. This observation suggests that commitment to a specific developmental pathway(s) at the blastomere level may not occur as early in the human as it apparently does in other mammals (15), or that early development may be sufficiently plastic to accomodate for a subset of aberrant cells. Therefore, the presence of abnormal blastomeres may not necessarily preclude normal preimplantation embryogenesis if such cells do not progress in development, are eliminated from the embryo, enter the extraembryonic lineage, or, as indicated by pseudopronuclear eggs, exhibit a developmentally benign anomaly. However, it should be noted that although such embryos may appear able to progress through the preimplantation stages, the lethality of a chromosomal abnormality (e.g., hypohaploidy) may be manifested prior to or after implantation and, consequently, may contribute to the high frequency of preimplantation loss and spontaneous abortions that can be attributed to genetic causes.

One of the foremost challenges in the study of early human development may well be the development of the ability to recognize abnormal cytoplasmic and genetic states in living embryos, and to distinguish those perturbations that are of no developmental consequence from those that are, or will become developmentally lethal. The types of real and apparent abnormalities discussed in this chapter may well represent only a small portion of the molecular and cellular problems that may lead to fertilization and developmental failure in the human. Increased experience with human IVF, as well as highly focused experimentation, should provide a more complete picture of the origin and causes of developmental failure and whether or not they can be recognized and corrected.

Several major developmental processes that have been well defined in other mammalian species, such as for the preimplantation mouse embryo, have not been presented in this chapter because comparable information is either presently not available in the same detail for the human or is discussed elsewhere in this volume. These critical events include temporal and spatial aspects of the following developmental processes: (1) establishment of intercellular communication during cleavage, (2) generation, maintenance, and developmental function of polar and apolar blastomeres in the morula-stage embryo (see 21), and (3) development and functional differentiation of inner cell mass and trophectoderm in the peri-implantation stage blastocyst (see Chapter 7, this volume). The continued analysis of human preimplantation embryos will indicate both the extent to which information derived from other species is applicable to the human, and whether early reproductive failure can be attributed to perturbations in these developmental processes that are known in other mammals to be essential for the normal progression of embryogenesis.

Acknowledgments

To no small degree, deep appreciation is expressed to all those patients who allowed their oocytes and abnormally fertilized/developing eggs to be examined. I thank Dr. Cathy Van Blerkom, Dr. Michael W. Klymkowsky, and Mr. Hobart Bell and Miss Virginia Fonte for their comments on the manuscript. To a large extent this work was made possible by funds made available by Reproductive Genetics In Vitro. Some of the studies of living unfertilized oocytes were supported by grants from the National Institutes of Health (HD 21582).

References

1. Leese HJ: Analysis of embryos by non-invasive methods. *Hum Reprod* 2: 37–40, 1987.
2. Van Blerkom J, Bell H, Henry GH: The occurrence, recognition and developmental fate of pseudomultipronuclear eggs after in vitro fertilization of human oocytes. *Hum Reprod* 2: 217–225, 1987.
3. Gifford DJ, Fleetham JA, Mahadevan MM, Taylor PJ, Schultz GA: Protein synthesis in human unfertilized eggs. *Gamete Res*, in press.
4. Edwards RG: Current status of human conception in vitro. *Proc R Soc Lond B* 223: 417–448, 1985.
5. Edwards RG: Causes of pregnancy loss. *Human Reprod* 1: 185–198, 1986.
6. Testart JR: Evidence of uterine inadequacy to egg implantation in stimulated in vitro fertilization cycles. *Fertil Steril* 47: 855–857, 1987.
7. Edwards RG, Fishel SB, Cohen J, Fehilly CB, Purdy JM, Slater JM, Steptoe PC, Webster JM: Factors affecting the success of in vitro fertilization for alleviating human infertility. *J In Vitro Fertil Embryo Transfer* 1: 3–23, 1984.

8. Fishel SB, Cohen J, Fehilly C, Purdy JM, Eurof Walters D, Edwards RG: Factors influencing human embryonic development in vitro. *NY Acad Sci* 442: 342–356, 1985.
9. Buster JE, Busillo M, Rodi IA, Cohen SW, Hamilton M, Simon JA, Thorneycroft IH, Marshall SR: Biologic and morphologic development of donated human ova recovered by non-surgical uterine lavage. *Am J Obstet Gynecol* 153: 211–217, 1985.
10. Van Blerkom J, Motta PM: The Cellular Basis of Mammalian Reproduction. Baltimore and Munich: Urban and Schwarzenberg, 1979.
11. Van Blerkom J, Janzen R, Runner MN: The patterns of protein synthesis during foetal and neonatal organ development are remarkably similar. *J Embryol Exp Morph* 72: 97–116, 1982.
12. Van Blerkom J: The structural relationship and post-translational modification of stage-specific proteins synthesized during early preimplantation embryogenesis in the mouse. *Proc Natl Acad Sci USA* 78: 7629–7633, 1981.
13. Pratt HPM, Bolton VN, Gudgeon KA The legacy from the oocyte and its role in controlling early development in the mouse embryo. In: Molecular Biology of Egg Maturation: CIBA Symposium 98. London: Pitmann, p 197–277, 1983.
14. Wassarman PM: Oogenesis: Synthetic events in the developing mammalian egg. In: Mechanism and Control of Animal Fertilization. JF Hartmann (ed), New York: Academic Press, p 1–54, 1983.
15. Johnson MH, McConnell J, Van Blerkom J: Programmed development in the mouse embryo. *J Embryol Exp* Morph 83 (suppl): 197–231, 1984.
16. Van Blerkom J: Posttranslational modification of proteins during resumption of arrested meiosis and early embryogenesis in the mammal. In: Control of Cell Growth and Proliferation. CM Veneziale (ed), Princeton, NJ: Van Nostrand, Reinhold, Co, p 67–86, 1985.
17. Van Blerkom J: Protein synthesis during oogenesis and early embryogenesis in the mammal. In Biology of Fertilization. Vol 3. A Monroy, C Metz (eds), New York, Academic Press, 1985.
18. Van Blerkom J: Extragenomic regulation and autonomous expression of a developmental program in the early mouse embryo. *Ann NY Acad Sci* 442: 58–72, 1985.
19. Maro B, Johnson MH, Webb M, Flach G: Mechanism of polar body formation in the mouse oocyte: an interaction between chromosomes, the cytoskeleton and the plasma membrane. *J Embryol Exp Morph* 92: 11 –32, 1986.
20. Van Blerkom J, Bell H: Regulation of development in the fully grown mouse oocyte: chromosome-mediated temporal and spatial differentiation of the cytoplasm and plasma membrane. *J Embryol Exp Morph* 93: 213–238, 1986.
21. Johnson MH, Maro B: Time and space in the mouse early embryo: cell biological approach to cell diversification. In: Experimental Approaches to Mammalian Embryonic Development. J Rossant, RA Pedersen (eds), Cambridge: Cambridge University Press, 1987.
22. Van Blerkom J: Molecular biology and cellular morphodynamics of preovulatory mammalian oocyte maturation: Implications for normal and abnormal development. In: Proceedings IVth World Congress of In Vitro Fertilization. I Johnston (ed), New York: Plenum Press, in press.
23. Wortham JWE, Veeck LL, Witmyer J, Sandow BA, Jones HW: Vital initiation of pregnancy (VIP) using human meopausal gonadotropin and human chorionic gonadotropin ovulation induction: Phase II-1981. *Fertil Steril* 40: 170– 177, 1983.
24. Testart J, Lefevre B, Castanier M, Belaisch-Allart J, Guillet-Rosso F, Frydman R: Comparative effect of clomiphene and clomiphene/hMG on the preovulatory follicles and fertilizability of the oocyte. *Ann NY Acad Sci* 442: 128–139, 1985.
25. Rogers P, Molloy D, Healy D, McBain J, Howlett D, Bourne H, Thomas A, Wood C, Trounson A: Crossover trial of superovulation protocols from two major in vitro fertilization centers. *Fertil Steril* 46: 424–431, 1986.
26. Baccetti B: The human spermatozoon. In: Ultrastructure of Human Reproduction. J Van Blerkom, P Motta (eds), Boston: Martinus Nijhoff, p 110–126, 1984.
27. Wramsby H, Fredga K: Chromosome analysis of human oocytes failing to cleavage after insemination in vitro. *Hum Reprod* 2: 137– 142, 1987.
28. Wramsby H, Fredga K, Liedholm P: Chromosome analysis of human oocytes recovered from preovulatory follicles in stimulated cycles. *N Engl J Med* 316: 121–124, 1987.
29. Wramsby H, Fredga K, Liedholm P: Ploidy in human cleavage stage embryos after in vitro fertilization. *Hum Reprod* 2: 233–236, 1987.
30. Schultz RM: Molecular aspects of oogenesis. In Experimental Approaches to Mammalian Embryonic Development. J Rossant, RA Pederson (eds) New York: Academic Press, p 195–237, 1987.
31. Soupart P, Strong PA Ultrastructural observations of human oocytes fertilized in vitro. *Fertil Steril* 25: 11–44, 1974.
32. Dunbar BS: Morphological, biochemical and immunochemical characterization of the mammalian zona pellucida. In: Mechanism and Control of Animal Fertilization. JF Hartmann (ed), New York: Academic Press, p 140–175, 1983.
33. Sathananthan AH, Trounson AO: Ultrastructural observations on cortical granules in human follicular oocytes cultured in vitro. *Gamete Res* 5: 191–198, 1982.
34. Sathananthan AH, Trounson AO Ultrastructure of cortical granule release and zona interaction in monospermic and polyspermic human ova fertilized in vitro. *Gamete Res* 6: 225–234, 1982.
35. Schmell ED, Gulyas BJ, Hedrick JL Egg surface changes during fertilization and the molecular mechanisms of the block to polyspermy. In: Mechanisms and Control of Animal Fertilization. J Hartmann (ed), p 365–413, 1983.
36. Hartmann JF: Mammalian fertilization: Gamete surface interactions in vitro. In: Mechanism and Control of Animal Fertilization. JF Hartmann (ed), New York: Academic Press, p 325–364, 1983.
37. Wolf DE, Ziomek CA: Regionalization and lateral diffusion of membrane proteins in unfertilized and fertilized

mouse eggs. *J Cell Biol* 96: 1786–0000, 1983.

38. Johnson MH, Eager D, Muggleton-Harris AL, Graves HM: Mosaicism in the organization of concanavalin A receptors on surface membrane of mouse eggs. *Nature* (London) 257: 321–322, 1975.
39. Eager DD, Johnson MH, Thurley KW: Ultrastructural studies on the surface membrane of the mouse egg. *J Cell Sci* 22: 345–353, 1976.
40. Longo FJ, Chen DY: Development of cortical polarity in mouse eggs: Involvement of the meiotic apparatus. *Devel Biol* 107: 382–394, 1985.
41. Maro B, Johnson MH, Pickering S, Flach G: Changes in actin distribution during fertilization of the mouse egg. *J Embryol Exp Morph* 81: 211–237, 1984.
42. Moore GPM, Lintern-Moore S: Transcription of the mouse oocyte genome. *Biol Reprod* 18: 865–870, 1978.
43. Wassarman PM, Letourneau GE: RNA synthesis in fully-grown mouse oocyte. *Nature* (London) 261: 73–74, 1976.
44. Van Blerkom J, Runner MN: Mitočhondrial reorganization during resumption of arrested meiosis in the mouse. *Am J Anat* 171: 335–355, 1984.
45. Fleming TP, Pickering SJ: Maturation and polarization of the endocytotic system in ouside blastomeres during mouse preimplantation development. *J Embryol Exp Morph* 89: 175–208, 1985.
46. Fleming TP, Cannon PM, Pickering SJ: The cytoskeleton, endocytosis and cell polarity in the mouse preimplantation embryo. *Devel Biol* 113: 406– 419, 1986.
47. Albertini DF: Cytoplasmic reorganizations during the resumption of meiosis in cultured preovulatory rat oocytes. *Devel Biol* 120: 121–131, 1987.
48. Schatten G, Simerly C, Schatten H Microtubule configurations during fertilization, mitosis, and early development in the mouse and the requirement for egg micorotuble-mediated motility during mammalian fertilization. *Proc Natl Acad Sci USA* 82: 4152–4156, 1985.
49. Lopata, A: Concepts in human in vitro fertilization and embryo transfer. *Fertil Steril* 40: 289–301, 1983.
50. McGaughey RM, Van Blerkonm J: Patterns of polypeptide synthesis of porcine oocytes during maturation in vitro. *Devel Biol* 56: 241–254, 1977.
51. Van Blerkom J McGaughey RW: Molecular differentiation of the rabbit ovum. I. During oocyte maturation in vivo and in vitro. *Devel Biol* 63: 139–150, 1978.
52. Van Blerkom J, McGaughey RW: Molecular differentiation of the rabbit ovum. II. During the preimplantation development of in vivo and in vitro matured rabbit oocytes. *Devel Biol* 63: 151–164, 1978.
53. Schultz RM, Wassarman PM Specific changes in the pattern of protein synthesis during meiotic maturation of mammalian oocytes in vitro. *Proc Natl Acad Sci USA* 74: 538–541, 1977.
54. Van Blerkom J: The molecular biology of mammalian oogenesis and oocyte maturation. In: The Biology of the Ovary. P Motta, ESE Hafez (eds), Boston: Martinus Nijhoff, p 179–190, 1980.
55. Osborn JC, Moor RM: Time-dependent effects of alpha-amanitin on nuclear maturation and protein synthesis in mammalian oocytes. *J Embryol Exp Morph* 73: 317–338, 1983.
56. Schultz RM Letourneau GE, Wassarman PE Meiotic maturation of mouse oocytes in vitro: Protein synthesis in nucleate and anucleate oocyte fragments. *J Cell Sci* 30: 251–264, 1978.
57. Van Blerkom J: Molecular differentiation of the rabbit ovum. III. Fertilization-autonomous polypeptide synthesis. *Devel Biol* 72: 188–194, 1979.
58. Petzholdt U, Hoppe PC, Illmensee K: Protein synthesis in enucleated fertilized and unferilized mouse eggs. *Wilhelm roux's Arch Dev Biol* 189: 215–219, 1980.
59. Wolgemuth DJ: Molecular changes at fertilization. In: Mechanism and Control of Animal Fertilization. J Hartmann (ed). New York: Academic Press, p 415–452, 1983.
60. Thibault C: In vitro maturation and fertilization of rabbit and cattle oocytes. In: The Regulation of Mammalian Reproduction. SJ Segal, R Crozier, PA. Corfman, PG Condliffe (eds), Spring Field: Charles C. Thomas, p 231–240, 1973.
61. Thibault C, Gerard M: Cytoplasmic and nuclear maturation of rabbit oocytes in vitro. *Ann Biol Anim Biochim Biophys* 13 (suppl.): 145 –156, 1973.
62. Petrov P, Raitchera E, Tusanev R: Nucleosomes and non-ribosomal RNA transcription in early mouse embryo. An electron microscopic study. *Eur J Cell Biol* 22: 708–713, 1980.
63. Kopecny V, Pavlok A: Incorporation of arginine-3H into chromatin of mouse eggs shortly after sperm penetration. *Histochemistry* 45: 341–345, 1975.
64. Krzanowska H: Toludine blue staining reveals changes in chromatin stabilization of mouse spermatozoa during epididymal maturation and penetration of ova. *J Reprod Fertil* 64: 97–101, 1982.
65. Van Blerkom J, Manes C: The molecular biology of preimplantation embryogenesis. In: Concepts in Mammalian Embryogenesis. MI Sherman (ed), Cambridge: MIT Press, p , 1977.
66. Richter JD, McGaughey RW Patterns of polypeptide synthesis in mouse oocytes during germinal vesicle breakdown and during maintenance of the germinal vesicle state by dibutryl cAMP. *Devel Biol* 83: 188–192, 1982.
67. Schultz RM, Montgomery R, Belanoff JR Regulation of mouse oocyte meiotic maturation: Implication of a decrease in oocyte cAMP and protein dephosphorylation in commitment to resume meiosis. *Devel Biol* 97: 264–273, 1983.
68. Calarco PG, Donahue RP, Szollosi D: Germinal vesicle breakdown in the mouse oocyte. *J Cell Sci* 10: 369–385, 1972.
69. Wassarman PM, Fujiwara K: Immunofluorescent anti-tubulin staining of spindles during meiotic maturation of mouse oocytes in vitro. *J Cell Sci* 29: 171–188, 1978.
70. Ezzell RM, Szego CM: Luteinizing hormone-accelerated redistribution of lysosome-like organelles preceding dissolution of the nuclear envelope in rat oocytes maturing in vitro. *J Cell Biol* 82: 264–277, 1979.
71. Ducibella T, Ukena T, Karnovsky M, Anderson E: Changes in cell surface and cytoplasmic organization

during early embryogenesis in the preimplantation mouse embryo. *J Cell Biol* 74: 153–167, 1977.

72. Van Blerkom J, Henry GP: Cytogenetic analysis of living human oocytes: cellular basis and developmental consequences of perturbations in chromosomal organization and complement. In press.
73. Janzen R, Van Blerkom J, Runner MR: Coordination of pronuclear DNA synthesis with cytoplasmic and molecular events in the early mouse embryo. Submitted, 1987.
74. Donahue RP: Fertilization of the mouse oocyte: Sequence and timing of the two-cell stage. *J Exp Zool* 180: 305–318, 1972.
75. Abramczuk J, Sawicki W: Pronuclear synthesis of DNA in fertilized and parthenogenetically activated mouse eggs. *Exp Cell Res* 92: 361–372, 1975.
76. Siracusa G, Coletta M, Monesi V Duplication of DNA during the first cell cycle in the mouse embryo. *J Reprod Fertil* 42: 395–398, 1975.
77. Wentz AC, Repp JE, Maxson WS, Pittuay DE, Torbit CA: The problem of polyspermy in in vitro fertilization. *Fertil Steril* 40: 748–754, 1983.
78. Van Blerkom J, Henry GH, Porreco R: Preimplantation human embryonic development from polypronuclear eggs after in vitro fertilization. *Fertil Steril* 41: 686–696, 1984.
79. Van Der Ven HH, Al-Hasani A, Diedrich K, Hamerich U, Lehmann F, Krebs D: Polyspermy in in vitro fertilization of human oocytes: Frequency and possible causes. *Ann NY Acad Sci* 442: 88–95, 1985.
80. Ahlgren M Boxtrom K, Malmkvist R: Sperm transport and survival in women with special reference to the fallopian tube. In: Biology of Spermatozoa. ESE Hafez, C Thibault (eds), Basel: Karqer, p 63–73, 1975.
81. Englert Y, Puissant F, Camus M, Degueldre M, Leroy F: Factors leading to tripronucleate eggs during human in vitro fertilization. *Human Reprod* 1: 117–119, 1986.
82. Diamond MP, Rogers JB, Webster BW, Vaughan WK, Wentz AC: Polyspermy: effect of varying stimulation protocols and inseminating sperm concentrations. *Fertil Steril* 43: 777–780, 1985.
83. Ben-Rafael Z, Meloni F, Strauss JF, Blasco L Mastroianni L, Flickenger GL: Relationships between polypronuclear fertilization and follicular fluid hormones in gonadotropin-treated women. *Fertil Steril* 47: 284–288, 1987.
84. Webster BW, Wentz AC, Osteen KG, Rogers JB Vaughan WK: Hormonal correlates with polyspermy. *Ann NY Acad Sci* 442: 332–335, 1985.
85. Testart JR, Frydman R, De Mouzon J, Lassalle B, Belaisch JC: A study of factors affecting the success of human fertilization in vitro. I. Influence of ovarian stimulation upon the number and condition of oocytes collected. *Biol Reprod* 28: 415–431, 1983.
86. Zamboni L: Ultrastructure of mammalian oocytes and ova. *Biol Reprod* 2 (suppl): 44–63, 1970.
87. Sathananthan AH, Ng SC, Chia CM, Lay HY, Edirisinghe WR, Ratnam SS The origin and distribution of cortical granules in human oocytes with reference to Golgi, nucleolar, and microfilament activity. *Ann. NY Acad Sci* 442: 251–264, 1985.
88. Longo FJ: Correlative ultrastructural and electro-physiological studies of sperm–egg interactions of the sea urchin, Lytechinus variegatus. *Devel Biol* 118: 155–166, 1986.
89. Michelmann HW, Bonhoff A, Mettler L: Chromosome analysis of polypoid human embryos. *Hum Reprod* 1: 243–246, 1986.
90. Plachot M, Junca AM, Mandlebaum J, de Grouchy J, Salat-Baroux J, Cohen J Chromosome investigations in early life. II. Human preimplantation embryos. *Hum Reprod* 2: 29–35, 1987.
91. Werteleki W, Graham JM, Sergovich FR: The clinical syndrome of triploidy. *Obstet Gynecol* 47: 69–73, 1976.
92. Tesarik J, Kopecny V, Plachot M, Mandelbaum J: Ultrastructural and autoradiographic observations on multinucleated blastomeres of human cleaving embryos obtained by in vitro fertilization. *Hum Reprod* 2: 127–136, 1987.
93. O'Rand MG, Herman J, Diguiseppi J, Halme J, Hammond MG, Talbert LM: Analysis of deoxyribonucleic acid distribution in non-cleaving oocytes from patients undergoing in vitro fertilization. *Fertil Steril* 46: 452–460, 1986.
94. Plachot M, Junca AM, Mandlebaum J, de Grouchy J, Salat-Baroux J, Cohen J: I. Human oocytes recovered in and IVF programme. *Hum Reprod* 1: 547–551, 1986.
95. Plachot M, Mandlebaum J, Junca A–M, Salat-Baroux J, Cohen J: Impairment of human development after abnormal in vitro fertilization. *Ann NY Acad Sci* 45: 442: 336–341, 1985.
96. Boue J, Boue A, Lazar P: The epidemiology of human spontaneous abortions with chromosomal anomalies. In: Aging Gametes, Their Biology and Pathology. RJ Blandau (ed), Basel: Karger, p 330–357, 1975.
97. Hassold TJ, Chen N, Funkhauser J, Jooss T, Manuel B, Matsuura J, Matsuyama A, Wilson C, Yamane JA, Jacobs PA: A cytogenic study of 1000 human abortions. *Ann Hum Genet* 44: 151–178, 1980.
98. Kajii T, Ferrier A, Niikawa N, Takahara H, Ohama K, Avirachin S: Anatomic and chromosomal anomalies in 639 spontaneous abortions. *Hum Genet* 55: 67, 1980.
99. Rudak E, Dor J, Mashiach S, Nebel L, Goldmann B Chromosome analysis of multipronuclear human oocytes fertilized in vitro. *Fertil Steril* 41: 538–545, 1984.
100. Rudak E, Dor J, Mashiach S, Nebel L, Goldman B Chromosome analysis of human oocytes and embryos fertilized in vitro. *Ann NY Acad Sci* 442: 476–486.
101. Jacobs PA, Angell RR, Buchanan IM, Matsuyama AM, Manuel B: The origin of human triploids. *Ann Hum Genet* 42: 49–57, 1978.
102. Robinson A, Puck TT Studies on chromosomal nondisjunction in man. *Am J Hum Genet* 19: 112–118, 1967.
103. Kalousek DK, Dill FJ: Chromosomal mosaicism confined to the placenta of human conceptions. *Science* 221: 665–669, 1983.
104. Patillo R, Shigeru S, Katayama P, Roesler M, Mattingly RF: Genesis of 46, XY hydatiform mole. *Am J Obstet Gynecol* 141:104–106, 1981.
105. Surani MAH: Evidence and consequences of differences

between maternal and paternal genomes during embryogenesis in the mouse. In: Experimental Approaches to Mammalian Embryonic Development. J Rossant, R Pederson (eds), New York: Academic Press, p 410–435, 1987.

106. Van Blerkom J, Manes C: Development of preimplantation rabbit embryos in vivo and in vitro. II. A comparison of qualitative aspects of protein synthesis. *Devel Biol* 40: 40–51, 1974.
107. Van Blerkom J, Brockway G: Qualitative patterns of protein synthesis in the preimplantation mouse embryo. I. Normal pregnancy. *Devel Biol* 44: 148–157, 1975.
108. Levinson J, Goodfellow P, Vadeboncoeur M, McDevitt H: Identification of stage-specific polypeptides synthesized during murine preimplantation development. *Proc Natl Acad Sci USA* 75: 3332–3336, 1978.
109. Cascio SM, Wassarman PM: Program of early development in the mammal: Posttranscriptional control of a class of proteins synthesized by mouse oocytes and early embryos. *Devel Biol* 89: 397–408, 1982.
110. Howlett SK, Bolton VN: Sequence and regulation of morphological and molecular events during the first cell cycle of mouse embryogenesis. *J Embryol Exp Morph* 87: 175–206, 1985.
112. Dvorak M, Cech S, Stastna J, Tesarik J, Travnik P: The Differentiation of Preimplantation Mouse Embryos. Brno, Czechoslovakia: Acta Fac. Med. Univ. Brno., 1985.
113. Biggers JD, Borland RM: Physiological aspects of growth and development of the preimpantation mammalian embryo. *Ann Rev Physiol* 38: 95–119, 1976.
114. Mills RM, Brinster RL: Oxygen consumption of preimplantation mouse embryos. *Exp Cell Res* 47: 337–344, 1967.
115. Brinster RL: Carbon dioxide production from glucose by the preimplantation mouse embryo. *Exp Cell Res* 47: 271–277, 1967.
116. Van Blerkom J, Manes C, Daniel J: Development of preimplantation rabbit embryos in vivo and in vitro. *Devel Biol* 35: 262–282, 1973.
117. Dvorak M, Tesarik J: Differentiation of mitochondria in the human preimplantation embryo grown in vitro. *Scripta Medica* 58: 161–170, 1985.
118. Gardner DK, Leese HJ: Non-invasive measurement of nutrient uptake by single cultured preimplantation mouse embryos. *Hum Reprod* 1:25–27, 1986.
119. Barbehenn EK, Wales RG, Lowry OH: Measurements of metabolites in single preimplantation embryos: a new means to study metabolic control in early embryos. *J Embryol Exp Morph* 43: 29–46, 1978.
120. Magnusson C, Hillensjo T, Hamberger L, Nilsson L: Oxygen consumption by human oocytes and blastocysts grown in vitro. *Hum Reprod* 1: 183–184, 1986.
121. Barbehenn EK, Wales RG, Lowry OH: The explanation for the blockade of glycolysis in early mouse embryos. *Proc Nat Acad USA* 71: 1056–1060, 1974.
122. Nilsson L, Magnusson C, Widehn S, Hillensjo T: Correlation between blastocyst oxygen consumption and trophoblast cytochrome oxidase reaction at initiation of implantation of delayed mouse blastocysts. *J Embryol exp Morph* 71: 75–82, 1982.
123. Leese HJ, Barton AM: Pyruvate and glucose uptake by mouse ova and preimplantation embryos. *J Reprod Fertil* 71: 9–13, 1984.
124. Wales RG Wittingham DG Hardy K, Craft IL: Metabolism of glucose by human embryos. *J Reprod Fertil*, in press.
125. Anderson E, Condon W, Sharp D: A study of oogenesis and early embryogenesis in the rabbit, Oryctolagus cuniculus, with special reference to the structural changes of mitochondria. *J Morphol* 130: 67–92, 1971.
126. Johnson LV, Walsh ML, Chen LB: Localization of mitochondria in living cells with rhodamine 123. *Proc Natl Acad Sci USA* 77: 990–994, 1980.
127. Dvorak M, Travnik P, Stankova J: A quantitative analysis of the incidence of certain cytoplasmic structures in the ovum of the rat during cleavage. *Cell Tiss Res* 179: 429–437, 1977.
128. Manes C: Genetic and biochemical activities in preimplantation embryos. In: The Developmental Biology of Reproduction. CL Market, J Papaconstantinou (eds), New York: Academic Press, p 133–163, 1975.
129. Johnson MH: The molecular and cellular basis of preimplantation mouse development. *Biol Rev* 56: 463–498, 1981.
130. Tesarik J, Kopecny V, Plachot M, Mandelbaum J, Da Lage C, Flechon J–E: (1986) Nucleoglogenesis in the human embryo developing in vitro. *Devel Biol* 115: 193–203, 1976.
131. Karp GC, Manes C, Hahn WE: Ribosome production and protein synthesis in the preimplantation rabbit embryo. *Differentiation* 2: 65–75, 1973.
132. Manes C: Nucleic acid synthesis in preimplantation rabbit embryos. II. Delayed synthesis of ribosomal RNA. *J Exp Zool* 176: 87–96, 1971.
133. Van Blerkom J, Runner MN: The fine structure of parthenogenetically activated preimplantation mouse embryos. *J Exp Zool* 196: 113–124, 1976.
134. Geuskens M, Alexandre H: Ultrastructural and autoradiographic studies of nucleolar development and rDNA transcription in preimplantation mouse embryos. *Cell Differ* 14: 125–134, 1984.
135. Flach G, Johnson MH, Braude PR, Taylor RS, Bolton VN: The transition from maternal to embryonic control in the 2-cell mouse embryo. *The EMBO J* 1: 681–686, 1982.
136. Braude P, Bolton V, Moore S: Human gene expression first occurs between four- and eight-cell stages of preimplantation development. *Nature* 332:459–461, 1988.
137. Marin RH, Mahadevan MM Taylor PJ, Hildebrand K, Long-Simpson L, Peterson D, Yamamoto J, Fleetham J: Chromosomal analysis of unfertilized human oocytes. *J Reprod Fert* 78: 673–678, 1986.
138. Jagiello G, Ducayen M, Fang J, Graffeo J: Cytogenetic observations in mammalian oocytes. In: Chromosomes Today. Vol. 5. PK Pearson, KR Lewis (eds), New York: John Wiley and Sons, p 43–63, 1976.
139. Wramsby H, Liedholm P: A gradual fixation method for chromosomal preparations of human oocytes. *Fertil Steril* 41: 736–738, 1984.

intracellular components which cannot be repaired. These effects can often be observed at both the light and ultrastructural level.

2. Effects of culture on oocytes and embryos

2.1. Culture media and embryo viability

The media used to culture human oocytes and embryos bear very little resemblance to the composition of the oviduct and uterine luminal fluids (7). There is also a great variability in the ionic and nonionic composition of the various media used for embryo culture and differences in the source and quantity of protein used to supplement these media. The most widely used culture media include Ham's F10, Whittingham's T6, Earle's solution (8) and the French B2 or B3 medium (9). These media are frequently supplemented with the patient's own serum, human cord serum, or fetal calf serum, or contain either bovine serum albumin or human serum albumin. More recently a medium which relates more closely with the ionic composition of human oviductal fluid has been introduced (10) and promising trials of culture in human amniotic fluid have been reported (11). However, at present embryos grown in synthetic media, with or without serum supplementation, cleave more slowly than those in vivo and have reduced viability when compared to those in vivo.

Some of the most serious defects observed in culture can occur when unsuitable materials are used. These may include the type of plastics or oil in which embryos are cultured which result in the complete cessation of embryo development (12). Exposure of oocytes or embryos to osmolarity and pH outside the normal physiological range may also have dramatic effects on the structure and function of oocytes and embryos.

2.2. Morphological effects on oocytes and embryos

The abnormalities that are frequently seen in human oocytes and embryos which appear to derive primarily from suboptimal culture conditions have been described extensively (3,6). These may be summarized as follows:

(a) Progessive swelling of vesicular elements of smooth endoplasmic reticulum (SER) culminating in the fusion of these vesicles to form larger vacuoles.
(b) Rearrangement of mitochondria around these vacuoles and the clumping of mitochondria.
(c) Increased electron-density of mitochondria and appearance of dense oval bodies within the mitochondria.
(d) Centripetal migration of cortical granules from the plasma membrane surface into the ooplasm.
(e) Loss of microvilli and pinocytotic activity at the plasma membrane surface or the gradual appearance of rounded, stumpy microvilli indicating the loss of plasma membrane function.
(f) Clumping or scattering of chromosomes, or the movement of metaphase chromosomes of the oocyte away from the normal peripheral position near the plasma membrane.

Some of these changes could also be attributed to ageing of oocytes and prolonged exposure of embryos to suboptimal culture conditions, which eventually culminates in arrested development and degeneration of cells. The most vulnerable component of oocytes and embryos to adverse culture conditions appears to be the vesicular SER. Swelling of vesicular SER is seen almost invariably in oocytes and embryos cultured under conditions known to reduce embryo viability and during ageing of unfertilized oocytes in vitro. The function of the vesicular SER in human oocytes and early cleavage-stage embryos has not been determined. They are apparently involved in the formation of the pronuclear envelope membranes of the fertilized oocyte and appear also to be involved in secretory processes in oocytes and early cleavage-stage embryos (3).

3. Effects of cryopreservation on human oocytes and embryos

The development of methods to cryopreserve successfully human oocytes and embryos (13) has resulted in the widespread use of embryo freezing in association with routine IVF. The freezing and thawing techniques used and the type of cryoprotectant employed may have major effects on the survival and normality of oocytes and embryos. For example, glycerol was found to have fusogenic properties on plasma membranes when used as a cryoprotectant to freeze early cleavage-stage human embryos (8). The damage observed to embryos during cryopreservation may range from complete lysis of all blastomeres (14) to damage of some individual blastomeres without any major effect on embryo survival (15) to more subtle effects on the normality of embryo development (14) and cell morphology at the light and ultrastructural levels (16).

3.1. Cryopreservation of human oocytes

3.1.1. Sensitivity of human oocytes to cooling

Ooctyes and early cleavage-stage embryos of many species do not survive cooling to temperatures around 0°C (17). In order to determine whether the human oocyte has intracellular components that are sensitive to cooling to temperatures at which ice nucleation may be initiated (-5 to -7°C), four mature oocytes were cooled to 0° or to -6°C at 2°C min^{-1} and held at these temperatures for 30 minutes before rewarming to room temperature. The oocytes were fixed and processed for electron microscopy (3). These oocytes showed little change in fine structure when compared to nonfrozen culture controls (3) except for elongtion of vesicular SER and cracks in the zona pellucida in one instance. This is in contrast to observations on early cow embryos cooled to 4°C (17) which showed distortion of blastomeres and uneven distribution of organelles but no change in their ultrastructure. Cell contacts between blastomeres were also preserved. Globular vesicles, some containing lipidlike material, were distorted or had fused. It has been shown that sensitivity to cooling is associated with the presence of these inclusions in pig and early-stage cow embryos (17), whereas cow blastocysts, which have no inclusions, are more resistant to cooling. Such lipid-containing vesicles are not found in human oocytes.

3.1.2. Freezing methods and survival of human oocytes

The freezing of human oocytes by the method used to freeze mouse oocytes (18) resulted in very low survival rates (13). The procedure involves addition of the cryoprotectant dimethyl sulphoxide (DMSO) at 0°C and slow cooling to low subzero temperatures (−50 to −80°C). The oocytes are warmed slowly to prevent solution effects on the dehydrated cell. If the oocytes do survive the freeze–thaw process, we have shown that they fertilize and develop normally to at least the eight-cell stage. The chromosomal normality of mouse oocytes frozen by this procedure has been assessed by counting the chromosomes of embryos that develop after fertilization (19). No increase was observed in aneuploidy but there was a significant increase in polyploidy that included a doubling of both the maternal and paternal chromosome complements.

A modification of this technique for freezing human oocytes has been reported by Chen (20). Oocytes were frozen in 1.5M DMSO, but the oocytes were cooled slowly to a relatively high subzero temperature of −36°C and were then rapidly cooled to −196°C. The oocytes were thawed rapidly to prevent crystallization and growth of any intracellular ice. High survival rates were obtained using this technique and in addition the birth of twins was reported following fertilization of frozen–thawed oocytes and embryo replacement (20). There has been no report of the assessment of oocytes frozen by this method for ultrastructural or genetic normality.

Other methods have also been examined for their suitability in human oocyte cryopreservation. A modified method of the vitrification technique reported by Rall and Fahy (21) has given high survival rates of human oocytes which may be fertilized and develop to at least the eight-cell stage (13). The vitrification technique involves the placement of oocytes in a solution containing high concentrations of DMSO and propanediol at 0°C and ultrarapid cooling by direct transfer to liquid nitrogen. The solution vitrifies forming glass rather than ice. Rapid thawing prevents ice formation and possible damage during warming. The vitrification solution is relatively toxic to cells and necessitates exposure of oocytes to high concentrations needed to achieve vitrification, to be limited to low temperatures. Assessment of oocytes after vitrification by genetic and ultrastructural methods are presently underway to ascertain the possible adverse effects of the procedure on normal embryo development and viability.

The freezing of oocytes in propanediol as the cryoprotectant has also been studied following reports (22) of high survival rates of early cleavage-stage embryos and pronuclear oocytes. The methods of freezing and thawing used are similar to those used for freezing in DMSO. Propanediol is related structurally to ethanol which is a potent parthenogenetic activator, and propanediol has been found to induce parthogenetic development of unfertilized mouse oocytes in our laboratory. This property is of concern in the use of propanediol for freezing of human oocytes. The viability of human oocytes frozen in propanediol has also been disappointing.

A variation of the freezing methods that have been described for vitrification and freezing in DMSO has also been studied. This method is termed ultrarapid freezing (13,23) and involves the brief exposure of oocytes to high concentrations of 2 to 4M DMSO and 0.25M sucrose before very rapid cooling by plunging into liquid nitrogen. The oocytes are thawed rapidly and placed in 0.25M sucrose for 30 minutes to remove the DMSO. This method has been very successful for freezing two-cell mouse embryos (23) which show excellent preservation of fine structure except for occasional zona damage.

3.1.3. Oocyte ultrastructure

Human oocytes frozen by the various methods described have been evaluated by electron microscopy immediately after thawing and removal of the cryoprotectant and following insemination to assess the fine structural normality of the intracellular components of the oocytes, the corona cells attached to the zona pellucida, and the integrity of the zona pellucida (33).

The human oocytes studied were derived from two sources:

1. Human oocytes cultured for 24–72 hours that failed to fertilize in the clinical IVF program. These oocytes were studied to assess the immediate effects of various freezing and thawing methods and cryoprotectants.
2. Mature oocytes provided by consenting IVF patients specifically for research to develop oocyte cryopreservation techniques. These oocytes were inseminated after thawing to assess their viability.

The procedures used for oocyte freezing have been described in detail previously (13) and in summary are

1. Slow cooling to low subzero temperatures (−50 to −80°C) in 1.5M DMSO or 1.5M propanediol.
2. Slow cooling to high subzero temperatures (−35 to −36°C) in 1.5M DMSO or 1.5M propanediol.
3. Vitrification using a modification of the method reported by Rall and Fahy (21).
4. Ultrarapid freezing in 2 to 4M DMSO and 0.25M sucrose (23).

The best preservation of ultrastructure so far has been obtained by the DMSO slow method followed by the DMSO ultrarapid method, while vitrification produced the worst results. Only few oocytes have been examined by the propanediol slow method (33).

Figures 6-1–6-14 show the structure of oocytes and their vestments, including the events of sperm penetration and fertilization after thawing and insemination.

The general organization was reasonably well preserved in most of the oocytes examined (Figs. 6-1,6-2), irrespective of the cryopreservation technique used. The remaining oocytes showed evidence of cryoinjury involving disruption of the cell membrane (oolemma) and extensive disorganization of the ooplasm, especially after vitrification (Fig. 6-9). It has been speculated that cell injury is due primarily to intracellular freezing or solution effects (24). The cell membrane in particular seems to be more vulnerable to freeze–thawing (16) and its integrity is vital for the survival of the oocyte, which is a single cell. Embryos, however, can survive if more than half their cells are intact (16,28). On the whole, the aged oocytes that failed to fertilize were more resistant to freezing and thawing compared to mature donated oocytes that were cultured for shorter periods (6–12 hours) before cryopreservation. Hence it may not be advisable to rely on these oocytes or embryos kept too long in culture for critical evaluation of the effects of freezing and thawing, as we have done with our initial studies on oocyte freezing. Clumping of chromosomes in mature oocytes and disruption of the meiotic spindle with loss of microtubules (Fig. 6-2) may be caused by freeze–thawing, although this needs to be confirmed. Damage to the spindle and consequent scattering of chromosomes may lead to the formation of micronuclei and may result in chromosomal aberrations (33).

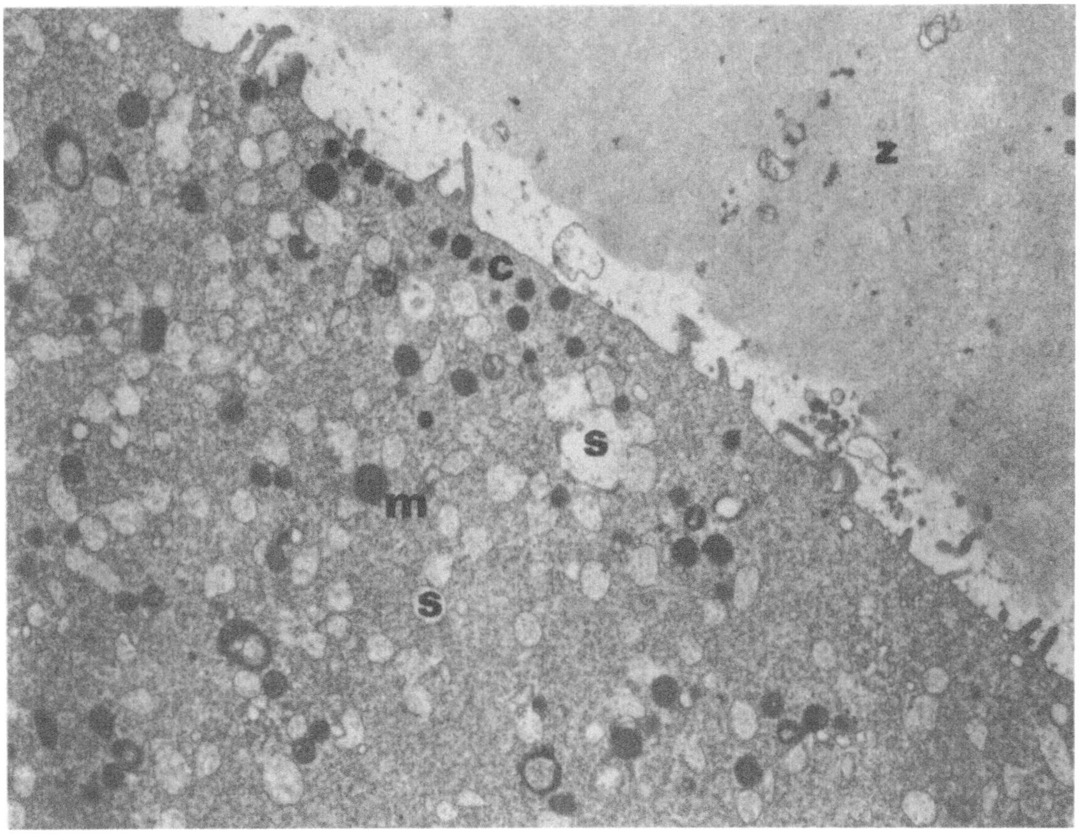

Figure 6-1. Surface of mature vitrified oocyte showing fairly good preservation of structure. Microvilli, cortical granules (C) and mitochondria (m) are evident. The swelling of vesicular elements of smooth endoplasmic reticulum (S) could be due to culture effects (frozen after 30 hours in culture). Z: zona pellucida. X 11,900.

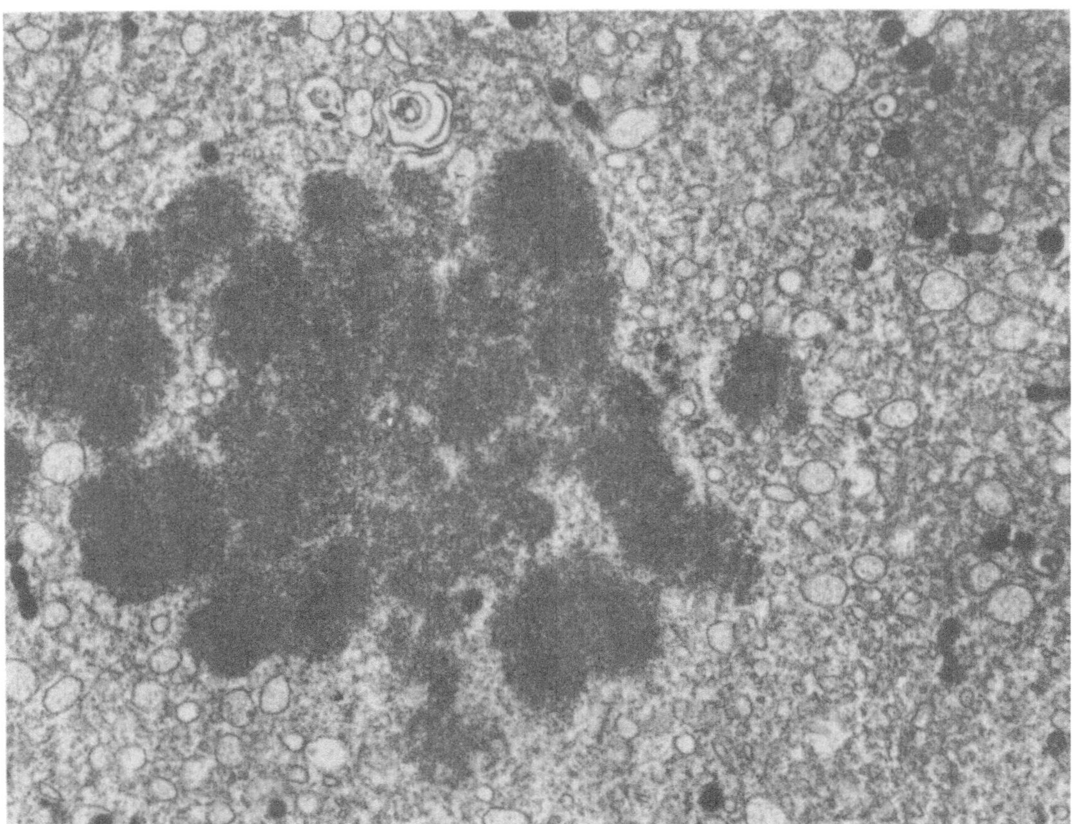

Figure 6-2. A clump of deep-seated metaphase II chromosomes of a mature oocyte frozen by the DMSO slow method. This egg was not fertilized and was frozen–thawed 48 hours after insemination and fixed immediately. Spindle microtubules are not discernible. X 15,400.

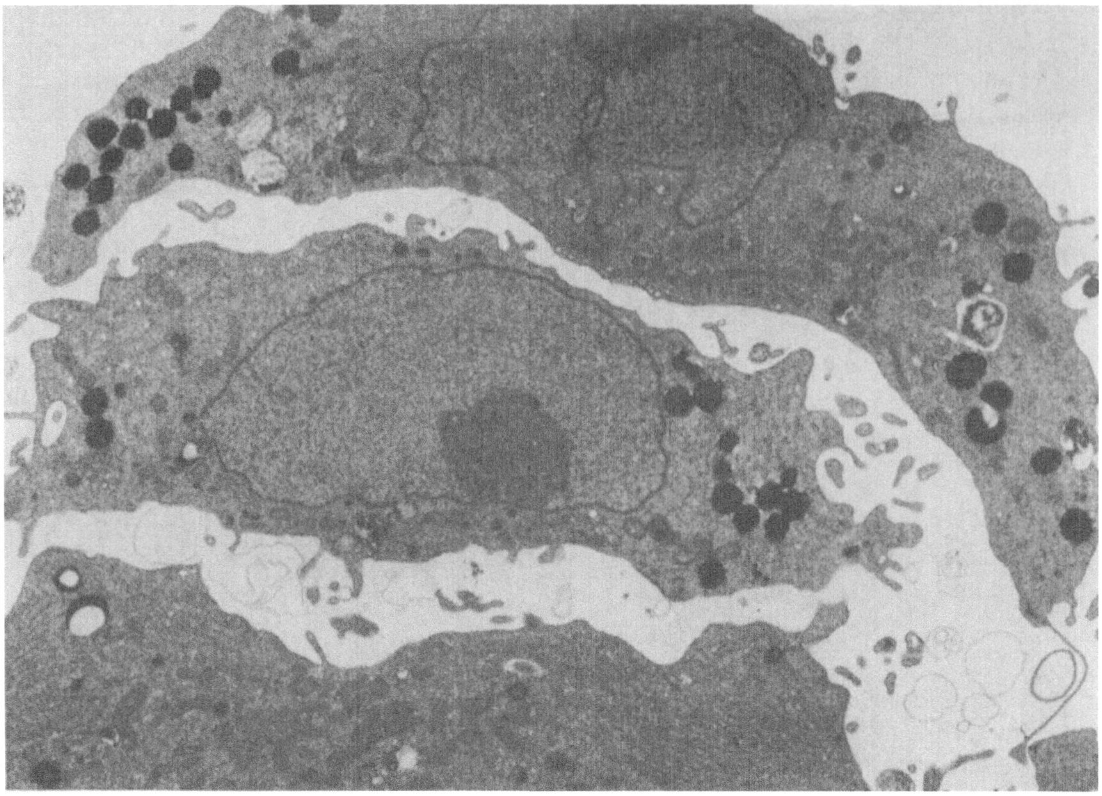

Figure 6-3. Crescent-shaped cumulus cells of a morula (90 hours in culture) frozen–thawed by the DMSO ultrarapid method. The fine structure of the cells appears to be well preserved. X 7,000.

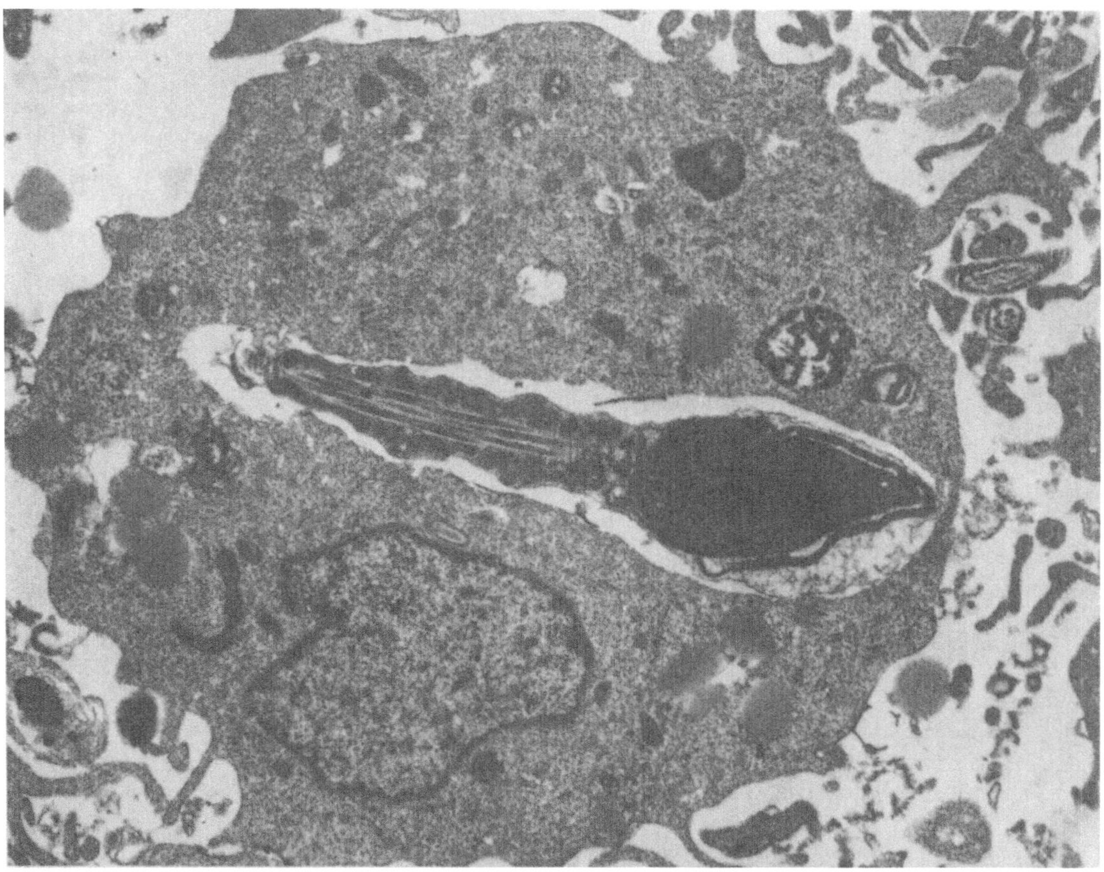

Figure 6-4. Cumulus cell of a vitrified oocyte showing a phagocytosed acrosome-intact sperm (three hours postinsemination after thawing). X 15,400.

Most cumulus and corona cells seem to be well preserved and retain their function after freeze–thawing (Figs. 6-3,6-4). Cells tend to become crescent-shaped and develop wider intercellular spaces. Cumulus cell processes were still intact and phagocytosis of sperm was evident after insemination (Fig. 6-4). Both acrosome-intact and acrosome-reacted sperm were sequestered by cumulus cells three hours after insemination, as reported in the case of fresh nonfrozen eggs (3). A few cumulus cells showed depletion of their cytosol, as observed in embryos.

Rupture of the zona is sometimes encountered in oocytes (Fig. 6-5), as well as in embryos. Supernumerary sperm are seen in the perivitelline space of these oocytes. Fertilization of zona-punctured oocytes has resulted in polyspermy (25), as the zona reaction and block to polyspermy is ineffective (26). Cumulus cells, too, were occasionally observed in the perivitelline space of zona-ruptured oocytes showing evidence of their pseudopodial activity. They have presumably migrated under the zona through its cracks and their processes were closely associated with the oolemma (Fig. 6-6).

Zona damage has been reported in 10–20% of frozen–thawed embryos, cleaving embryos being more susceptible than blastocysts (27). It is obvious from our studies that a higher incidence of polyspermy may be expected during fertilization of frozen–thawed oocytes. Fine cracks in the zona are not always visible with the light microscope, and zona damage may be minimized by slow cooling from −80°C to −110°C using DMSO as the cryoprotectant (28).

3.1.4. Fertilization of frozen oocytes

All stages of sperm penetration through the cumulus and zona pellucida are evident in oocytes inseminated after freeze–thawing (Figs. 6-7–6-9), as observed in nonfrozen oocytes (3). Both acrosome-reacted and unreacted sperm penetrate freely between cumulus cells, and sperm receptivity at the surface of the zona, evidenced by sperm binding (Fig. 6-7), seems to be largely unaffected by cryopreservation. In oocytes with damaged zonae, sperm were never seen binding to its inner surface, suggesting that receptors for human sperm are exclusively located on its outer surface. The acrosome reaction takes place in the usual way and the reacted sperm penetrate the zona (Fig. 6-8) to enter the perivitelline space (Fig. 6-9). Sperm that had already penetrated the vestments of unfertilized oocytes and embryos prior to freezing appear to be well preserved.

Sperm–oocyte membrane fusion occurs if the oolemma is intact. Fusion begins between the oolemma and plasma membrane of predominantly the postacrosomal region of the sperm head, and the sperm is incorporated into the ooplasm by phagocytosis (Fig. 6-10). These events have been documented during normal monospermic fertilization of nonfrozen oocytes (29) and during polyspermic interactions (25). It is remarkable that the posterior region of the sperm head is often incorporated first (Fig. 6-10), as was observed during normal fertilization when the zona was present (29). In the absence of the zona, however, sperm are incorporated head first, held perpendicularly or tangentially to the egg surface (25), but the fusogenic region of the sperm remains much the same. Sperm chromatin decondensation occurs in the usual way preceded by progressive inflation of the nuclear envelope (25), whilst its inner membrane forms a series of folds directed towards the expanding chromatin (Fig. 6-11). If the oolemma does not survive freeze–thawing, membrane fusion does not eventuate and sperm merely penetrate the damaged ooplasm and their chromatin remains unexpanded (Fig. 6-9). This phenomenon was also reported during penetration of nonfrozen degenerated oocytes (3).

Finally, normal fertilization was confirmed 17–18 hours after postthaw insemination, fulfilling all the criteria necessary (26), including pronuclear formation (Fig. 6-12), cortical granule exocytosis (Fig. 6-13) and extrusion of the second polar body, followed by activation of the egg and development to the morula stage. Pronuclear ova, fertilized after vitrification, showed monospermic and polyspermic penetration (Fig. 6-14) and had multiple micronuclei in their ooplasm, although they showed good preservation of fine structure. Polyspermy may have resulted through damage to the zona pellucida during freeze–thawing, while micronuclear formation may be attributed to scattering of chromosomes caused by damage to the meiotic spindle of the mature oocyte or incomplete incorporation of chromatin during pronuclear for-

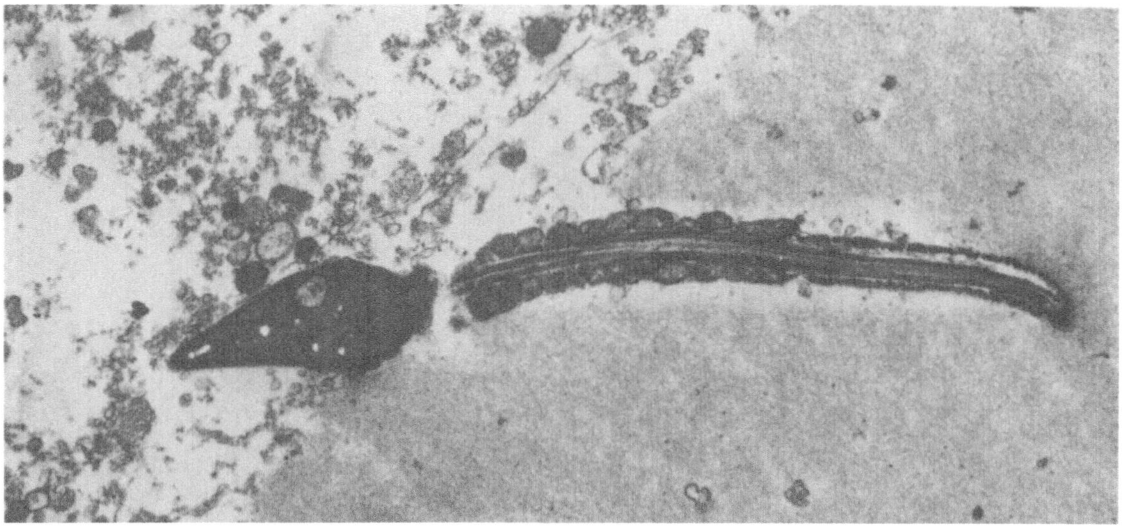

Figure 6-9. Acrosome-reacted sperm entering the perivitelline space of a highly disorganized vitrified egg (three hours postinsemination after thawing). Since the oolemma has disrupted, sperm–egg membrane fusion has not occurred. X 15,400.

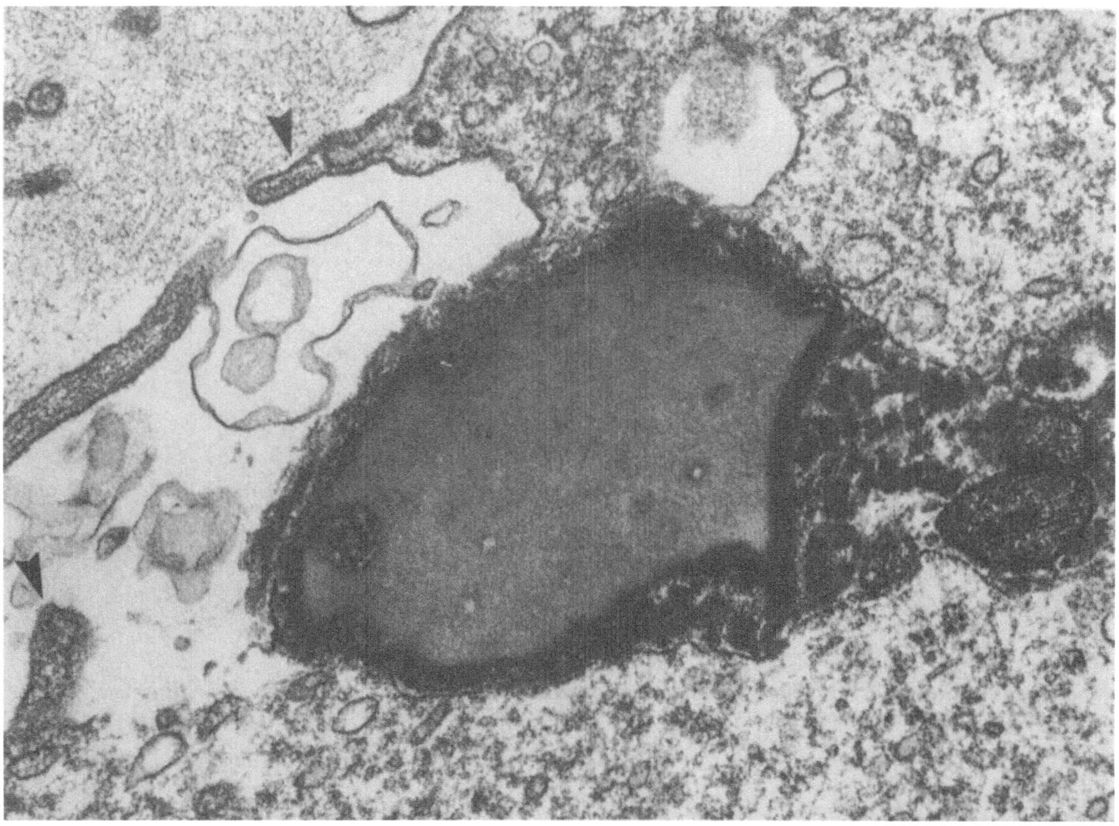

Figure 6-10. Sperm–egg fusion observed in a polyspermic ovum derived from a vitrified egg (18 hours postinsemination after thawing). The postacrosomal region and midpiece have already been incorporated into the ooplasm, while the anterior region of the spermhead is being phagocytosed by tongues of ooplasm (arrows). X 54,600.

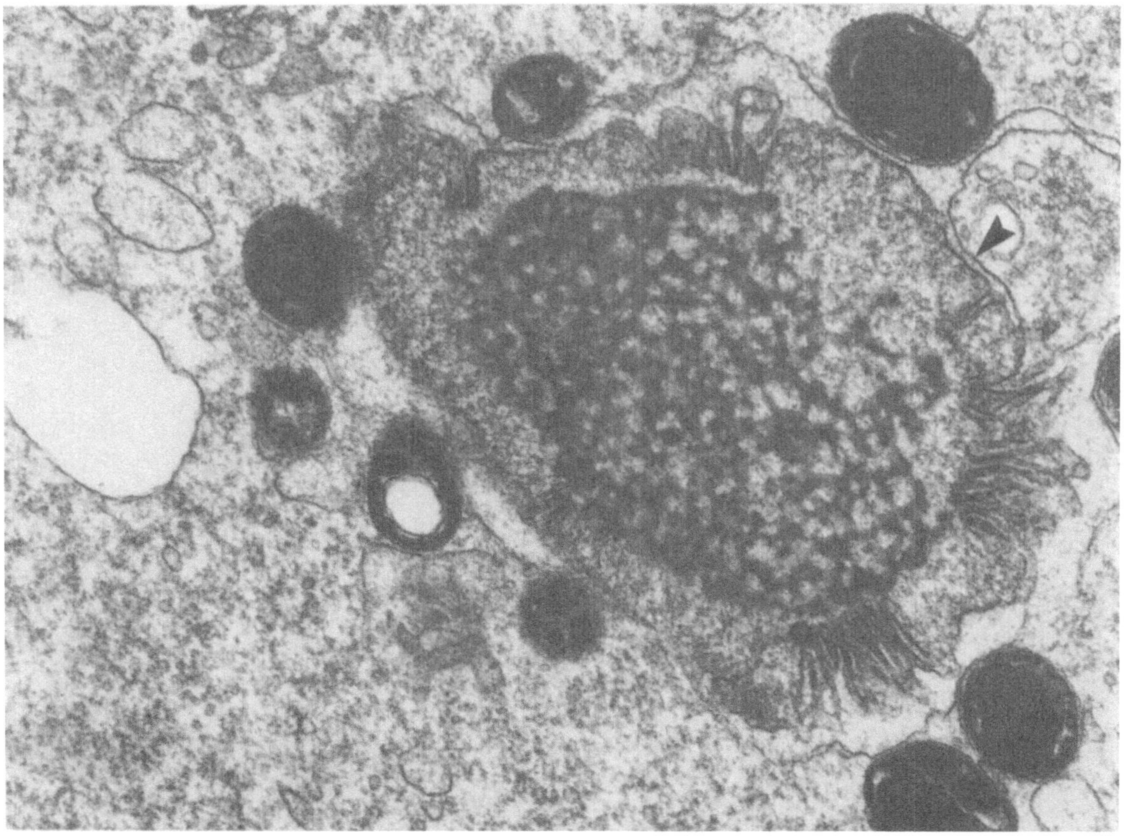

Figure 6-11. Decondensing spermhead in the ooplasm of the polyspermic ovum in Figure 6-10. The sperm nuclear envelope has inflated (arrow) while the chromatin has partially expanded. Multiple folds originating from the inner nuclear membrane are seen in the posterior and other regions of the sperm nucleus. X 35,700.

mation (3). An increased incidence of polyploidy has been reported in embryos developed from frozen–thawed mouse eggs (19).

3.1.5. Embryo development from frozen oocytes

A few embryos that developed from vitrified oocytes fertilized in vitro have been assessed by electron microscopy. These range from eight-cell stages to morulae obtained 60–120 hours after thawing and insemination. Such embryos showed some abnormal features such as enucleate fragments and multinucleated crescent-shaped cells curling around other cells (Fig. 6-15), in addition to normal blastomeres. Attempted internalization of a cell by another at the normal time of compaction is often observed when few viable cells remain in the embryo. Some embryonic fragmentation is inevitable during routine IVF and may affect embryo viability if the degree of fragmentation is excessive (28). Embryos with multinucleated cells or fragments can originate from polyspermic ova (3,32).

3.2. Cryopreservation of human embryos

Unlike many other animal species, human embryos may be cryopreserved at all preimplantation stages. The only other mammalian species in which embryos of all these stages have been successfully cryopreserved is the mouse. The extensive damage observed when early cleavage-stage bovine embryos are cooled and frozen has been described by Mohr and Trounson (17). In addition, following differentiation of cells of the bovine blastocyst into trophectoderm at the time of

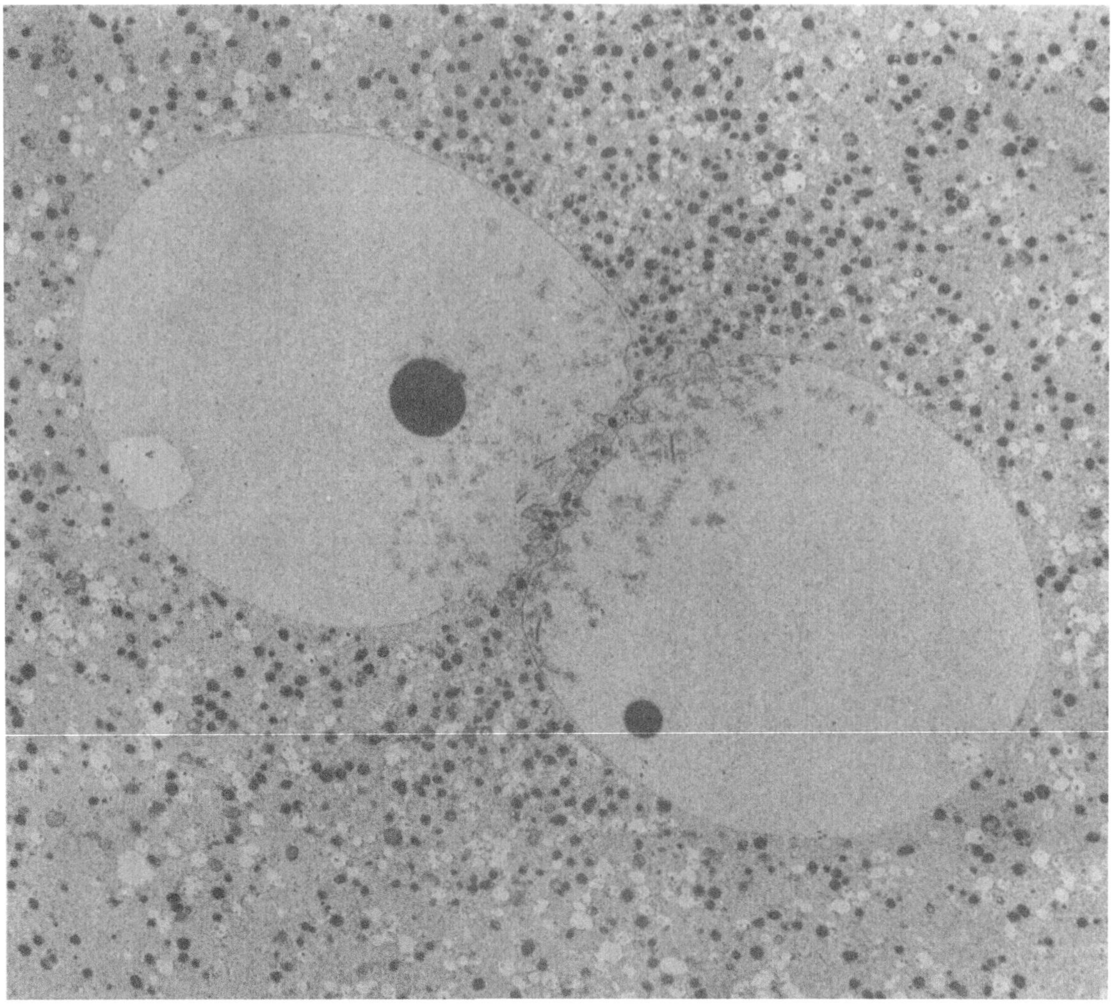

Figure 6-12. Normal pronuclear ovum obtained from a frozen oocyte using propanediol (17 hours postinsemination after thawing). The two pronuclei are closely associated with each other in the central ooplasm. This ovum showed all features of normal fertilization including a second polar body. X 2,940. From Sathananthan et al. (33).

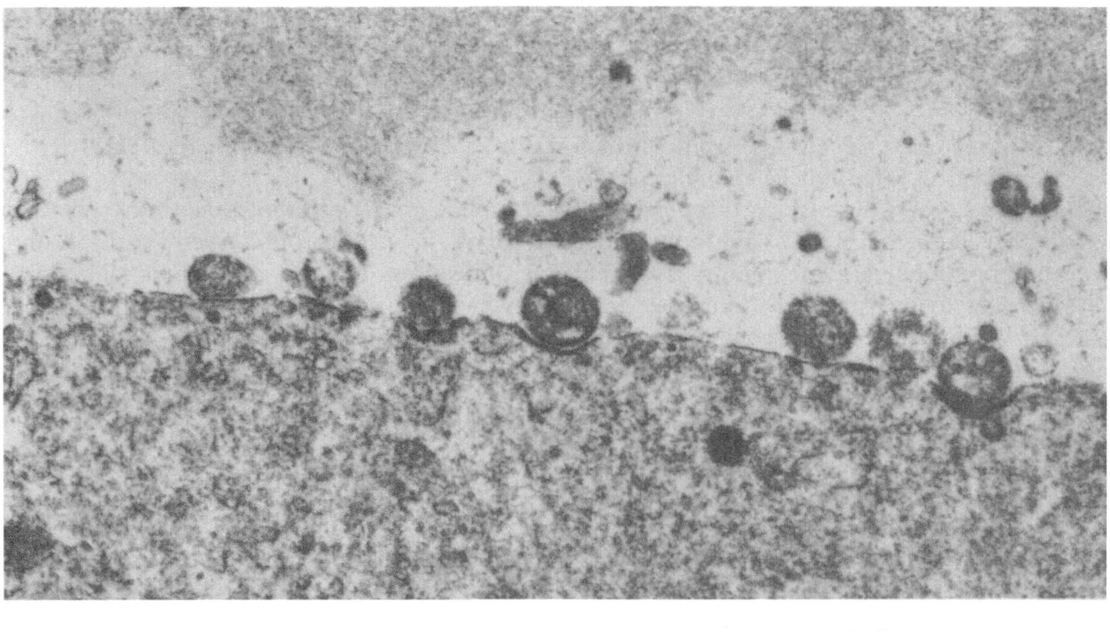

Figure 6-13. Cortical granule exocytosis at the surface of the ovum depicted in Figure 6-12. X 27,300.

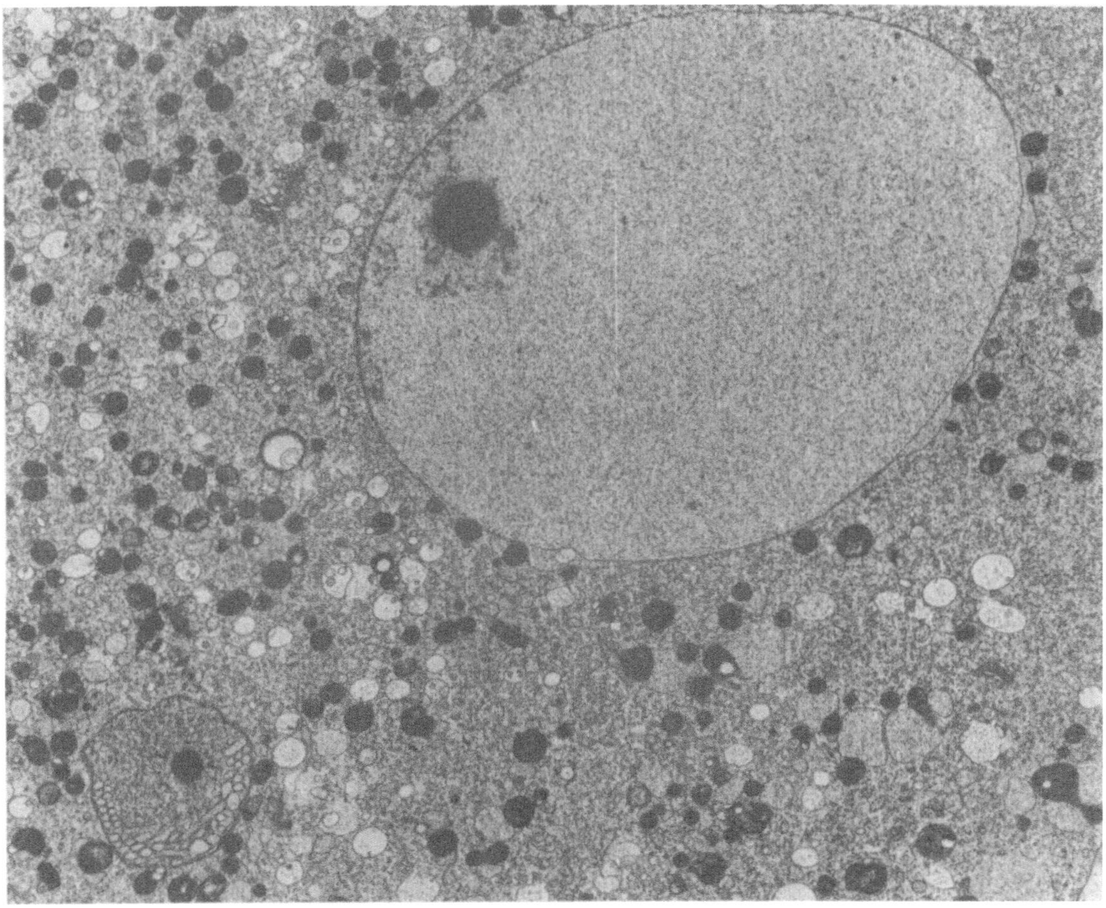

Figure 6-14. Polyspermic ovum obtained from a vitrified oocyte (18 hours postinsemination after thawing) showing a micronucleus adjacent to a pronucleus. Membranous profiles are seen within the micronucleus which also has a nucleolus. X 7,000.

hatching from the zona pellucida, these cells are severely damaged by cryopreservation while the undifferentiated inner cell mass cells are well preserved (17). There are clearly temperature-sensitive subcellular components present in embryos of some species and in some stages of embryonic development. Intracellular lipoprotein is considered to be one of the most unstable vesicular components of embryos sensitive to cooling (17). Human preimplantation embryos appear to be remarkably resistant to freezing damage and appear to contain little of the temperature-sensitive material that exists extensively in preblastocyst stages of the pig, horse, cow and to a lesser extent in sheep and goats.

3.2.1. Freezing methods for embryos

The choice of cryoprotectants for human embryos depends to some extent on the developmental stage. Pronuclear and two-cell embryos may be frozen successfully in propanediol (22), whereas four-cell to 16-cell embryos are frozen successfully in DMSO, when cooled slowly to low subzero temperatures and thawed slowly (15,16). The blastocyst stage may be frozen very successfully in glycerol, cooling slowly to high subzero temperatures before plunging into liquid nitrogen and rapid thawing (27,30).

Research in our own laboratory has shown that freezing early cleavage-stage embryos in glycerol

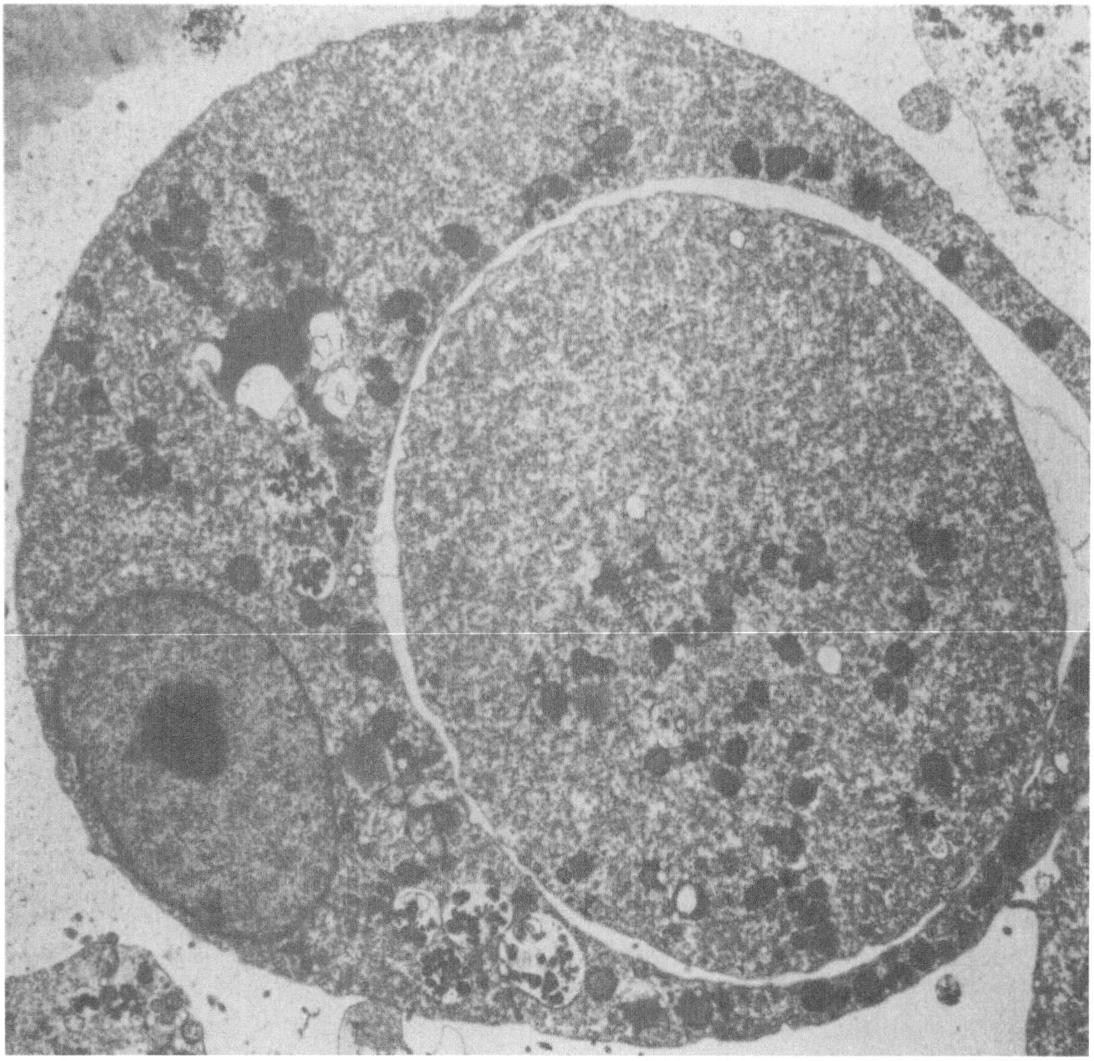

Figure 6-15. Two blastomeres of an abnormal morula developed from a vitrified oocyte (72 hours postinsemination after thawing). One cell has curled around the other and both cells had multiple nuclei in serial sections. X 7,000.

results in fusion of adjacent blastomeres (8), the development of abnormal blastocysts with multiple blastocoeles following cryoinjury to embryos at the morula stage (14), and damage to many intracellular organelles and other cellular components (16). However, provided that half or more of the original blastomeres are intact after freeze–thawing, embryos are capable of continued development and pregnancy when transferred in utero (13,28).

3.2.2. *Ultrastructure of frozen embryos*

Human embryos considered to have survived cryopreservation are transferred to patients in the IVF program. Embryos that have more than half their blastomeres lysed have been studied by electron microscopy to assess the extent and type of damage caused by cryopreservation. These embryos were usually frozen in 1.5M DMSO, cooled slowly to −80°C and thawed slowly (16,28). A small

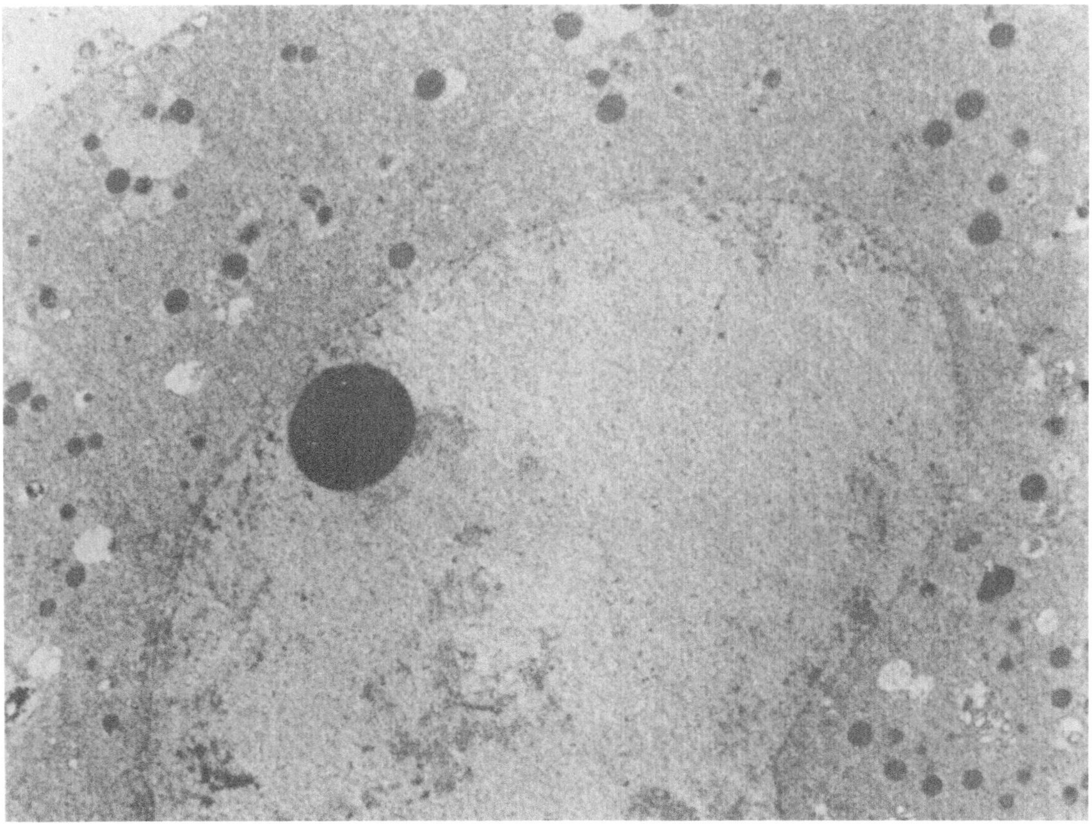

Figure 6-16. Blastomere of a two-cell embryo frozen–thawed by the DMSO ultrarapid method (54 hours after insemination). The cytoplasm and nucleus seem to be well preserved. X 7,000.

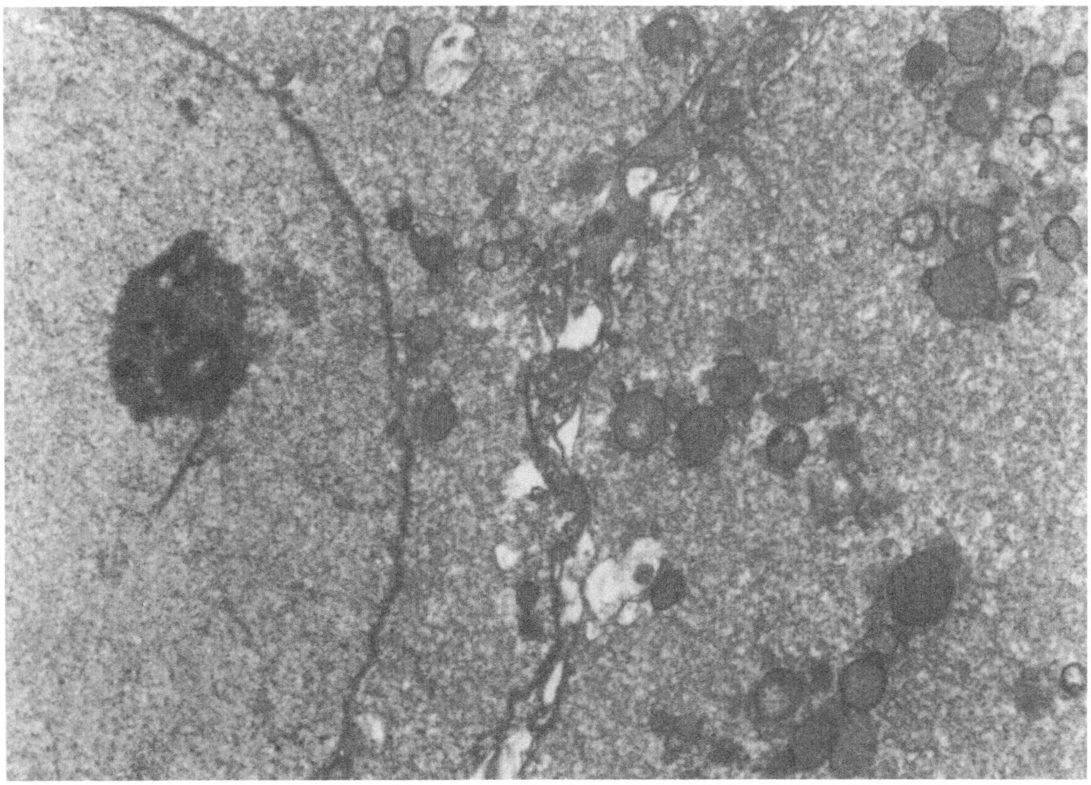

Figure 6-17. Two intact blastomeres of a fragmented eight-cell embryo frozen–thawed by the DMSO slow method (60 hours posinsemination). X 15,400.

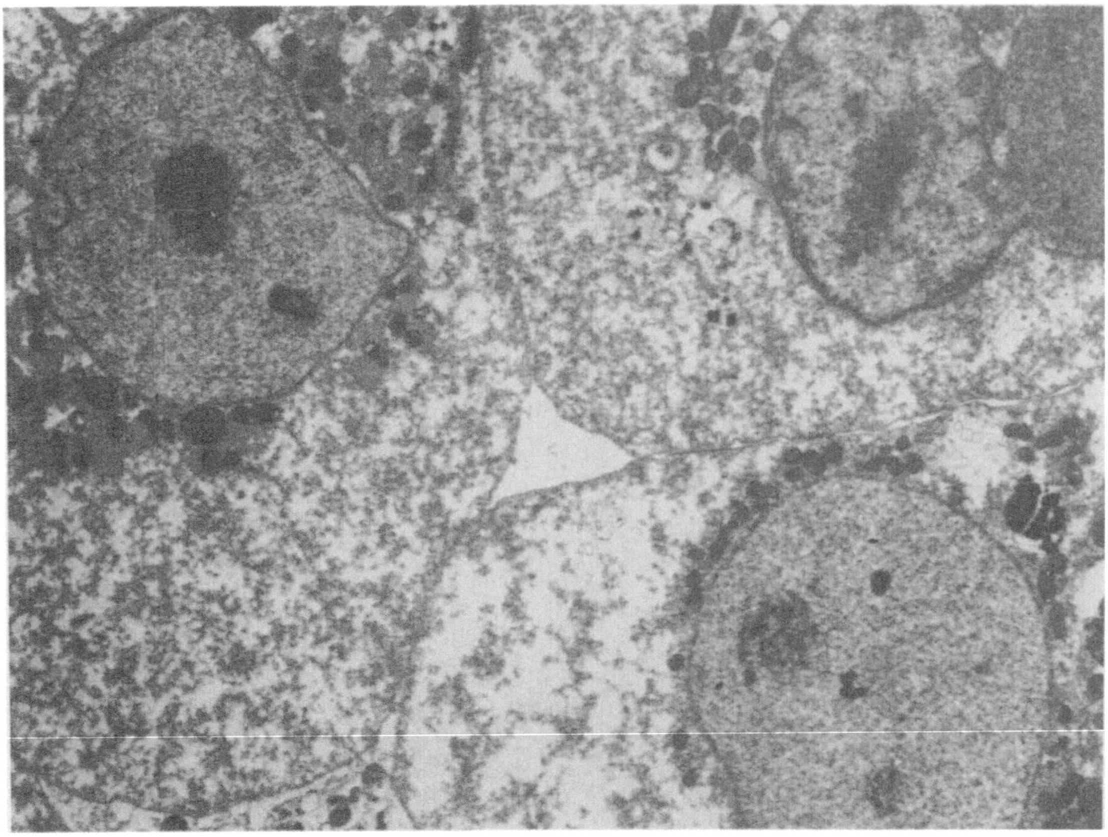

Figure 6-18. Three intact blastomeres of a fragmented 16-cell morula frozen–thawed by the DMSO ultrarapid method (72 hours after insemination and examined 24 hours postthaw culture). Varying degrees of disorganization of the ground cytoplasm (clear areas) are evident. Note cell junctions are still intact. X 7,000.

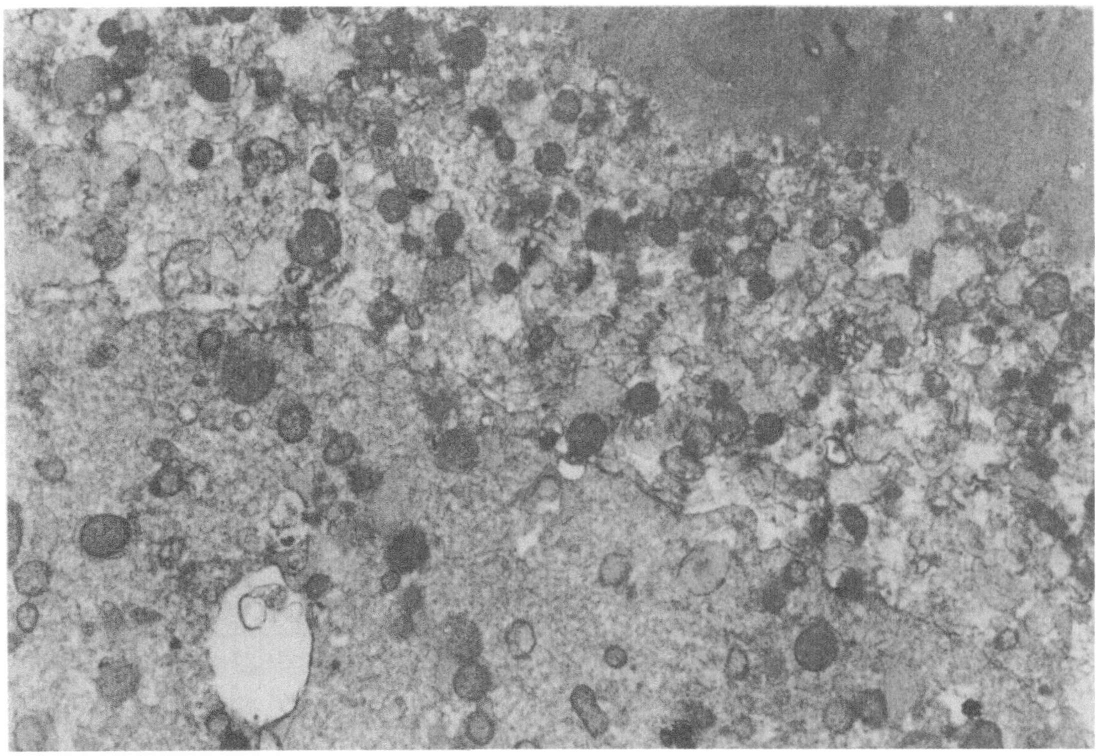

Figure 6-19. Disorganized blastomere lacking a cell membrane seen adjacent to an intact blastomere of the embryo in Figure 6-17. X 15,400.

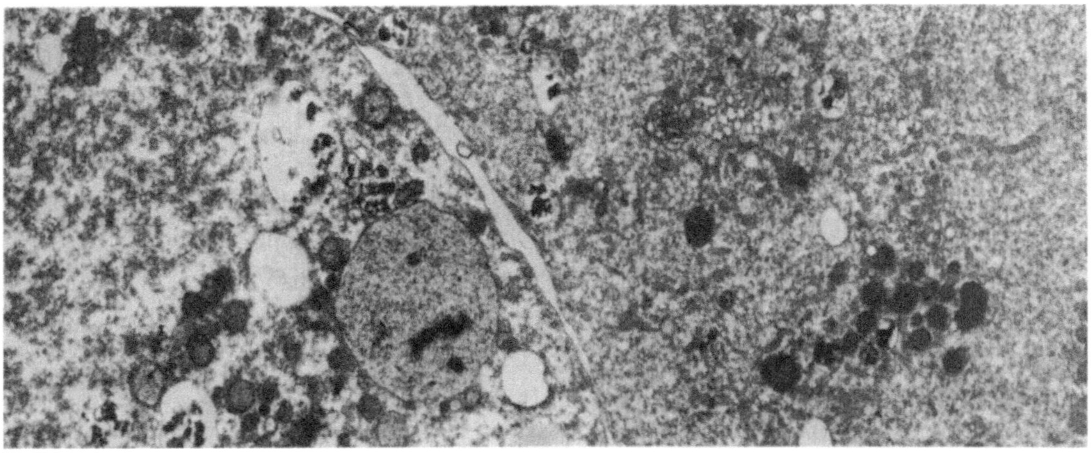

Figure 6-20. Micronucleus in a blastomere of the morula shown in Figure 6-18. The ground cytoplasm of the cell on the left has been damaged. X 9,100.

number of embryos have been frozen ultrarapidly or frozen in 1.5M propanediol, cooling slowly to low subzero temperatures. The other category of embryos examined by electron microscopy were those that had already partially fragmented or were retarded in development before freeze–thawing. Figures 6-16–6-20 portray the ultrastructure of embryos after freeze–thawing. As in the case of oocytes, over 50% of the blastomeres of the retarded embryos examined were intact and had survived freezing and thawing (Figs. 6-16,6-17). This has been reported previously with the DMSO slow method (16,28). The intact cells showed fairly good preservation of structure and still maintained their intercellular junctions. Subtle changes in fine structure, possibly caused by ice crystallisation, were noted in otherwise intact blastomeres that could not be detected by light microscopy (Fig. 6-18) where varying degrees of disorganization of the ground cytoplasm (cytosol) of cells was evident. The cytosol predominantly consists of water and soluble substrates such as sugars, enzymes, and salts. It seems plausible that intracellular freezing of water can lead to formation of minute ice crystals, which is believed to be a lethal event (24). Disruption of the cytosol can also cause disorganization of the microtrabecular architecture and cytoskeletal elements, particularly the microtubules that are universal components of cells (31). The cells that did not survive freezing had no cell membranes and were highly disorganized (Fig. 6-19). Whether fragmented or retarded embryos should be used at all in freeze–thaw studies is questionable, although it is possible to evaluate some of the affects of freezing in the remaining intact blastomeres. Many of these blastomeres are well preserved in spite of their initial fragmentation. Other embryos developed for further periods after freeze–thawing may have micronuclei in some blastomeres (Fig. 6-20), as was observed in pronuclear ova (Fig. 6-14). The origin of micronuclei has been already discussed.

4. Conclusions

In conclusion it is evident that oocyte and embryo freezing are in their infancy, although pregnancies and births have been obtained (15,16,20,27,28). The techniques used at present need to be optimized and standardized using critical methods of evaluation such as electron microscopy, genetic analysis, and developmental studies parallel with experiments on animal models such as mice. Mouse eggs are used to standardize culture media in IVF and should be the obvious choice. Appropriate clinical trials need to be carefully planned and it would seem prudent to ensure the safety of new techniques by assessing embryos before embarking on clinical trials. The results of oocyte freezing and postthaw studies of fertilization and embryonic development are at the stage in which further evaluation by electron microscopy and chromosomal analysis should accompany the improvement in survival as new procedures are described.

References

1. Rogers PAW, Milne BJ, Trounson AO: A model to show human uterine receptivity and embryo viability following ovarian stimulation for in vitro fertilization. *J Vitro Fert Embryo Transfer* 3: 93–98, 1986.
2. Asch RH, Balmaceda JP, Ellsworth LR, Wong PC: Preliminary experiences with gamete intrafallopian transfer (GIFT). *Fertil Steril* 45: 366–371, 1986.
3. Sathananthan AH, Trounson, AO, Wood C: Atlas of fine structure of human sperm penetration, eggs and embryos cultured into vitro. Philadelphia, : Praeger Scientific, 1986.
4. Trounson AO, Mohr LR, Wood C, Leeton JF: Effect of delayed insemination on in-vitro fertilization, culture and transfer of human embryos. *J Reprod Fert* 64: 285–294, 1982.
5. Sathananthan AH, Trounson AO: Ultrastructural observations on cortical granules in human follicular oocytes cultured in vitro. *Gamete Res* 5: 191–198, 1982.
6. Trounson AO, Sathananthan AH: The application of electron microscopy in the evaluation of 2–4 cell human embryos cultured in vitro for embryo transfer. *J Vitro Fert Embryo Transfer* 1: 153–165, 1984.
7. Lippes J, Enders RG, Pragay DA, Bartholomew WR: The collection and analysis of human fallopian tubal fluid. *Contraception* 5: 85–87, 1972.
8. Trounson A: In vitro fertilization and embryo preservation. In: In Vitro Fertilization and Embryo Transfer. A Trounson C Wood (eds), Edinburgh: Churchill Livingstone, p 111–130, 1984.
9. Menezo Y, Testart J, Perrone D: Serum is not neccessary in human in vitro fertilization, early embryo culture and transfer. *Fertil Steril* 42: 750–755, 1984.
10. Quinn P, Kerin JF, Warnes GM: Improved pregnancy rate in human in vitro fertilization with the use of medium based on the composition of human tubal fluid. *Fertil Steril* 44: 493–498, 1985.
11. Gianaroli L, Seracchioli R, Ferraretti AP, Trounson A, Flamigni C, Bovicelli L: The successful use of human amniotic fluid for mouse embryo culture and human in vitro fertilization, embryo culture and transfer. *Fertil Steril* 46: 907–911, 1986.
12. Mohr LR, Trounson AO: The use of fluorescein diacetate to assess embryo viability in the mouse. *J Reprod Fert* 58: 189–196, 1980.
13. Trounson A: Preservation of human eggs and embryos. *Fertil Steril* 46: 1–12, 1986.
14. Mohr LR, Trounson AO, Leeton JF, Wood C: Evaluation of normal and abnormal human embryo development during procedures in vitro. In: Fertilization of the Human Egg In Vitro. HM Beier, HR Lindner (eds), Berlin: Springer Verlag, p 211–221, 1983.
15. Trounson A, Mohr L: Human pregnancy following cryopreservation, thawing and transfer of an eight-cell embryo. *Nature* 305: 707–709, 1983.
16. Mohr LR, Trounson A, Freemann L: Deep-freezing and transfer of human embryos. *J Vitro Fert Embryo Transfer* 2: 1–10, 1985.
17. Mohr LR, Trounson AO: Structural changes associated with freezing of bovine embryos. *Biol Reprod* 25: 1009 –1025, 1981.
18. Whittingham DG: Fertilization in vitro and development to term of unfertilized mouse oocytes previously stored at −196°C. *J Reprod Fertil* 49: 89–94, 1977.
19. Glenister PH, Wood MJ, Kirby C, Whittingham DG: The incidence of chromosome anomalies in first-cleavage mouse embryos obtained from frozen–thawed oocytes fertilized in vitro. *Gamete Res* 16: 205–216, 1987.
20. Chen C: Pregnancy after human oocyte cryopreservation. Lancet 1: 884–886, 1986.
21. Rall WF, Fahy GM: Ice-free cryopreservation of mouse embryos at -196°C by vitrification. *Nature*: 573–575, 1985.
22. Lasalle B, Testart J, Renard JP: Human embryo features that influence the success of cryopreservation with the use of 1,2 propanediol. *Fertil steril* 44: 645–651, 1985.
23. Peura A, Trounson A, Freemann L: Ultra-rapid embryo freezing (Abstr). Presented at the Fourth World Congress on In Vitro Fertilization, Melbourne, Australia, 1985, p 244.
24. Schneider U: Cryobiological principles of embryo freezing. *J Vitro Fertil Embryo Transfer* 3: 3–9, 1986.
25. Sathananthan AH, Ng SC, Edirisinghe R, Ratnam SS: Sperm–oocyte interaction in the human during polyspermic fertilisation in vitro. *Gamete Res* 15: 317–326, 1986.
26. Sathananthan AH: Ultrastructural morphology of fertilization and early cleavage in the human. In: In Vitro Fertilization and Embryo Transfer. A Trounson, C Wood (eds), London: Churchill Livingstone, p 131–158, 1984.
27. Cohen J, Simons RF, Fehilly CB, Edwards RG: Factors affecting survival and implantation of cryopreserved human embryos. *J Vitro Fertil Embryo Transfer* 3: 46–52, 1986.
28. Freemann L, Trounson, A, Kirby C: Cryopreservation of human embryos: progress on the clinical use of the technique in human IVF. *J Vitro Fert Embryo Transfer* 3: 53–61, 1986.
29. Sathananthan AH, Chen C: Sperm–oocyte membrane fusion in the human during monospermic fertilization. *Gamete Res* 15: 177–186, 1986.
30. Cohen J, Simons RF, Edwards RG, Fehilly CB, Fishel SB: Pregnancies following the frozen storage of expanding human blastocysts. *J Vitro Fertil Embryo Transfer* 2: 59–64, 1985.
31. Fawcett DW: Bloom and Fawcett: A Textbook of His-

tology. Philadelphia: Saunders, 1986.
32. Van Blerkom J, Henry GP, Porreco RP: Preimplantation human embryonic development from polynuclear eggs after in vitro fertilization. *Fertil Steril* 41: 686–696, 1984.
33. Sathananthan AH, Trounson AO, Freemann L: Morphology and fertilizability of frozen human oocytes. *Gamete Res* 16: 343–354, 1987.

Author's address:
A. Henry Sathananthan
Lincoln School of Health Sciences
LaTrobe University
625 Swanston St.
Carlton, Victoria 3053
Australia

CHAPTER 7

In vitro studies of the peri-implantation phase of human embryos

SVEND LINDENBERG & POUL HYTTEL

Since Hertig and Rock (1,2) described human embryonic development from fertilization until the late implantation stage, very few morphological studies on the human peri-implantation embryo have been reported (3,4,5). Hertig and Rock's material detailed the normal light-microscopic morphology of the human embryo until the blastocyst stage, and from the implanted blastocyst until the 17th day of gestation. Unfortunately, they were unable to include the implanting blastocyst stage. Human embryos have been studied in detail at the ultrastructural level during the immediate preimplantation (3) and early postimplantation (days 10–11) stages. Thus, morphological information about the process of implantation itself is still lacking for the human.

In the study of implantation, it is useful to separate events into the following series of developmental phases: 1) the phase of apposition and adhesion of trophoblast cells to the uterine surface; 2) the phase of penetration of trophoblast into the uterine epithelium; and 3) the phase of placental organization. In this chapter, only the apposition–adhesion phase and the very early stages of trophoblast penetration of the uterine epithelium will be discussed.

Penetration of the trophoblast into the epithelium has been extensively studied in different mammalian species. These studies have revealed three primary types of penetration: 1) a fusional type, such as observed in the rabbit (6), where the trophoblast cells fuse with the epithelial cells of the endometrium; 2) an intrusive type, such as in the ferret (7), where trophoblast cells extend cytoplasmic protrusions between individual epithelial cells without primary phagocytosis; and 3) a displacement type, such as that observed in the rat and mouse (8,9), where trophoblast cells simply displace the epithelial cells, i.e., lifting them away from the stroma.

To determine how epithelial penetration occurs in the human, an in vitro implantation system has been developed. This approach has been taken owing to the obvious difficulties in access to properly staged specimens. Implantation in vitro admittedly does not precisely reflect the in vivo situation, since the endometrial stroma and maternal hormonal environment are absent in culture. Nevertheless, from our results (10), and from those of others (11), it seems reasonable to consider that some of the peri-implantation characteristics of embryonic development in vitro may reflect the in vivo situation. However, the problem of artifact associated with in vitro conditions should always be kept in mind.

Nearly all in vitro implantation experiments have employed mammalian species with a haemochorial type of in vivo implantation (12–15). Haemo-chorial implantation is characterized by an invasive trophoblast that penetrates the uterine epithelium and results in direct haemo-chorial contact, i.e., the uterine epithelium is disrupted by penetrating cells. In vitro studies of this type of implantation reveal a trophoblast penetration and disruption of the uterine monolayer cultures. The epithelial-chorial implantation observed in the pig (16) and cow (17) is characterized by trophoblast outgrowth upon the epithelium, but without disruption and penetration of the uterine epithelium. It is therefore of interest to compare the in vitro correlates of chorial-epithelial implantation (i.e., cow) to the findings derived from human embryo implantation in vitro. Such a comparative study could reveal whether or not the displacement of epithelial cells that accompanies trophoblast outgrowth in vitro is an artifact of the culture environment, or whether it is caused by some mechanism inherent to this type of implantation.

1.1. Methodology of in vitro implantation studies

Details of the methods have previously been described (10,18,19) but the basics of the protocols will be summarized.

Human blastocysts were developed from in vitro-fertilized oocytes that were recovered by ultrasound-guided percutaneous follicle aspiration from women whose ovulatory cycles were controlled with clomiphene citrate, hMG and hCG as pre viously described (10). Mature oocytes were inseminated and cultured in either Ham's F-10 medium or Earle's balanced salt solution supplemented with maternal serum. Preovulary oocytes were washed in follicular fluid and allowed to complete maturation for five hours before insemination with approximately 0.5×10^6 motile sperm per ml. Eighteen hours after insemination, eggs containing two pronuclei were transferred to new medium for the duration of culture. Hatched and hatching blastocysts were transferred in the same medium to uterine epithelial monolayer cultures prepared as previously described (10,18). *Cow blastocysts* were obtained from superovulated dairy cows by means of uterine lavage performed seven days after artificial insemination. These blastocysts were kept frozen until usage. After thawing, the blastocysts were cultured in Alfa-modified Earle's balanced salt solution supplemented with 20% fetal calf serum until expansion and were subsequently transferred to cow uterine epithelial monolayer cultures. These cultures were prepared from cyclic cow uterine biopsies by the same method as for the human uterine monolayer preparations.

Co-cultures of blastocysts and endometrial cells were observed daily in an inverted phase microscope. Adhesion was operationally defined as tandem motion of the blastocyst and monolayer when gently tapping the bottom of the culture dish. The human implantation sites were fixed for microscopy 24 hours after adhesion was observed. The bovine implantations were fixed 24 to 72 hours after adhesions were seen. Fixation was performed in 70% Karnowsky fixative for 20 minutes at 4°C. After repeated washing in 0.1M cacodylate buffer at pH 7.3, postfixation in 1% osmium tetroxide in 0.1M cacodylate buffer for 2 hours at 4°C was performed. Subsequently, the specimens were dehydrated and embedded in Epon. Each implantation site was sectioned into semithin and ultrathin sections. Ultrathin sections were obtained after each 5–10 semithin sections. The semithin sections were stained with toluidine blue, while the ultrathin sections were contrasted with uranyl acetate and lead citrate.

The findings discussed in this chapter were derived from 15 women where 45 mature cumulus-oocyte complexes were retrieved, resulting in five hatched blastocysts on day 6–7 after insemination. Four blastocysts implanted within 24 hours after adhesion were observed on day 7–8. Also 15 bovine thawed blastocysts were included. Six blastocysts hatched in vitro and five implanted within two days after hatching.

2. Implantation of the human blastocyst in vitro

2.1. Hatching — appositional–adhesional phase

Hatching in vitro appears to initiate at one region of the zona pellucida where the mural trophoblast cells protrude through a crack in the zona. With successive expansions of the blastocoel cavity, progressively more of the trophoblast protrudes. Apposition and adhesion of the embryo apparently occurs within the next 24 hours. All five hatched blastocysts apposed and adhered to the monolayer by their *polar trophoblast area* (embryonic pole), indicating that a polarity of the implanting blastocyst already exists at the phase of apposition–adhesion

2.2. The implanting blastocyst

In the implanting blastocysts, *mural* and *polar* trophoblast cells, and *embryonic* cells, i.e., the developing inner cell mass including endodermal cells, could be recognized (Fig. 7-1). However, in two blastocysts the inner cell mass showed degenerative changes. An incomplete germinal disc, composed of developing inner cell mass, i.e., embryonic cells, and delaminating endodermal cells was also observed (Fig. 7-1).

2.3. *The mural trophoblastic cells*

The mural trophoblast cells form the wall of the blastocoele and elaborate abundant microvilli on their outer surfaces. By contrast, the surface facing the blastocoele was relatively smooth and devoid of microvilli. Adjoining cells were associated by means of junctional complexes and exhibited relatively large intercellular spaces basal to the complexes. Numerous microfilaments extended from the junctional complexes into the cytoplasm (Fig. 7-2).

2.4. *The polar trophoblast cells*

The polar trophoblast cells, which cover the inner cell mass, are formed by cytotrophoblast. These cells are situated partially under the germinal disc and partially organize the peripheral outgrowth. Adjoining cytotrophoblast cells are associated by junctional complexes. The contact between cytotrophoblast and embryonic cells involved desmosomes and long slender projections from the cytotrophoblast that invaginated into the embryonic cells. Often, these projections terminated at the perinuclear region of the embryonic cells (Fig. 7-3). In the central portion of the implantation site, polar trophoblast cells, which at this location were in contact with the culture vessel, formed an early syncytium which appeared to have at least two nuclei (Fig. 7-4). The syncytium contained electron-dense nuclei and abundant free polyribosomes.

2.5. *The embryonic cells*

Embryonic cells were located above the polar trophoblast and formed an incomplete germinal disc, of which the endodermal cells constituted the outer layer (Fig. 7-1). Intercellular junctions included desmosomes and structures resembling gap junctions. In the cytoplasm, crystallike inclusions and annulate lamellae were seen. The nucleus was centrally situated and contained a prominent nucleolus.

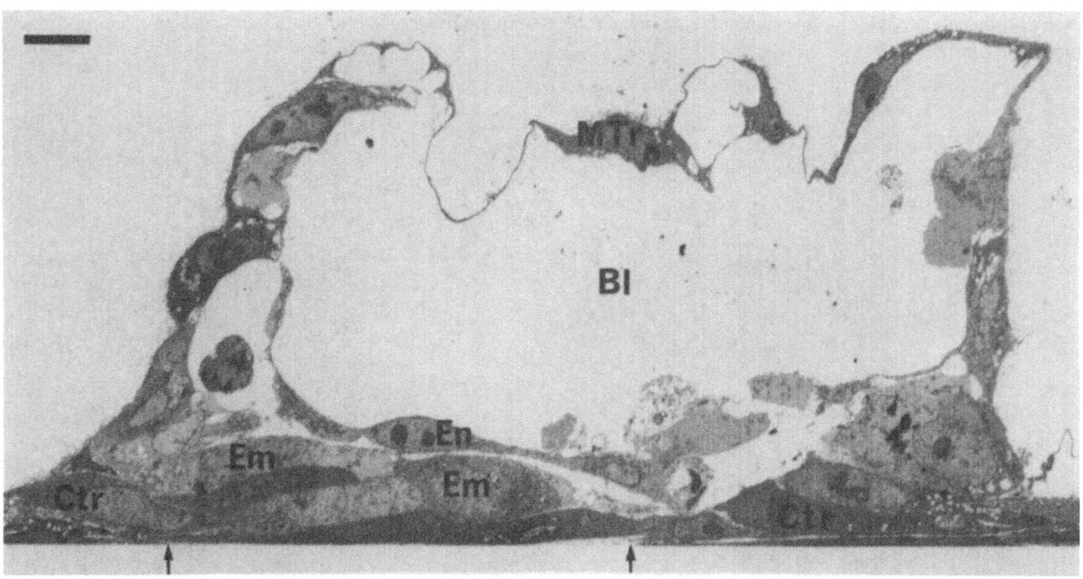

Figure 7-1. Human implantation showing the blastocoel (Bl) and the delaminating mural trophoblast cells (MTr). In the basal portion of the implantation site, embryonic cells (Em), including endodermal (En), and cytotrophoblast (Ctr) are recognized. The arrows indicate to which level the endometrial cells could be traced. Bar indicates 10 μm.

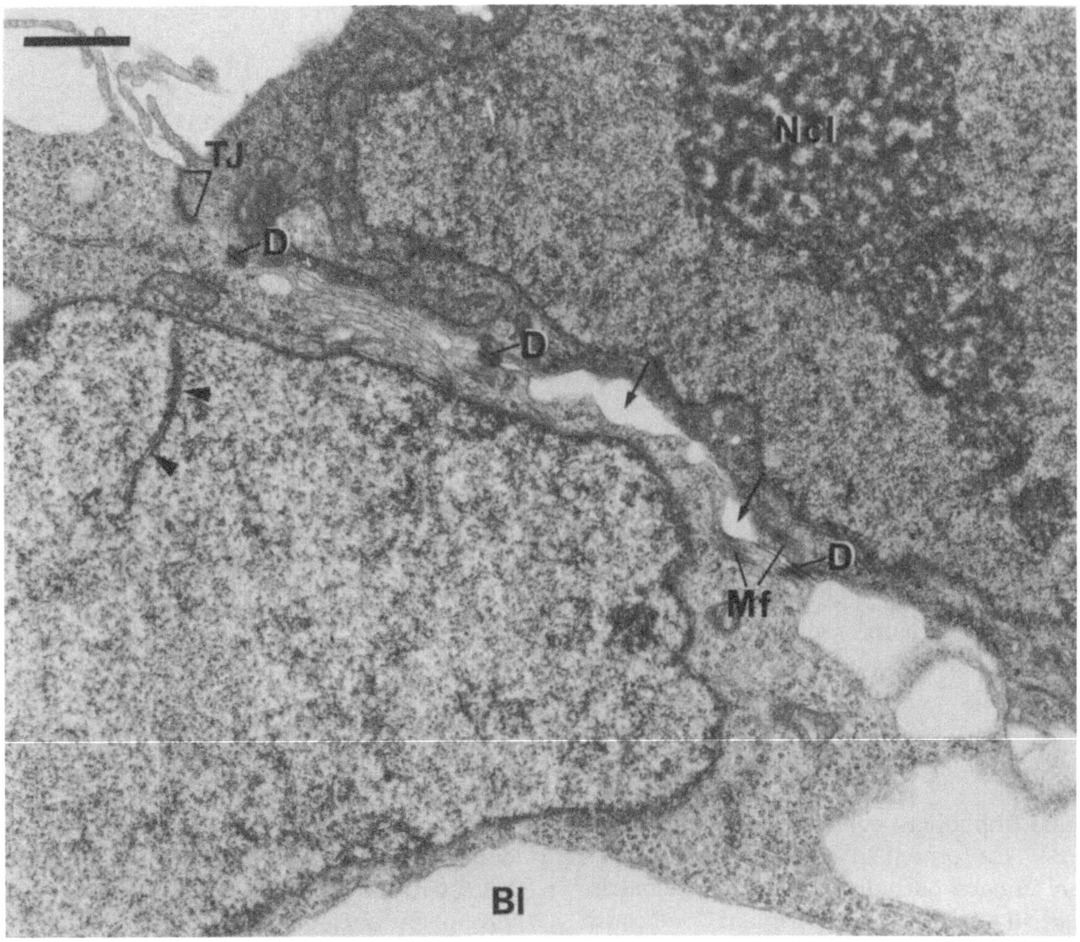

Figure 7-2. Detail from a human implantation indicating part of the blastocoel (B1) and communication between two mural trophoblast cells. The two cells form a tight junction (TJ) and several desmosomes (D) with associated microfilaments (Mf). Note the widened intercellular space (arrows) between desmosomes. In the nucleus of one of the trophoblast cells, infoldings of the inner membrane of the nuclear envelope (arrowheads) are present. A prominent nucleolus (Ncl) is present in the nucleus of the other cell. Bar indicates 1 μm.

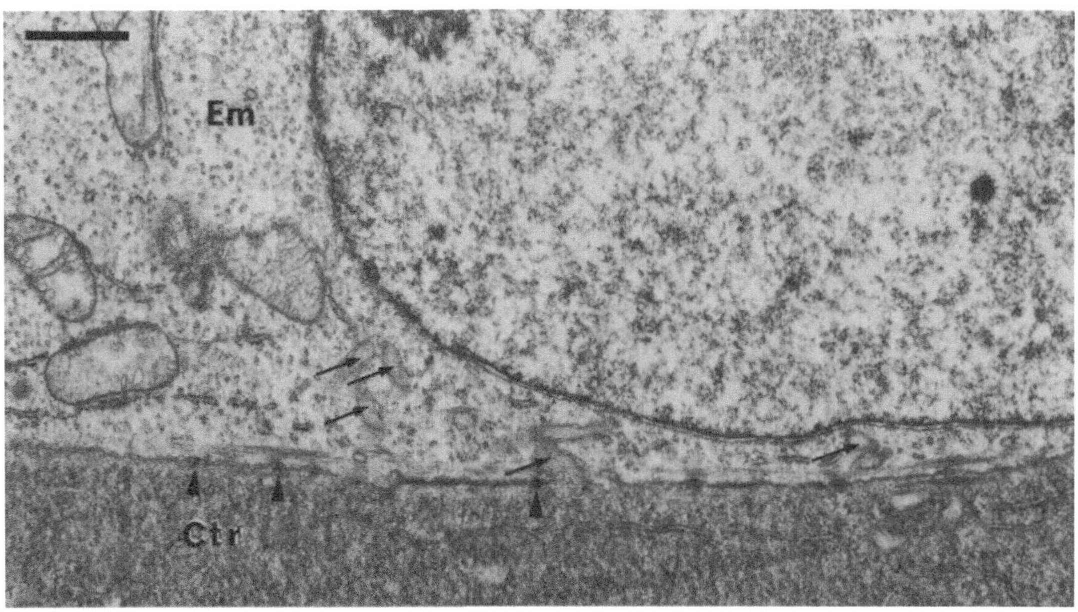

Figure 7-3. Detail from the basal portion of a human implantation site showing the communication between a cytotrophoblast (Ctr) and an embryonic cell (Em). Microvillilike projections (arrows) from the cytotrophoblast invaginate the plasmalemma of the embryonic cell. Several desmosomes (arrowheads) are also evident. Bar indicates 1 μm.

2.6. The endodermal cells

Endodermal cells constituted the innermost layer of the incomplete germinal disc and tended to form a lining of the inner surface of the trophoblast cells. Apart from a more flattened appearance, the ultrastructure of these cells resembled that of the remaining embryonic cells. Crystallike inclusions were also observed in the cells (Fig. 7-5).

2.7. The endometrial cells

The epithelial monolayer was composed of flattened cells which elaborate microvilli, cilia and cytoplasmic protrusions. Prominent cytoplasm components included desmosomes, cisternae of rough endoplasmic reticulum containing electron-dense material, polyribosomes and microfilaments. The microfilaments were characteristically situated in the periphery of the cells. An elongated nucleus was usually located in the center of the cell, often with an indented nuclear membrane, a prominent nucleolus, and peripherally located heterochromatin.

2.5. The endometrial–trophoblastic interphase

In all specimens, the polar trophoblast cells had displaced the endometrial cells in the central portion of the implantation site, thereby making direct contact to the culture vessel (Fig. 7-1). In the peripheral area around the implanting blastocyst, the endometrial cells were stacked 3–4 cells thick. The cytotrophoblast formed long slender cytoplasmic protrusions which penetrated the intercellular space of adjacent endometrial cells (Fig. 7-6). No regular junctional complexes were seen between the two cell types, but focally, the cytotrophoblast formed close appositions with the endometrial cells (Fig. 7-7). Neither phagocytosis of the endometrial cells by the cytotrophoblasts nor any degenerated endometrial cells were seen. The endometrial cells could be distinguished from the trophoblastic cells by their content of glycogen, rough endoplasmic reticulum, peripherally located tonofilaments, dark mitochondria and polyribosomes. By contrast, trophoblast cells contained few polyribosomes and cisternae of the rough endoplasmic reticulum. The trophoblast was characterised by the presence of elements of the smooth endoplasmic retriculum, tubular mitochondria and bundles of filaments dispersed in the cytoplasm. Nevertheless, the morphology of the endometrial cells adjacent to the cytotrophoblast had changed when compared to endometrial cells distal to the implantation site. These endometrial cells had fewer microvilli and appeared to be polarized, as indicated by the presence of numerous invaginations (100 nm) of the plasmalemma in the areas facing the cytotrophoblastic protrusions (Fig. 7-8).

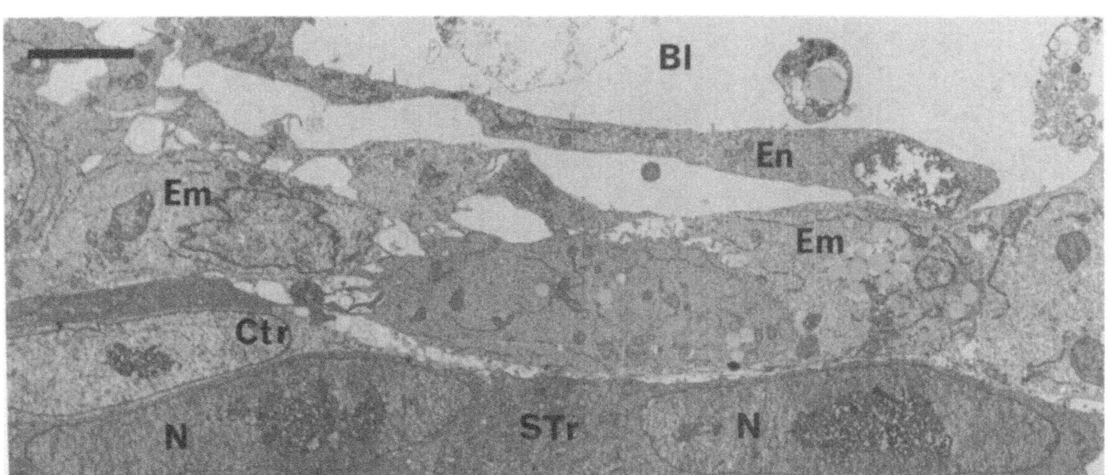

Figure 7-4. Detail from the basal part of a human implantation site showing parts of the blastocoel (Bl), embryonic cells (Em) including endodermal cells (En), cytotrophoblast (Ctr) and an early formation of syncytial trophoblast (STr) with two apparent nuclei (N). Bar indicates 5 μm.

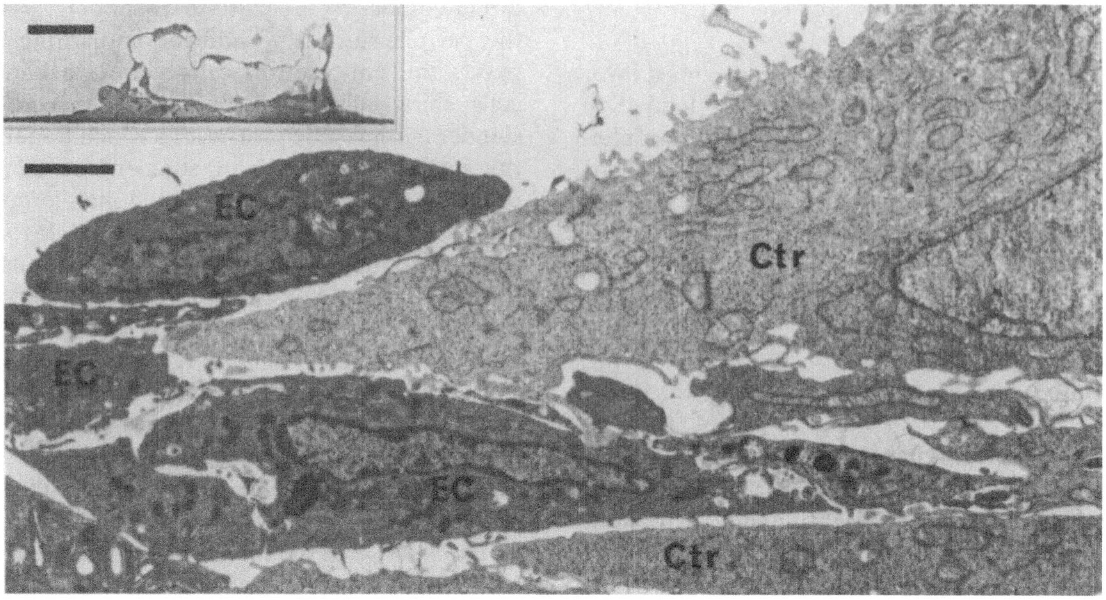

Figure 7-5. Detail from a human implantation site showing parts of the blastocoel (Bl), and an endodermal cell (En) located on other embryonic cells (Em). In the endodermal cell, a crystallike inclusion (Cl) is seen. Bar indicates 1 μm. Inset: the crystallike inclusion at a larger magnification. Bar indicates 0.25 μm.

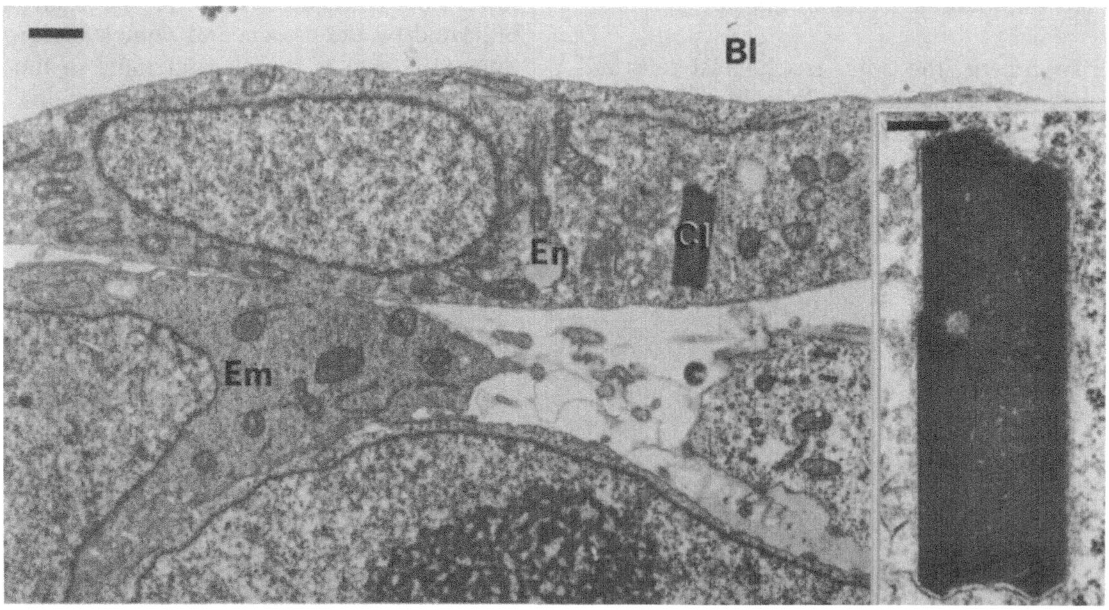

Figure 7-6. Detail from a human implantation site showing the trophoblastic-endometrial interphase where cytotrophoblast (Ctr) insinuate ectoplasmic protrusions between the stacked endometrial cells (EC). Bar indicates 2 μm. Inset: the same implantation indicating location within the confluent monolayer of endometrial cells. Bar indicates 50 μm.

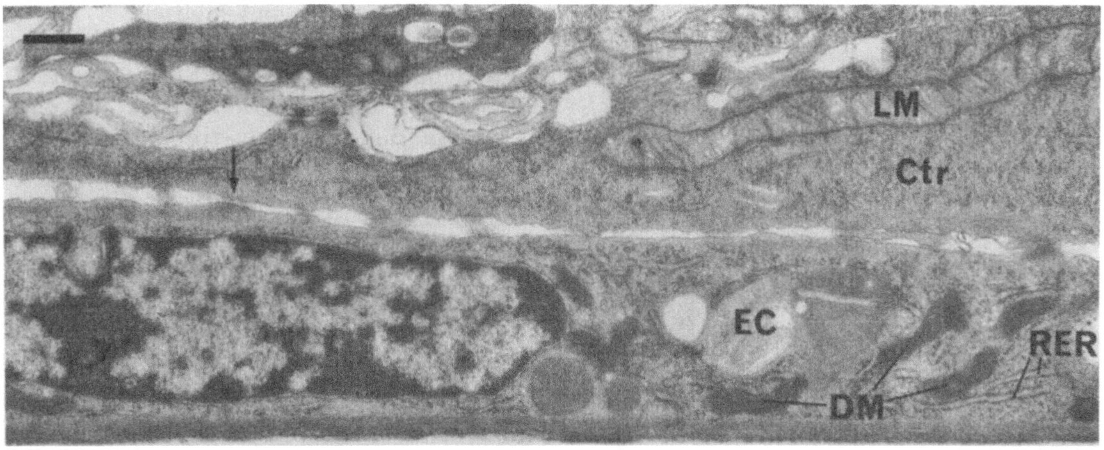

Figure 7-7. Detail from the basal portion of a human implantation site showing the communication between a cytotrophoblast (Ctr) and an endometrial cell (EC). Focal appositions (arrow) with increased density of the adjacent cytoplasm are present. In the cytotrophoblast, a long mitochondrion (LM) of low electron density is seen while the endometrial cell contain mitochondrial (DM) of high electron density and cisternae of rough endoplasmic reticulum (RER). Bar indicates 0.5 μm.

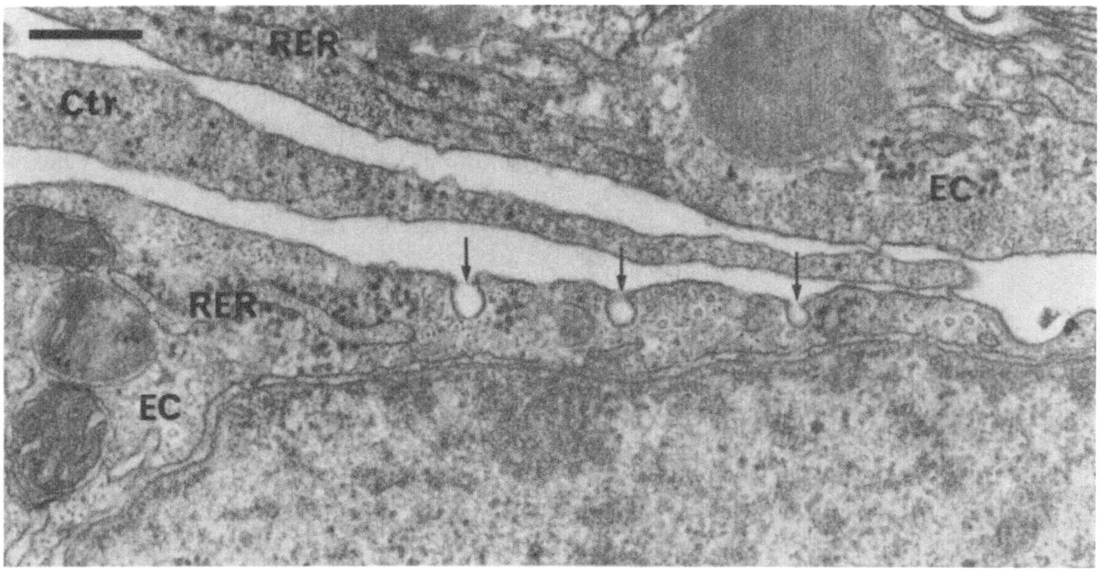

Figure 7-8. Detail from the trophoblastic–endometrial interphase of a human implantation site showing two endometrial cells (EC) and a cytoplasmic protrusion from a cytotrophoblast cell (Ctr). In the endometrial cells, cisternae of rough endoplasmic reticulum (RER) and numerous invaginations of the plasma membrane (arrows) are evident. Bar indicates 0.25 μm.

3. Implantation of the bovine blastocyst in vitro

3.1. Hatching — appositional and adhesional phase

The hatching of bovine in vitro-cultured blastocysts occurred in the same manner as observed for the human species, i.e., through a crack in the zona pellucida that was progressively expanded by the protruding trophoblast cells. In contrast to the human, the bovine zona was firmly attached before hatching by means of long slender projections of endometrial origin (Fig. 7-9). These projections contacted the zona and apparently phagocytosed parts of it.

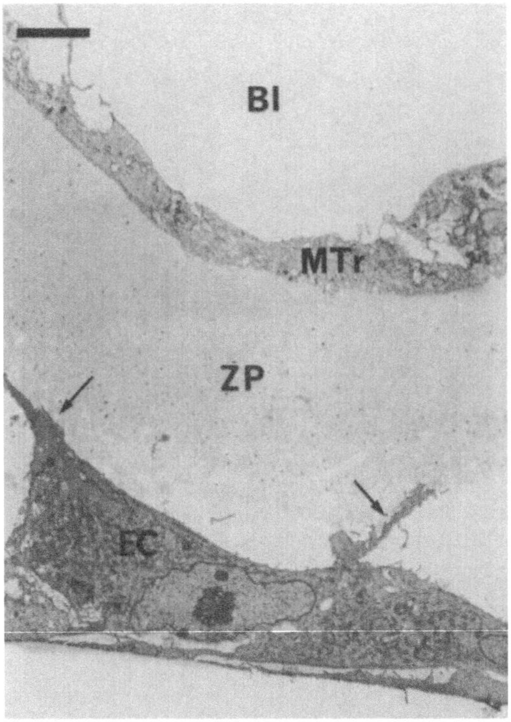

Figure 7-9. Detail from a bovine implantation showing the blastocoel (Bl) delinated by mural trophoblastic cells (MTr). Note the activated endometrial cell (EC) which elaborates projections (arrows) that contact the zona pellucida (ZP). Bar indicates 5 μm.

3.2. The endometrial–trophoblastic interphase

Endometrial cells were not displaced by the outgrowing trophoblast cells. By contrast, the trophoblast cells covered the surface of the endometrial monolayer (Fig. 7-10). Again, a remarkable polarity of the endometrial cells was induced: the plasmalemma of the endometrial cells facing the outgrowing trophoblast had numerous invaginations. There were no signs of phagocytosis of endometrial cells by the trophoblast since degenerated endometrial cells were not observed.

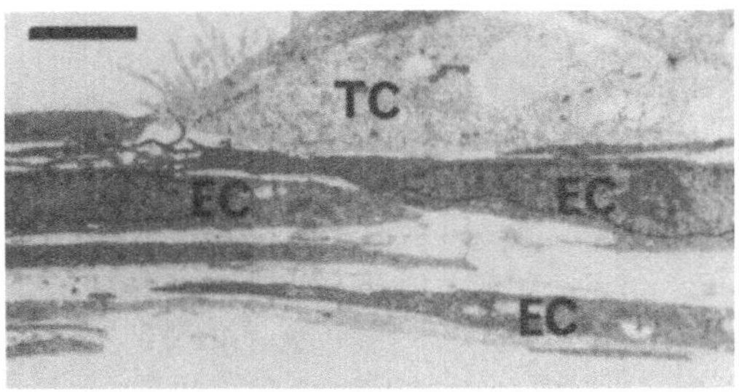

Figure 7-10. Detail from a bovine implantation site showing the trophoblastic–endometrial interphase. Endometrial cells are marked EC, and trophoblast TC. Bar indicates 2.5 μm.

3.3. The in vitro implantation model

The cytological evidence of hatching, adhesion and further development of the endometrial–trophoblast interphase is of particular significance for the understanding of the mechanism of implantation. Hatching of the cultured human blastocyst (from in vitro fertilization) was approximately one day delayed as compared to in vivo observations (1,20). This delay may be due to a hardening of the zone pellucida in vitro, an event which has been reported for the mouse (21). Other factors may also be important. For example, Mohr and Trounson (3) observed the hatching of a morphologically normal human blastocyst occurred on day 6 after fertilization and culture in vitro. The delayed hatching observed in the present in vitro system could be associated with suboptimal media or with the building of toxic agents (22). This may also explain the extensive and progressive loss of viable human embryos during culture (Table 7-1).

Differences between the implantation of the human blastocyst in vivo and in vitro are apparent. For example, not only are the growth rates of embryos different but also the mechanism of hatching is different. In descriptions of embryos (in vivo fertilization) flushed from the human uterine cavity (1,20), most blastocysts found were either with or without the zona pellucida. How-

Table 7-1. Results following in vitro fertilization and culture.

	Number
Patients	15
Mature oocytes	45
Ova with 2 pronuclei	30
Morulae	22
Blastocysts	12
Hatched blastocysts	5
Adhesive blastocysts	5
Implantations	4

ever, empty zonae were never reported. This is in contrast to the situation in vitro, where empty human zonae were present in the cultures days after hatching. These findings suggest that enzymatic activities are important in vivo for the zona dissolution. Such activities may be absent in vitro. Only the trophoblast cells covering the inner cell mass (i.e., polar trophoblast) were involved in the initial phases of apposition and adhesion, as well as the subsequent penetration of the epithelium. The orientation and polarization of the implanting blastocyst during the initial phases of implantation in vitro seem to be the same as observed in vivo for later implantations (1,4,5). It is possible that an interaction between the inner cell mass and the trophoblast cells is responsible for the polarity of the implanting blastocysts. Cytological evidence of such an interaction is the presence of polar trophoblast cell projections that invaginate the plasmalemma of the embryonic cells, and the fact that signs of syncytial formation of the trophoblast were only recognized in relation to the inner cell mass.

Although degenerative changes were observed in the inner cell mass of two embryos, the trophoblast outgrowth of these blastocysts was indistinguishable from that of implanting blastocysts with an apparently normal inner cell mass (composed of embryonic cells and developing endoderm) and an expanded blastocoele. The degeneration of the inner cell mass may be reflected in the often-encountered biochemical pregnancies seen in IVF programs (23), or may be the in vitro correlate of the blighted ovum syndrome.

The initial penetration of the epithelium by the cytotrophoblast involves two separate components, namely the trophoblast cell outgrowth and the response of the endometrial cells. Both bovine and human endometrial cells in vitro responded to the presence of trophoblast cells by membrane activity. Whether this activity represents endo- or exocytosis remains to be determined. However, in cyclic cows (24), and during the peri-implantation phases of rat (25) and mouse development in vivo (26), similar membranous activity of the endometrial epithelium has been shown to be endocytotic in nature.

The membrane activity of the endometrial cells might be morphological evidence of local embryo maternal signaling. Substances secreted by the embryo as it lies free in the uterine lumen during the initial penetration of the endometrium are involved in signaling the endometrium to prepare for implantation (27,28). Thus, possible localized pinocytic activity in the endometrial cells in vitro may be morphological signs of such an interaction. Another observation that characterizes the endometrial cells in vivo is the reduced number of microvilli (29) in areas adjacent to the cytotrophoblastic cells. This phenomenon is also seen during the human implantation in vitro.

An intriguing feature of human implantation in vitro was the stacking of endometrial cells around the implantation site. Whether this is a result of pushing from the trophoblast, as suggested by Larsen (4,5) on the basis of in vivo observations, or a local redistribution of the endometrial cells themselves, as reported for the rat (32), is not clear. The stacking of the epithelial cells around the implanting blastocyst may also be the in vitro correlate of plaque formation — i.e., epthelial hypertrophy or hyperplasia as observed in vivo during the initial stages of implantation in the rhesus monkey (30). With respect to trophoblast cells, the mural trophoblast remained unchanged during the initial phases of in vitro implantation, whereas the polar trophoblast cells changed in two areas. In the central part of the implantation site and adjacent to the inner cell mass, an early syncytial formation was observed. The morphology of the syncytium resembled that described for very early implantation sites in the ferret (7). Moreover, our observations probably relate to the first stages of syncytial formation, since Mohr and Trounson did not find any syncytial formation in a hatched but unimplanted human blastocyst (3). In the periphery of the implantation site, the cytotrophoblast insinuates, between the endometrial cells, long slender ectoplasmic protrusions. In this intimate interphase between trophoblast and endometrial cells, the latter cell type becomes polarized. The cytotrophoblasts were not cytolytic during the initial stages of implantation, since no degenerating endometrial cells were seen in proximity to the outgrowing cytotrophoblast. No junctions were seen between adjoining trophoblast and endometrial cells, but close apposition, by others referred to as primitive junctions (32), were detected.

The in vitro observations of human implantation correlate surprisingly well with the intrusive type of implantation observed in the Rhesus monkey and described in detail by Enders (31). Implantation in the Rhesus monkey involves plaque formation of the epithelium and a noncytotoxic but intrusive trophoblast migration of long slender ectoplasmic protrusions through the uterus epithelium. Syncytial formation in the human in vitro implantation model is also situated in the vicinity of the inner cell mass and is not observed until the blastocyst is actually implanting. With respect to the bovine species, the trophoblast cells begin placentation by covering the endometrial cells rather than by penetration of the monolayer. These results indicate that while the human and bovine culture systems have been prepared according to identical protocols, these two species behave significantly differently in the in vitro implantation system. This observation corresponds very closely to the in vivo situation (33). Therefore, the penetration of the human cytotrophoblast may not be an in vitro artifact, but rather appears to be a species-related phenomenon. Our findings indicate that the initial stages of implantation in the human are of the intrusive type.

4. Conclusions

Morphological features of human in vitro fertilized and cultured blastocysts during the initial stages of in vitro implantation on uterine monolayers were studied by transmission electron microscopy. After hatching, the blastocysts attached to the uterine cells within 24 hours and during the next 24 hours trophoblast outgrowth was recognized. The implanting blastocyst had differentiated into mural and polar trophoblast and embryonic cells including endoderm. Trophoblast cells migrated through the endometrial monolayers by displacing the endometrial cells. These cells elaborated long, slender ectoplasmic protrusions into the intercellular space which disrupted endometrial cell junctions. The endometrial cells were multilayered at the periphery of the implantation site, allowing the trophoblast to come into contact with the culture dish. This situation may parallel the invasive haemo-chorial implantation process seen in vivo. In the trophoblast–endometrial interphase, local endo- or exocytotic processes were encountered in the endometrial cells. The findings are compared with the bovine in vitro implantation system. The results correlate well with what is observed in vivo, i.e., an epithelial-chorial, noninvasive type of implantation. Findings from the human implantation model system suggest that the initial phase of implantation may be of the intrusive type. This mode involves epithelial plaque formation and trophoblast protrusions that insinuate between endometrial cells.

In the analysis of human in vitro implantation, which in vivo represents a haemo-chorial placentation, we have included controls in the form of a chorial-epithelial implantation (cow) to elucidate some of the possible artifacts that may be associated with the in vitro system.

References

1. Hertig AT, Rock J, Adams EC: A description of 34 human ova within the first 17 days of development. *Am J Anat* 98: 435–491, 1956.
2. Hertig AT, Rock J, Adams EC, Mulligan WJ: On the preimplantation stages of the human ovum: a description of four normal and four abnormal specimens ranging from the second to the fifth day of development. *Contrib Embryol* 240: 201–220, 1954.
3. Mohr LR, Trounson AO: Comparative ultrastructure of hatched human, mouse and bovine blastocysts. *J Reprod Fert* 66: 499–504, 1982.
4. Larsen JF, Knoth M: Ultrastructure of the anchoring villi and trophoblastic shell in the second week of placentation. *Acta obstet Gynec Scand* 50: 117–128, 1971.
5. Knoth M, Larsen JF: Ultrastructure of a human implantation site. *Acta Obstet Gynec Scand* 51: 385–393, 1972.
6. Schlafke S, Enders AC: Cellular basis of interaction between trophoblast and uterus at implantation. *Biol Reprod* 12: 41–65, 1975.
7. Enders AC, Schlafke S: Implantation in the ferret: epithelial penetration. *Am J Anat* 133: 291–316, 1972.
8. Tachi S, Tachi C, Lindner HR: Ultrastructural features of blastocyst attachment and trophoblastic invasion in the rat. *J Reprod Fert* 21: 37–56, 1970.
9. Finn CA: Cellular changes in the uterus during the establishment of pregnancy in rodents. *J Reprod Fert Suppl* 31: 105–111, 1982.
10. Lindenberg S, Lenz S, Hyttel P, Holmes P: Ultrastructure

of the early human implantation in vitro. In press.
11. Sherman MI, Wudl LR: The implanting mouse blastocyst. *Cell Surface Rev* 1: 81–125, 1976.
12. Morris JE, Potter SW, Buckley PM: Mouse embryos and uterine epithelia show adhesive interactions in culture. *J Exp Zool* 222: 195–198, 1982.
13. Blerkom JV, Chavez DJ: Morphodynamics of outgrowths of mouse trophoblast in the presence and absence of a monolayer of uterine epithelium. *Am J Anat* 162: 143–155, 1981.
14. Sherman MI: Implantation of mouse blastocysts in vitro. In: Methods in Mammalian Reproduction. New York, San Francisco, London: Academic Press, Inc., p 247–257, 1978.
15. Salomon DS, Sherman MI: Implantation and invasiveness of mouse blastocysts on uterine monolayers. *Exp Cell Res* 90: 261–268, 1975.
16. Dantzer V: Electron microscopy of the initial stages of placentation in the pig. *Anat Embryol* 172: 281–293, 1985.
17. Wathes DC, Wooding FBP: An electron microscopic study of implantation in the cow. *Am J Anat* 159: 285–306, 1980.
18. Lindenberg S, Lauritsen JG, Nielsen MH, Larsen JF: Isolation and culture of human endometrial cells. *Fertil Steril* 41(4): 650–652, 1984.
19. Lindenberg S, Nielsen MH, Lenz S: In vitro studies of human blastocyst implantation. *Ann NY Acad Sci* 442: 368–374, 1985.
20. Buster JE, Bustillo M, Rodi IA, Cohen SW, Hamilton M, Simon JA, Thorneycroft IH, Marshall JR: Biologic and morphologic development of donated human ova recovered by nonsurgical uterina lavage. *Am J Obstet Gynecol* 153: 211–217, 1985.
21. De Felici M, Salustri A, Sicracusa G: "Spontaneous" hardening of the zona pellucida of mouse oocytes during in vitrol culture. II. The effect of follicular fluid and glycosaminoglycans. *Gamete Res* 12: 227–235, 1985.
22. Quinn P, Warnes GM, Kerin JF, Kirby C: Culture factors in relation to the success of human in vitro fertilization and embryo transfer. *Fertil Steril* 41: 202–209, 1984.
23. Lopata A, Martin M, Oliva K, Johnston I: Embryonic development and blastocyst implantation following in vitro fertilization and embryo transfer. *Fertil Steril* 38(6): 682–687, 1982.
24. Hyttel P: The epithelium of uterine biopsies from cyclic dairy cattle: ultrastructure and endocytotic activity. *Acta Anat* 123: 93–100, 1985.
25. Parr MB, Parr EL: Uterine luminal epithelium: protrusions mediated endocytosis, nor apocrine secretion, in the rat. *Biol Reprod* 11: 220–233, 1976.
26. Parr MB, Parr EL: Endocytosis in the uterine epithelium of the mouse. *J Reprod Ferti* 50: 151–153, 1977.
27. Bitton-Casimiri V, Brun JL, Psychoyos A: Active release of material from rat blastocysts developing in vitro. *J Reprod Ferti* 27: 461–462, 1971.
28. Bazer FW, Roberts RM: Biochemical aspects of conceptus-endometrial interactions. *J Exp Zool* 228: 373–383, 1983.
29. Nilsson O. The morphology of blastocyst implantation. *J Reprod Ferti* 39: 187–194, 1974.
30. Enders AC. Anatomical aspects of implantation. *J Repr Fert Suppl* 25: 1–15, 1976.
31. Enders AC, Henoricks AG, Schlafke S. Implantation in the rhesus monkey: initial penetration of endometrium. *Am J Anat* 167: 275–298, 1983.
32. Enders AC, Schlafke S. Cytological aspects of trophoblast–uterine interaction in early implantation. *Am J Anat* 125: 1–30, 1968.
33. Lindenberg S, Hyttel P, Lenz S: In-vitro studies of an epitheliochorial and haemochorial implantation. Human Reproduction: Abstract of the First Meeting of the European Society of Human Reproduction and Embryology. IRL Press, 1985.

Author's address:
Dr. S. Lindenberg
Department of Obstetrics and Gynecology
Chromosome Lab 4041
Rigshospitalet
University of Copenhagen
Blegdamsvej 9, DK-2100
Denmark

Index

GPSR Compliance

The European Union's (EU) General Product Safety Regulation (GPSR) is a set of rules that requires consumer products to be safe and our obligations to ensure this.

If you have any concerns about our products, you can contact us on ProductSafety@springernature.com

In case Publisher is established outside the EU, the EU authorized representative is:

Springer Nature Customer Service Center GmbH
Europaplatz 3
69115 Heidelberg, Germany

Batch number: 10705930

Printed by Printforce, the Netherlands